THE CHEMISTRY OF PERFUMES

Academic and Industrial Perspectives

The Chemistry of Perfumes

Academic and Industrial Perspectives

Edited by

Sylvain Antoniotti • Véronique Michelet

Université Côte d'Azur, CNRS, France

World Scientific

NEW JERSEY • LONDON • SINGAPORE • BEIJING • SHANGHAI • HONG KONG • TAIPEI • CHENNAI • TOKYO

Published by

World Scientific Publishing Europe Ltd.
57 Shelton Street, Covent Garden, London WC2H 9HE
Head office: 5 Toh Tuck Link, Singapore 596224
USA office: 27 Warren Street, Suite 401-402, Hackensack, NJ 07601

Library of Congress Cataloging-in-Publication Data
Names: Antoniotti, Sylvain editor | Michelet, Véronique editor
Title: The chemistry of perfumes : academic and industrial perspectives /
edited by Sylvain Antoniotti, Université Côte d'Azur, France |
Véronique Michelet, Université Côte d'Azur, France.
Description: London ; Hackensack, NJ : World Scientific, 2026. |
Includes bibliographical references and index.
Identifiers: LCCN 2025020923 | ISBN 9781800617773 hardcover |
ISBN 9781800617780 ebook | ISBN 9781800617797 ebook other
Subjects: LCSH: Perfumes--Formulae | Chemistry, Technical
Classification: LCC TP983 .C534 2025 | DDC 668/.54--dc23/eng/20250710
LC record available at https://lccn.loc.gov/2025020923

British Library Cataloguing-in-Publication Data
A catalogue record for this book is available from the British Library.

For any available supplementary material, please visit
https://www.worldscientific.com/worldscibooks/10.1142/Q0523#t=suppl

Desk Editors: Nambirajan Karuppaiah/Gabriel Rawlinson/Shi Ying Koe

Typeset by Stallion Press
Email: enquiries@stallionpress.com

https://doi.org/10.1142/9781800617780_fmatter

Preface

The Chemistry of Perfumes: Academic and Industrial Perspectives has been designed to offer an integration of both the academic and industrial viewpoints on the dynamic fields of fragrance chemistry and olfaction. It is composed of 10 chapters collected from 2021 to 2024 contributed by scientists from academia and fragrance companies in an interdisciplinary and intersectoral approach.

Chapter 1, entitled "History of Important Perfumery Aromachemicals" by Olivier R. P. David, describes the evolution of synthetic aromachemicals in perfumery, starting from the use of natural ingredients to the development of artificial substances. It highlights key discoveries in the nineteenth and twentieth centuries, such as nitrobenzene, coumarin and vanillin, which revolutionised the perfume industry by allowing more affordable and diverse scent profiles. It further explores the introduction of synthetic musks, ionones and other odorants that mimic natural scents, as well as unique artificial molecules and how advances in chemistry, like the synthesis of macrocyclic musks and ambery woods, expanded the perfumer's palette.

Chapter 2, entitled "Sustainable Extraction of Perfumery Plants" by Xavier Fernandez *et al.*, deals with the emergence of a new generation of environmentally friendly solvents for eco-extraction processes. In particular, the use of solvents produced from bioresources, deep eutectic solvents and ionic liquids (ILs) will be discussed. Activation techniques, namely microwave-assisted extraction, ultrasound-assisted extraction, pressurised fluid extraction, enzyme-assisted extractions and other intensification techniques will then be presented.

Chapter 3, entitled "Modern or Renewed Techniques and Approaches for the Detection of Adulteration of Essential Oils" by Marissa Pierson and Sylvain Antoniotti, analyses the literature on the topic that appeared between 2015 and 2021, complementary to existing approaches such as enantioselective chromatography and isotopic analysis with mass spectrometry. Their developments such as in multidimensional gas chromatography (MDGC) or coupling of liquid chromatography with MDGC will be presented along with more recent approaches using techniques such as Raman, fluorescence and infrared spectroscopies, as well as global data analysis with metabolomics and deep learning.

Chapter 4, entitled "Full Complementarity of Three Aroma Extraction Methods and Two GC Detection Techniques for the Analysis of Natural Odorants in Fermentation Media: Focus on Terpenoids, Phenols, Esters and Polyfunctional Thiols" by Cécile Chenot and Sonia Collin, deals with specific analytical approaches, including sample preparation, applied to sensory molecules with thresholds sometimes as low as nanograms per litre. Solvent-assisted flavour evaporation, followed by GC-mass spectrometry, solid-phase extractions using mercury- or silver-based phases and pulsed flame photometric detectors, will be presented with applications to the analysis of thiols in beer.

Chapter 5, entitled "Manufacturing Natural Ingredients" by Bernard Blerot and Marine Goubet, contains a review of all the classic extraction methods currently used in the flavour and fragrance industry. Their principles, specificities and their advantages and disadvantages are discussed. In the first part, primary extraction methods, which are applied directly to the plant matter and which very often occur near the site of harvest, are presented (cold pressing for citrus fruits, steam distillation and hydrodistillation, organic solvent extraction and supercritical CO_2 extraction). In the second part, secondary extraction or purification processes, such as molecular distillation, fractional distillation and the refinement of absolutes, are detailed.

Chapter 6, entitled "Biocatalysis and Biotransformation for Optimised Natural Flavour and Fragrance Ingredients" by Mathilde Lecourt and Sylvain Antoniotti, reports on the use of biocatalysis and biotransformations to provide improved natural ingredients, either in terms of extraction or distillation yields, sensory properties or safety profile. The reader is provided with examples from the literature and from patents where enzymatic or cellular biosolutions have been developed either at the R&D level or at the production stage to access optimised natural flavour and fragrance ingredients.

Chapter 7, entitled "Rose Ketones: Past and Present History of This Major Class of Compounds for the F&F Industry" by Fabrice Robvieux, is dedicated to this iconic floral scent. Different strategies with their advantages and disadvantages for the synthesis of damascones are presented. With the increasing importance of green and sustainable chemistry, the chapter shows how damascones remain a vivid and active subject of research for the entire industry.

Chapter 8, entitled "Lewis Acid-Catalysed Cyclisations in the Field of Fragrances" by Raphaël Saget *et al.*, shows how the catalytic activity of Lewis superacids, in particular metal trifluoromethanesulfonates (triflates), can be used in several cyclisation reactions of interest in the field of fragrance compounds. The chapter reports examples of multiple reaction types, such as Diels–Alder, Friedel–Crafts, Prins, carbonyl-ene and related ene processes, as well as cyclisations involving unsaturated alcohols, carboxylic acids and thiols.

Chapter 9, entitled "Metal-Catalysed Synthesis of Fragrances" by Véronique Michelet *et al.*, focuses on the use of transition metals allowing access to challenging molecules with organoleptic properties. Some selected examples are proposed, including cycloadditions, reductions (C=C or C=O), syntheses of carbonyl compounds (oxidation, dehydrogenation, allylic oxidation and isomerisation), carbonylation reactions and metathesis reactions.

Chapter 10, entitled "Sensory Effects of COVID-19 on Perfume Creation and Perception" by Stephanie Hunter and Valentina Parma, explores the types of olfactory disorders that are associated with SARS-CoV-2 and their known biological bases, with specific reference to the damage caused by SARS-CoV-2 to the sustentacular cells in the olfactory epithelium. The available interventions and the implications of temporary and permanent alterations of the sense of smell will be discussed, with reference to the experience of and strategies available to professionals relying on their intact chemical sense for their careers.

Véronique Michelet and Sylvain Antoniotti
Editors

https://doi.org/10.1142/9781800617780_fmatter

About the Editors

Véronique Michelet is Professor of Chemistry at Université Côte d'Azur, where she heads the "Catalysis, Organic Synthesis Methods & Fine Chemistry" team within the Institut de Chimie de Nice. Her research focuses on the development of new catalytic methodologies, in particular gold catalysis, for the synthesis of heterocyclic molecules of interest in perfumery, pharmaceuticals and sustainable chemistry.

Sylvain Antoniotti is Director of Research at CNRS and Director of the Institute for Innovation and Partnerships Flavour, Fragrance, Cosmetics at Université Côte d'Azur. He heads the sustainable chemistry research group at the Institut de Chimie de Nice, specialising in catalysis, biocatalysis, natural complex substances and the synthesis of innovative fragrance ingredients.

https://doi.org/10.1142/9781800617780_fmatter

About the Contributors

Bernard Blerot is the R&D Director for Naturals at International Flavors & Fragrances (IFF) in Grasse, France, where he leads research at LMR Naturals, a division specialising in high-quality natural ingredients for perfumery. His work focuses on the scientific exploration of natural raw materials, combining data science and neuroscience to assess their sensorial and wellness benefits.

Jean-Baptiste Chagnoleau is currently a postdoctoral researcher at the Institut de Chimie de Nice (CNRS, UMR 7272). His research focuses on the sustainable extraction of natural compounds, particularly using deep eutectic solvents and ionic liquids.

Cécile Chenot is a postdoctoral researcher and guest lecturer at the Université Catholique de Louvain (UCLouvain), where she is affiliated with the Louvain Institute of Biomolecular Science and Technology (LIBST), specifically within the Unit of Brewery and Food Sciences.

Sonia Collin is professor at the Université Catholique de Louvain (UCLouvain) and leads the Unit of Brewery and Food Sciences (INBr) within the Louvain Institute of Biomolecular Science and Technology (LIBST). Her research focuses on the chemistry of flavour compounds in beer, wine and chocolate, particularly polyphenols and volatile sulphur compounds.

João A. P. Coutinho is professor in the Department of Chemistry at the University of Aveiro in Portugal and the director of CICECO — Aveiro Institute of Materials. His multidisciplinary research focuses on thermodynamics, ionic liquids, and deep eutectic solvents, with applications in the sustainable extraction of natural compounds and green chemistry.

Olivier R. P. David is a lecturer (Maître de Conférences) in organic chemistry at the Université de Versailles — Paris-Saclay, attached to the Institut Lavoisier de Versailles (ILV — UVSQ/CNRS). A specialist in perfume chemistry, his research is dedicated to the links between chemistry and olfactory creation, and the history and science of fragrance ingredients.

Elisabet Duñach is CNRS Director of Research Emeritus at the Institut de Chimie de Nice of the Université Côte d'Azur. A recognised specialist in organic chemistry and catalysis, she has contributed to the development of eco-compatible methods for the synthesis of ingredients used in flavours and fragrances.

Xavier Fernandez is a Professor of Chemistry at Université Côte d'Azur, based at the Institut de Chimie de Nice. He specialises in natural product chemistry, analytical techniques and the development of ingredients for the fragrance, cosmetics and food industries. He leads the "Metabolome and Valorisation of Plant Biodiversity" research group, focusing on sustainable extraction methods and the valorisation of aromatic plants.

Marine Goubet is a Research and Development Engineer at International Flavors & Fragrances (IFF) in Grasse, France. Her work focuses on the development of natural ingredients for perfumery, contributing to the creation of high-quality, sustainable fragrances.

Stephanie Hunter is a postdoctoral fellow at the Monell Chemical Senses Center in Philadelphia (USA), where she specialises in chemosensory science, focusing on the sense of smell.

Piotr Jaunky was a researcher in organic chemistry at the Centre de Recherche en Chimie Organique of V. MANE Fils, a family business based in Bar-sur-Loup, specialising in flavours and fragrances.

Romain Laher completed his Ph.D. in chemistry from Université Côte d'Azur, conducting his research at the Institut de Chimie de Nice on gold-catalysed synthesis of olfactory molecules. He is serving as the Operational Manager for dealcoholisation at Le Chai Sobre, a new production centre established by the cooperative group Vivadour.

Mathilde Lecourt completed her Ph.D. in chemistry at Université Côte d'Azur, conducting her research at the Institut de Chimie de Nice on sustainable chemistry and biocatalysis, in the context of fragrance and flavour applications. She is a Research and Development Engineer at Elicit Plant, an agri-biotech company specialising in the development of plant-based biostimulants aimed at enhancing crop resilience to abiotic stresses such as drought.

Christophe Marin is President of Expressions Parfumées, a fragrance composition house founded in 1982 in Grasse and integrated into the Givaudan group since 2018. Under his leadership, the company has strengthened its local roots while developing an international expansion strategy focused on innovation, sustainability and high-quality customer service.

Nicolas Papaiconomou is a lecturer (Maître de Conférences) in physical chemistry at Université Côte d'Azur, within the Institut de Chimie de Nice. His research focuses on the extraction of natural compounds from plant materials, using sustainable solvents such as ionic liquids, aqueous two-phase systems, eutectics and pressurised solvents.

Valentina Parma is a psychologist and Assistant Director at the Monell Chemical Senses Center in Philadelphia, where she leads research on human olfaction across the lifespan. Her work explores how the sense of smell influences social behaviour, health and quality of life, with a particular focus on chemosensory dysfunctions related to COVID-19 and long-COVID.

Marissa Pierson earned her Ph.D. in Analytical Chemistry from Université Côte d'Azur on natural product analysis, employing techniques such as liquid chromatography, gas chromatography, and mass spectrometry. She is a Senior R&D Analytical Chemist at VIQI Group in France, specialising in method development for HPLC, GC and MS.

Fabrice Robvieux is a chemist specialising in organic synthesis and fragrance innovation. He currently holds a position at DSM-Firmenich, a global leader in the fragrance and flavour industry, where he contributes to the development of new perfumery ingredients.

Raphaël Saget is a researcher in organic chemistry at the Centre de Recherche en Chimie Organique of V. MANE Fils, a family business based in Bar-sur-Loup, specialising in flavours and fragrances. His work focuses on the development of new fragrance molecules using innovative, environmentally friendly catalytic synthesis methods.

https://doi.org/10.1142/9781800617780_fmatter

Contents

https://doi.org/10.1142/9781800617780_0001

Chapter 1

History of Important Perfumery Aromachemicals

Olivier R. P. David

Institut Lavoisier, Versailles — UMR8180,
Laboratoire Synthèse Organique,
Université de Versailles SQY — Paris Saclay, Versailles, France

olivier.david@uvsq.fr

Abstract

The chapter describes the evolution of synthetic aromachemicals in perfumery, starting from the use of natural ingredients to the development of artificial substances. It highlights key discoveries in the nineteenth and twentieth centuries, such as nitrobenzene, coumarin and vanillin, which revolutionised the perfume industry by allowing more affordable and diverse scent profiles. It further explores the introduction of synthetic musks, ionones and other odorants that mimic natural scents, as well as unique artificial molecules and how advances in chemistry, like the synthesis of macrocyclic musks and ambery woods, expanded the perfumer's palette. The development of synthetic materials has continually shaped modern perfumery, allowing for new artistic expressions and more sustainable practices.

1.1 Introduction

Since antiquity, the art of perfumery relied on the use of natural ingredients. The odoriferous substances were employed after expression, oil infusions, then by distillation and solvent extraction. The use of pure, man-made substances appeared with the first progresses of modern chemistry during the nineteenth century. This chapter deals with the stories of the major discoveries in this field, following a roughly chronological order and gathered in important olfactory families [1–5].

To commence forthwith, the very first artificial substance employed for its odorant properties was nitrobenzene, prepared by Eilhardt Mitscherlich in 1834 by nitration of benzene with fuming "azotic acid". The very harsh odour of almond of this "essence of Mirbane" confined its use to soaps and low-priced toiletries; moreover, a rapid recognition of skin sensitisation problems led to its disappearance from the market [6]. In consequence, the first synthetic compound that has genuinely been used in a fine perfume is coumarin.

1.2 First "Artificial Perfumes"

While studying Tonka beans (*Dipteryx odorata*, called *coumarou* in its growing place), Friedrich Wöhler and Justus Liebig succeeded in isolating its odorant principle in 1856 and christened it coumarin. Without precise knowledge of its structure, William H. Perkin realised its synthesis from salicylic aldehyde in 1868. Ferdinand Tiermann and Hermann Herzfeld devised a more efficient preparation of coumarin in 1877, allowing for its industrial production; hence, in 1882, Paul Parquet could employ synthetic coumarin to create *Fougère royale* for Jean–François Houbigant [7]. Noteworthily, industrial optimisation reduced the price of coumarin by a factor of eight between 1880 and 1900, allowing for its use in larger quantities and in inexpensive goods [8].

While studying natural extracts, chemists also discovered odorant ingredients like heliotropin (also called piperonal) that was produced for the first time in 1869 by the oxidation of pepper aromatic principle by Wilhelm Rudolph Fittig and W.H. Mielck. It was commercialised by *Schimmel & Co*. in 1879 as *heliotropin* at a price of 3,790 francs per kilo, but it decreased to 15 francs in 1914. Similarly, aubepine or 3-methoxybenzaldehyde was obtained by oxidation of anethol by Frédéric Landolph in 1875 [9]. These compounds were used by perfumers as *heliotropin* and *aubepine*, respectively, and both are present in *Après l'Ondée*, created by Jacques Guerlain in 1906, and contribute to its bucolic feel (Figure 1.1).

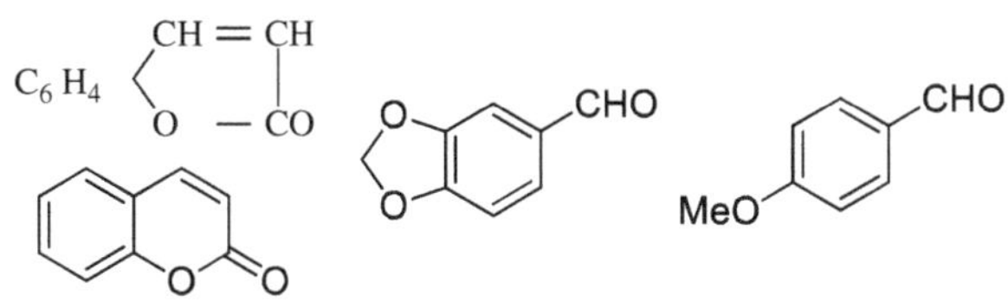

Figure 1.1. Perkin's depiction of coumarin and its actual structure, along with heliotropin and aubepine (3-methoxybenzaldehyde).

1.3 Vanillin

The true nature of vanillin remained a mystery for almost 60 years after the first study by Bucholz and Vogel in 1816 on "vanilla frost". Several authors analysed these crystals of nearly pure vanillin forming on high-quality vanilla pods, but all failed to identify its correct structure (Gobley (1858), Amédée Vée (1858), Stokkeby (1865) and Carles (1872)). Only in 1874, Ferdinand Tiemann and Wilhelm Haarmann obtained the correct formula of vanillin and were able to produce it artificially from coniferin.

The process was low-yielding but was however applied on an industrial scale in the newly created company *Haarmann's Vanillinfabriek in Holzminden* (1875). Later, in 1876, Karl Reimer and Tiemann showed that the action of chloroform on guaiacol under strong basic conditions delivers vanillin in a much more economical way, leading to the renaming of the company to *Haarmann & Reimer*. In close contact with his friend Haarmann, George de Laire in France was patenting the same year a preparation of vanillin from acetyleugenol, derived from cloves. From that day on, both companies, *H&R* and *de Laire*, worked together and exploited patents in parallel. Here again, continuous progress in the making of artificial vanillin spectacularly reduced its price, from 8,750 francs per kilo in 1876 to 873 francs when Aimé Guerlain launched *Jicky* in 1889 — with high dosages of vanillin and coumarin — and down to 55 francs in 1908. This had a tremendous impact on modern perfumery since perfumers realised that the cheapness of synthetic ingredients reduced the price of their perfumes without affecting olfactory qualities, even permitting artistic effects unachievable with natural ingredients.

While preparing vanillin derivatives, Ernst Schering discovered the strong flavouring power of ethyl vanillin, which he patented in 1894. It was also adopted by perfumers and plays a crucial role in the soft-warm impression of *Shalimar* created by Jacques Guerlain in 1925 (Figure 1.2).

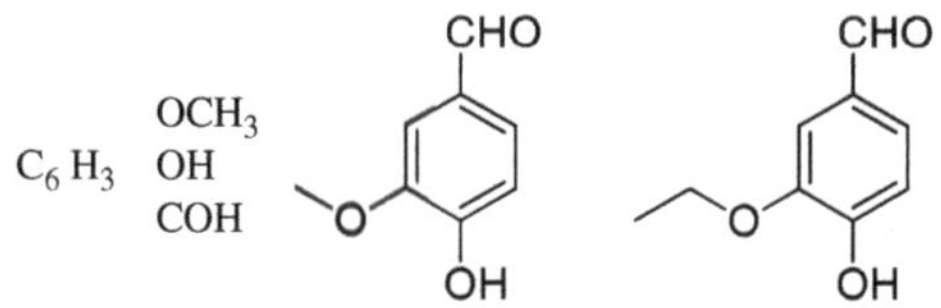

Figure 1.2. Tiemann's depiction of vanillin, vanillin and ethyl vanillin.

1.4 Nitromusks

A genuine revolution then occurred in perfumery with the release on the market of the first artificial substance having the smell of musk, when Albert Baur patented in 1888 *musk Toluene*, later known as *musk Baur*. Following a long lineage of pharmacists and chemists studying nitrated hydrocarbons with musky odours, which had its origin in the work of Andreas Sigismund Marggraf (1758), Baur was the first to bring to an industrial level the production of a substance not found in nature but reproducing the smell of musk Tonkin [10]. Just six months after Baur, two chemistry students, Emil Schnaufer (at the age of 22 in 1888) and Heinrich Hupfeld patented *musk Xylene*.

Then, Baur discovered musk Ambrette in 1891, and a Dutch entrepreneur, Willem Mallmann, patented musk Ketone in 1893. Eventually, all these nitromusks became the properties of *Fabrique de Produits Chimiques de Thann et Mulhouse* and were sold to perfumers by *Fabriques de Laire* until expiration of the patents. These nitromusks were immediate commercial successes, as they boosted the musky facets of natural Tonkin musk in fine perfumes for an affordable price. But nitromusks were also employed for scenting cosmetics, rice powders and talcum, and even pipe tobacco or beverages, testifying to their universal acclaim. Later investigations by Henri Barbier from *Givaudan–Vernier* brought *Moskene* (1930), and the work of Marion Scott Carpenter at *Givaudan–Delawanna* gave *Tibetene* (1935), closing the small family of nitromusks. These gave strong and sound base notes to all perfume creations and were considered excellent "fixatives". They worked beautifully in support of the natural animal musks like Tonkin and civet infusions; for instance, the musks complex of *Chanel N°5* by Ernest Beaux constituted no less than 36% of the formula. Photosensitising effects of musk ambrette and xylene were noted by dermatologists in 1977, leading to ever-increasing regulations of these ingredients, ultimately being totally banned by the European Union in 1998, except for musk ketone, which is recognised as safe (Figure 1.3).

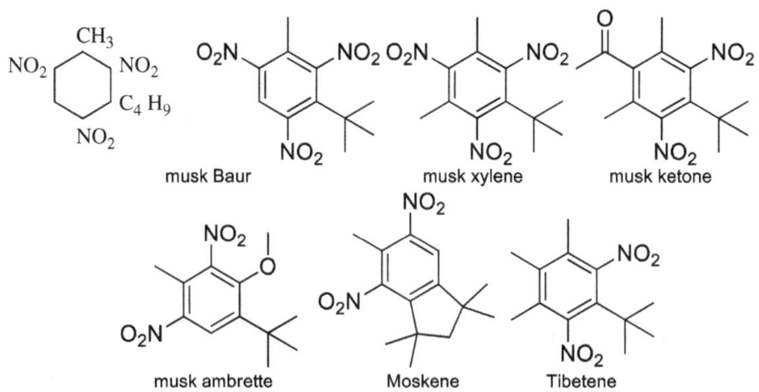

Figure 1.3. Baur's depiction of "musk Toluene", later musk Baur, along with all commercial nitromusks.

In short succession to the start of nitromusks, a second revolution took place in perfumery, now in the floral territories with ionones and methyl ionones.

1.5 Ionones

Since antiquity, the delicate smell of violets was appreciated and associated with distinction and sophistication. During the nineteenth century, the fragrance of Parma violets was available to perfumers through a violet essence distilled from natural flowers with an infinitesimal yield and, accordingly, a crazily exorbitant price. After his success with vanilla, Ferdinand Tiemann [11] searched for the odoriferous principle in the essence of violets with support from *Haarmann & Reimer* and *de Laire* firms [12]. These studies, started in 1880, were not fruitful, and he thus fell back on the "cheaper" orris butter and succeeded in isolating irone from it, whose fragrance he considered closely resembling that of violets.

An error in the brut formula of irone led him and his colleague, Paul Krüger, to synthesise what was in fact the genuine molecule of ionone that he never succeeded in isolating from violets. To realise it, he condensed citral with acetone under basic catalysis, obtaining pseudo-ionone, this one being cyclised by strong acids into a mixture of alpha and beta ionones. The process was patented on April 25, 1893, and two days later, the preparation of the homologous compounds, methyl ionones, was also protected for *H&R* in Germany and *de Laire* in France.

The first perfume employing these compounds was launched already in 1892, *Vera Violetta*, by *Roger & Gallet*. Interestingly, at first, hostile reactions ensued from the violet grower's part, fearing the extinction of their business. But soon, it was realised that the introduction of these inexpensive ionones and methyl ionones to the perfumer's palette increased the demand for violet essence. Perfumers were now less shy on violet perfumes, using large amounts of synthetic ionones, rounded by small hints of the expensive essence, thus resulting in an overall growth of the market. However, eventually the progress made in perfume compounding permitted the confection of fine violet perfumes without any natural essence in them, causing the vanishing of this traditional cultivation. During 1902–1906, Philippe Chuit, Martin Naef and Fred Firmenich (*Chuit & Naef*, to become *Firmenich*) worked with the chemist Jules Bolle and realised the preparation of pure ionones, alpha and beta, sold under the names *Violettones A* and *B*. They also developed an optimised process for methyl ionones and were able to sell *Iralia*, a mixture of isomers particularly appreciated by perfumers. Ionones and methylionones gained a prominent place within the perfumer's organ because, beyond the sole reconstitution of the violet scent, they assumed a crucial role of linking the other notes together by bridging floral and woody materials. They constitute the flesh of a perfume; conversely, since their release, and even today, almost no fragrance is compounded without them.

Concerning irone, the compound isolated initially by Tiemann in orris butter, its synthetic preparation was far more arduous and was only reported in 1944 by Yves–René Naves from *Givaudan*. This was actually the case of a primacy struggle between him and Ruzicka from *Firmenich*, both groups investigating irone synthesis but protecting their discoveries under sealed envelopes. After unsealing from all parties, Naves was assured to have the precedence [13] (Figure 1.4).

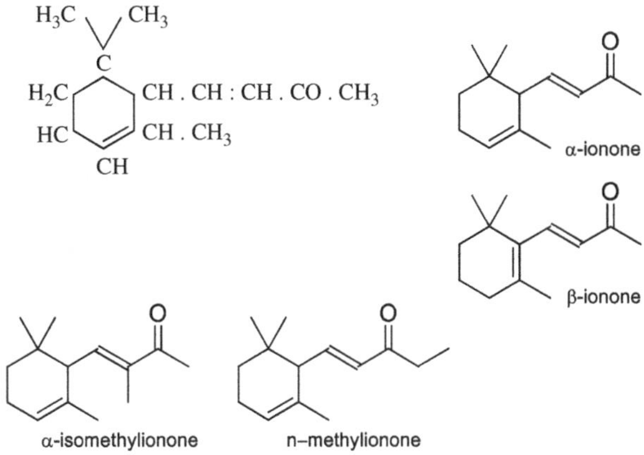

Figure 1.4. Tiemann's (erroneous) depiction of ionone, isomers of ionones and methylionones.

With all the previous perfumery materials, we have seen only natural compounds or artificial ones imitating a natural scent. At the turn of the nineteenth to twentieth centuries, artificial molecules with odours not found in nature started to emerge in perfumery.

1.6 Novel Odorants for the Dawn of the Twentieth Century

Aliphatic aldehydes are widespread in nature, especially the light ones of even numbers of carbon atoms (C4–C10) that are abundant in citruses, while unevenly numbered or branched aldehydes are extremely rare. In 1903, Auguste Darzens published a procedure to homologate carbonyl compounds into the higher aldehydes *via* a glycidic epoxide. This reaction was somewhat capricious on certain substrates, but allowed the industrial production of 2-methyl undecanal. It was employed by perfumers as aldehyde C12-mna, like in *Quelques Fleurs* by Robert Bienaimé for *Houbigant* in 1912. A reliable production of simple aliphatic aldehydes could be achieved thanks to the heterogeneous hydrogenation of acyl chlorides, reported by Karl Rosenmund in 1918. This process was applied by Léon Givaudan in his plant at Lavirotte, near Lyon, and could provide Ernest Beaux and others with C10, C11 and C12 aldehydes of constant quality. The famous *N°5* (1921) by the former for Gabrielle Chanel would have been inconceivable without this industrial process. This abstract odour of aldehydes was so successful that it gave rise to a specific class of "aldehydic" perfumes with countless representatives [14].

Another example of an unnatural molecule with an original odour is isobutyl quinoline. Zdenko Skraup developed in 1880 a synthetic path to quinolines from anilines and glycerol. A particular quinoline, isobutylated (IBQ), exerted an interesting and unprecedented olfactory profile of leather and was thus produced by the *Fabriques de Laire* for perfumers. But IBQ was initially not sold in pure form; it was incorporated in a small composition, a base, specifically for IBQ in the famous *Mousse de Saxe*, whose commercialisation started in around 1911. A prominent perfume built on *Mousse de Saxe* is *La Nuit de Noël* by Ernest Daltroff for *Caron* in 1922. Later on, pure IBQ could be employed, which allowed Germaine Cellier to overdose it (2.5%) in *Bandit* for *Robert Piguet* in 1944.

Phenylethanol (PEA) is an interesting case; it was first prepared by Bronislaw Radziszewski in 1876 while studying homologues of benzylic alcohol, but he did not note its particular odour. Despite studies by Tiemann on the constituents of rose absolute in 1896, phenylethanol was detected in this precious material only in 1900 by Dr. Rojan at *Heine & Co*, only six days before Walbaum at *Schimmel & Co* actually made the same discovery. Phenylethanol could be produced industrially using the procedure of ester reduction developed by Bouveault and Blanc at La Sorbonne in 1904; PEA was then employed by perfumers to convey the fresh rose facets to their creations.

Then, in 1908, two Russian chemists, Schukow and Schestakow, prepared γ-undecalactone for the first time, giving perfumers access to realistic reconstitutions of fruits. Jacques Guerlain employed this lactone (although sold as "aldehyde C-14"!) to round the austere aspects of his monumental chypre composition, *Mitsouko* (1919), with delicate notes of peach. Edmond Roudnistka later used the same undecalactone but oriented in the plum direction in *Femme* for *Marcel Rochas* in 1944 (Figure 1.5).

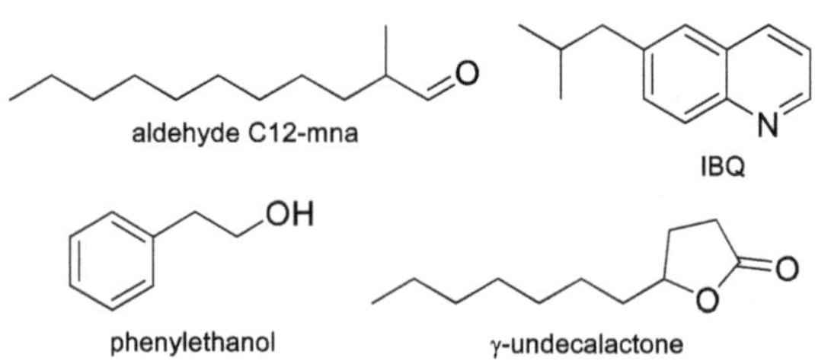

Figure 1.5. Structures of four important odorants.

The next tremendous step in aroma chemistry that helped perfumers' creativity concerned lily-of-the-valley odorants.

1.7 Lily-of-the-Valley Molecules

As muguet cannot be extracted to capture its natural scent, synthetic compounds reconstituting a lily-of-the-valley impression were much sought-after [15]. The first material to do so was hydroxycitronellal, initially obtained by Knoll & Co in 1905 by hydration of citronellal. *Givaudan* accordingly offered it as *Laurine* in 1906, a finer quality, made from enantiopure citronellal, was later offered by *Chuit & Naef* as *Cyclosia* (1908). This resulted in a proliferation of Muguet perfumes, the culmination of this family being *Diorissimo* by Edmond Roudnitska for *Christian Dior* in 1956 with 10% of "Hydroxy". Lily-of-the-valley molecules are of utmost importance in floral bouquets to convey freshness and volume to a composition. This fostered manifold efforts from chemists to extend the number of muguet odorants. Hence, other aldehydes enriched the family with Cyclamen aldehyde (*Givaudan*, 1919), *Lilial* (*Givaudan*, 1956), *Helional* (*IFF*, 1958), *Lyral* (*IFF*, 1958), *Bourgeonal* (*Givaudan* as Isolilial, 1959), etc. This field is still very active, with recent introductions of *Hivernal* (*Firmenich*, 1994), *Mimosal* (*Firmenich*, 2008), *Lilybelle* (*Symrise*, 2011), *Aquaflora* (*IFF*, 2011), *Mahonial* (*Givaudan*, 2013) and *Nympheal* (*Givaudan*, 2014) (Figure 1.6).

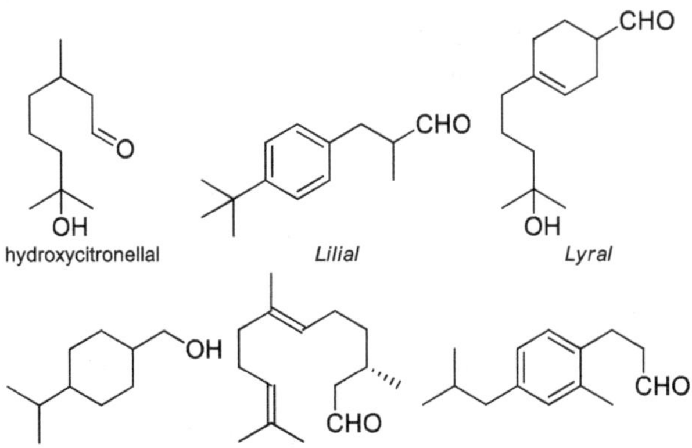

Figure 1.6. A selection of lily-of-the-valley odorants.

Interestingly, in 2012, *Takasago* patented an enantioselective process to synthesise (3*S*)-(6*E*)-2,3-dihydrofarnesal, a natural compound present in the headspace of lily-of-the-valley flowers, thus constituting the sole material on the market to be nature identical, sold under the name *Biomuguet*.

In 1973, a small revolution occurred in the muguet field, with the introduction of *Mayol* by *Firmenich*, a lily-of-the-valley alcohol, devoid of the stability and oxidation problems affecting aldehydic compounds. Conversely, other alcohols were discovered, like *Majantol* (*Haarmann & Reimer*, 1985), *Florol* (*Firmenich*, 1986) and *Lilyflore* (*Firmenich*, 1999) that can be safely used in corrosive media, like soaps and detergents. This perpetual quest for novel muguet material is stirred by successive restrictions and banishments of several very important molecules for skin sensitisation problems. Hence, hydroxycitronellal is highly limited, while *Lilial* and *Lyral* are now completely banned, stressing the need for substitutes which are benign for health and the environment.

While hydroxycitronellal was produced industrially, a laboratory discovery was the seed of a major change in the chemistry of musks.

1.8 Macrocyclic Musks

Heinrich Walbaum, chemist at *Schimmel & Co in Miltitz*, undertook the purification of the odorous principle of musk Tonkin, and in 1905, he patented the isolation of a particular ketone, which he named muscone, for which he determined the raw formula $C_{16}H_{30}O$. Further studies indicated that it was a saturated compound, but Walbaum could not ascertain the complete formula of muscone. In the same vein, Erwin Sack succeeded in isolating the odorous material of civet, which he christened civetone with a raw formula of $C_{17}H_{30}O$ and which contains one insaturation. This discovery was protected by a patent application in 1912, without the complete knowledge of its molecular structure. This work was done for both compounds by Leopold Ruzicka after deploying treasures of ingenuity to decipher the unprecedented structural features of civetone and muscone. Hence, after numerous and rational degradation studies, Ruzicka could ascribe to civetone a macrocyclic structure, with a ketone function and a *cis* carbon–carbon double bond, breaking with the belief that large molecular cycles were unstable and impossible to observe in nature. These spectacular studies were published in 1924 for civetone and in 1925 for muscone. During these investigations, Ruzicka synthesised several macroketones by thermal cyclisation of thorium salts of aliphatic diacids [16]. Although low-yielding, this procedure allowed for the preparation of a novel simple macroketone in C15 with a fine musk odour, which was patented in 1924 for *Naef & Co* and sold under the name *Exaltone*. Obviously, with this seminal production being inefficient (a 6% yield) and *Exaltone* being hugely expensive — 50,000 CHF/kg in 1930, more efficient macrocyclisation protocols were developed after WWII, and the price dropped down to 500 CHF/kg in 1947. In the meantime, Stevens and Erickson recognised the presence of *Exaltone* in secretions of the Louisiana muskrat, making it a natural substance. More complex, synthetic muscone had to wait until 1933 to be available through a Thorpe–Ziegler reaction employed in a process patented by *Neaf & Co* and *Schering–Kahlbaum* at the same time. Notably, in 2015, *Firmenich* was able to furnish the perfumer with enantiopure muscone, fully identical to the natural one, under the name *Muscone Laevo*.

The family of musky macrocyclic ketones was constantly extended with, for example, *Muscenone* (*Firmenich*, 1967) or *Cosmone* (*Givaudan*, 1998) which, with a low molecular weight, is able to operate in the early evaporation phase of a perfume, and more recently we can also mention *Aurelione* (*Symrise*, 2005) (Figure 1.7).

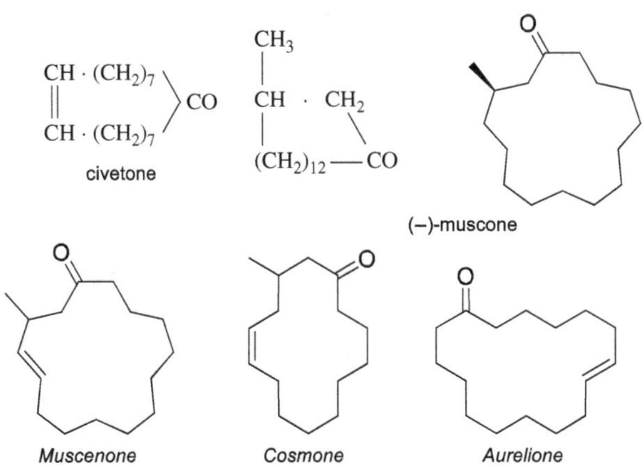

Figure 1.7. Ruzicka's structures for civetone and muscone, along with a selection of macrocyclic ketones.

After Ruzicka authenticated the macrocyclic nature of civetone and muscone, chemists unveiled more large-ring molecules from natural sources of muskiness. Hence, vegetal musks revealed the presence of macrolactones. Max Kerschbaum at *Haarmann & Reimer* could isolate 15-pentadecanolide from Angelica root oil in 1926. It was commercialised as *Exaltolide* in 1927, obtained through the Baeyer–Villiger oxidation of *Exaltone* at 100,000 CHF/kg. Again, process chemistry progress in macrolactonisation reactions reduced the price of *Exaltolide* to 60 CHF/kg in 2001. The same Kerschbaum also discovered an unsaturated macrolactone in ambrette seed oil in 1927, whose synthetic form was offered to perfumers as *Ambrettolide*. Later, structure-activity studies gave birth to artificial lactonic musks, like *Habanolide* (*Firmenich*, 1971) or *Nirvanolide* (*Givaudan*, 1997). But a genuine breakthrough took place in 1933 when Wallace Hume Carothers from *Dupont de Nemours* patented ethylene brassylate. This specialist of polymers discovered a very simple setup to distil macrocyclic dilactones from mixtures of polyesters of diacides and diols. Applying this process to ethylene glycol and brassylic acid gave a macrodilactone with a powerful musk odour in a very efficient and very economical way, which could then be commercialised as *Astratone* for a moderate price (Figure 1.8). He invented the polyamide *Nylon* in 1935, but, suffering from depression, he committed suicide in 1937, thinking that he hadn't achieved enough.

As we have seen in the previous paragraphs, during World War I and the interwar period, chemists were very active in developing novel

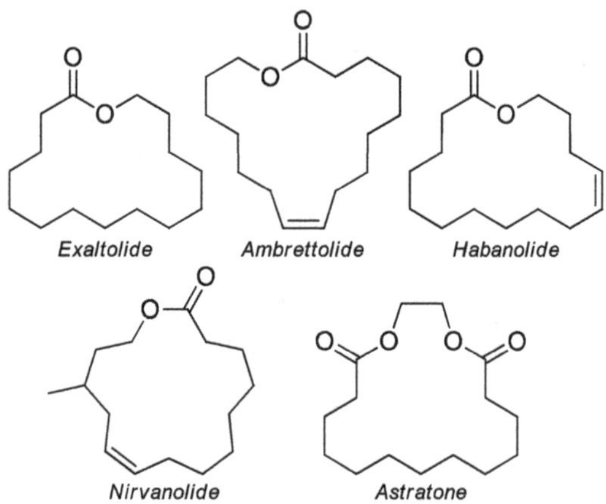

Figure 1.8. Commercial macrolactones used as musks.

compounds in known families of odorants. The progress made in analytical chemistry, in structural spectrometry and synthesis during World War II (WWII) then allowed aroma chemists to tackle complex natural ingredients. Hence, novel types of odorants could be detected in rose, jasmine, ambergris or sandalwood, opening new avenues in olfactory territories.

1.9 Artificial Sandalwood Odorants

Sandalwood is a unique and precious material that was first scientifically studied by Chapoteaut in 1882. Semmler pushed these studies further, but the complexity of the compounds responsible for the odour of sandalwood made necessary no less than 30 years of research to decipher their structures. In 1935, he could ascertain the molecular structure of α-santalol. The same year, Ruzicka, who got involved in almost all investigations concerning perfume chemistry, could give the exact formula of β-santalol, the most important odorant in sandalwood. The structure is so complex that its total synthesis had to wait decades before being achievable. Hence, in 1969, Kretschman et Erman at *Procter & Gamble* patented an unstereoselective preparation of β-santalol; the first synthetic route to pure (−)-(*Z*)-β-santalol was published by Krotz and Helmchen only in 1990. None of the procedures reported then could be scaled up to an industrial level. Fortunately for perfumers, artificial molecules were discovered that imitate the odour of sandalwood. The first was serendipitously found while preparing plastic modifiers by *Lech Chemie Gersthofen* in 1942. The process consisted of the condensation of phenols onto camphene under Lewis acid catalysis; then, hydrogenation of the complex mixture furnished a nicely odorant material evoking sandalwood. Studies conducted by Edouard Demole in 1964 at *Firmenich* determined that a single compound among a hundred and amounting to less than 1% of the mixture, was responsible for the sandalwood scent [17]. This artificial sandalwood was put on the market by *Givaudan* around 1960 under the name *Sandela*. More rational investigations led to the discovery of a very important class of artificial sandalwood materials, derived from (+)-campholenal, a building block easily accessed from (−)-α-pinene. Hence, in 1968, *VEB Chemisch Fabrik Miltitz* (DDR) patented a series of campholenic molecules appended with the side chain of santatol or close derivatives. Two were commercialised as *Sandacore* by *Kao* and *Bacdanol* by *IFF*. Structure-odour explorations later gave birth to a whole lineage

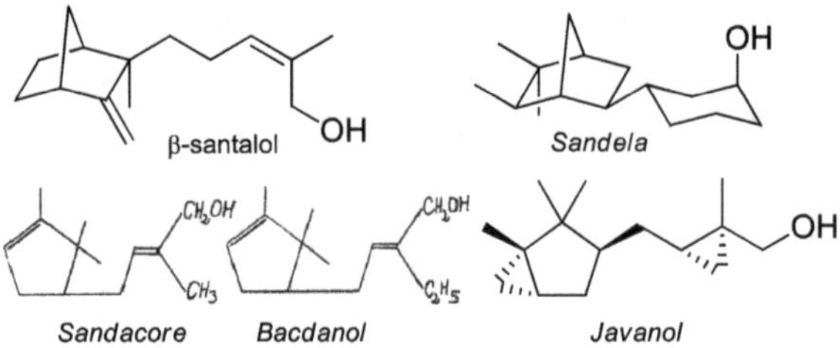

Figure 1.9. Natural β-santalol, the most potent isomer of *Sandela*, three examples of campolenic artificial sandalwood odorants, with the original patent drawings by Sandacore and Bacdanol.

of campholenic sandalwoods: *Sandalore* (*Givaudan*, 1976), *Brahmanol* (*Dragoco*, 1978), *Polysantol* (*Firmenich*, 1984), *Ebanol* (*Givaudan*, 1986) and *Firsantol* (*Firmenich*, 1994). Cyclopropanated derivatives of *Sandacore* brought very interesting compounds, marketed as *Javanol* (*Givaudan*, 1996) and *Pashminol* (*Givaudan*, 2006) (Figure 1.9).

Natural sandalwood experienced serious shortages, lifting its price dramatically; these artificial molecules were then used as back-up materials to convey the sandalwood impression while keeping perfume formulas within reasonable price limits. For instance, Jean-Paul Guerlain used large amounts of *Polysantol* to reinforce the already prodigious quantity of natural sandalwood employed in *Samsara*, 1989, both constituting 26% of the formula.

The next odorant substance to have been at the origin of a new range of perfume chemicals was discovered while studying ambergris in the mid-twentieth century.

1.10 Ambrox and Ambery Wood Odorants

Ambergris is one of the very few animal substances used in perfumery; considering its scarcity and exorbitant price, chemists were eager to identify the odorant molecules present in it [18]. Studies were initiated in 1934 by Ruzicka after *Firmenich* gifted his laboratory with ten kilograms of the highest quality of ambergris. The first relevant result arrived in 1946 with the identification of the main components, but none with a stunning perfuming power. However, the compounds identified recalled the molecular architectures found in labdanum and clary sage. Chemists

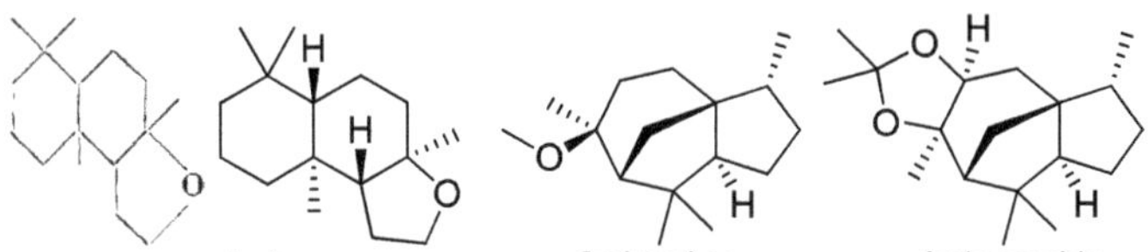

Figure 1.10. Stoll's depiction of Ambrox, Ambrox and two cedrene-derived ambery woods.

at *Firmenich* thus explored the field by effecting chemical modifications on sclareol, readily available by distillation of clary sage oil. Max Stoll prepared a tetrahydrofuran derivative of sclareol, which exerted an outstanding ambergris scenting power; this was patented in 1949 and sold as *Ambrox* [19]. More in-depth studies on ambergris in 1977 by Mookherjee revealed the occurrence of *Ambrox* in it, which could then be considered a natural substance. Other derivatives of sclareol were then discovered, such as *Amberketal* (*Firmenich*, 1953), *Superambrox* (*Firmenich*, 1992) and *Grisalva* (*IFF*, 2007).

More importantly, the discovery of *Ambrox* fostered systematic studies in R&D ventures of fragrance companies to convert cheap terpenic materials into powerful ambery odorants, some also exhibiting strong woody notes, giving birth to a major class of perfume chemicals, the ambery woods, or dry woods. For instance, explorations made on cedrene derivatives gave *Andrane* (*IFF*, 1962), *Cedramber* (*IFF*, 1966), or the more woody *Vertofix* (*Givaudan*, 1974). The most powerful representatives are the cedrene-derived ketals, *Ambrocenide* (*Dragoco*, 1997) and *Ambrostar* (Symrise, 2010) (Figure 1.10). Other terpenes were also good starting materials for discovering novel amber woods, like *Spirambrene* (*Roure*, 1986) from carene, *Bornafix* (*IFF*, 1989) from borneol or *Belambre* (*Givaudan*, 1995) from camphor; while variations on longifolene gave *Ysamber K* (*Dragoco*, 1991), *Symroxan* (*Dragoco*, 2000) and *Ambermax* (*Givaudan*, 2005).

We can now examine an interesting case of filiation in aroma chemistry, where several ambery woods could be invented from a totally unrelated family of odorants. As we will see in the following chapter on polycyclic aromatic musks, highly substituted indans are very valuable compounds in aroma chemistry. The first one was *Phantolid*, implying that the hydrocarbon core necessary for its production was synthesised on a large scale. John Hall at *IFF* had the idea to conduct the same type of exploration by chemical modifications on this compound, a fruitful

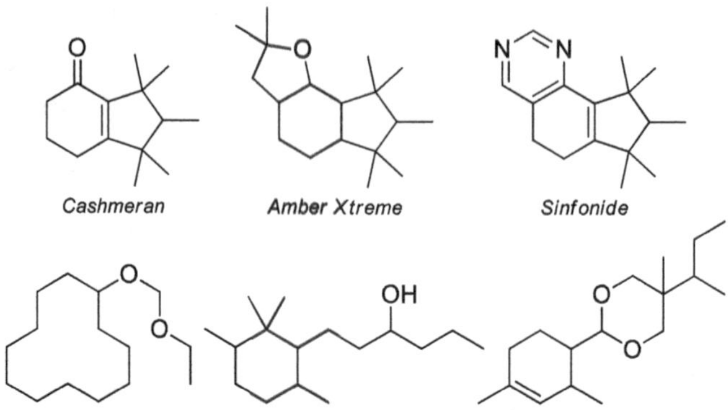

Figure 1.11. Indan-derived ambery woods, three examples of dry woods from known odorants.

initiative that brought him to discover *Cashmeran* in 1969 after a reduction/oxidation sequence. This very peculiar compound is fascinating for its blend of ambery, musky, powdery, mineral, vanillic and fruity characters. More importantly for our topic, Anubhav Narula and colleagues at *IFF* used *Cashmeran* again as a starting platform and could discover strong amber odorants, *Amber Xtreme* and *Trisamber* (2001), *Operanide* (2003), *Sinfonide* and *Ambertonic* (2012). Other isolated examples of ambery woods prepared from known perfumery chemicals are not rare: *Boisambrene* from a macrocyclic musk precursor (*Kao Co*, 1974), *Okoumal* from a polycyclic musk (*Givaudan*, 1989), *Timberol* from ionone (*Dragoco*, 1978), *Limbanol* from irone (*Firmenich*, 1983) and *Karanal* from *Triplal* (*Quest/Unilever*, 1987) (Figure 1.11).

This impressive list of compounds, all concentrated on a particular olfactory domain, stresses the importance it took for perfumers in modern creations and functional perfumery. All are characterised by a strong hydrophobicity and high molecular weights, assuring a good fixation on fabric and a very slow evaporation profile. Moreover, their extremely low detection thresholds imply that even minute quantities are efficient for scenting a composition, a feature that made them very popular in functional applications like washing powders and softeners.

While natural raw materials triggered the discovery of novel odours, fundamental research performed with the modern tools accessible to the chemist of the mid-twentieth century also brought very important new perfume chemicals.

1.11 Polycyclic Aromatic Musks

In the course of structural investigations on musk Ambrette in 1947 by Marion Scott Carpenter at *Givaudan–Delawanna*, it was realised that highly substituted benzenes appended with a simple carbonyl group could possess a very strong musk odour. This was a major breakthrough since only nitroaromatics and macrocyclic compounds were believed to have musky notes [20]. The compound discovered was however impossible to bring to industrial production. This discovery was made probably in parallel at *Polak's Frutal Works* by Kurt Fuchs, who was studying *Moskene*, a nitromusk from *Givaudan*'s portfolio. He could produce a powerful musk by simple acetylation of the hydrocarbon core of *Moskene*. This was thus the first polycyclic aromatic musk, patented in 1951 and released on the market as *Phantolid* the next year. The genuine structures of *Moskene* and *Phantolid* were revealed to be quite tedious to determine, an unexpected reaction occurring during the first step of the process.

The conundrum was resolved in 1955 after huge analytical efforts from *Givaudan* and *PFW*, and resulted in a deeper comprehension of the chemistry of indans and tetralins that helped in the development of additional polycyclic musks, all based on the same skeletons. Hence, *Versalide*, a tetralinic ketone, was patented in 1953 by Carpenter for *Givaudan*. Then, another acetylated tetralin is protected by *PFW* in 1954 and marketed as *Tonalid*. An original indanic ketone was discovered by Muus Beets at *Polak & Schwarz's Essencefabricken* (*P&SE*) in 1955 and sold as *Celestolide*. Beets proceeded with his investigations; with a very rational mind, he explored unconventional molecular structures guided by a structure–activity relationship theory that he forged on musky compounds. After a merger — *P&SE* becoming a part of *IFF* — Beets patented in 1962 for *IFF* a series of isochromane derivatives, from which one became the most successful musk in the whole history of musks, *Galaxolide*. The colossal commercial success of *Galaxolide* propelled the young *IFF* into the club of major companies in perfumery. Two other late members would join the polycyclic family, the indanic ketone, *Traseolide*, patented in 1977 by Traas and Boelens at *Naarden Int NV*, and the tetralinic aldehyde *Vulcanolide*, discovered by Fehr in 1989 at *Firmenich SA* (Figure 1.12). These polycyclic musks were almost ideal scenting materials, inexpensive, combining a powerful odorising capacity to an impressive level of persistence thanks to their high molecular weights. Like woody ambers, their high hydrophobicity grants optimal fixation on fabric, and they are

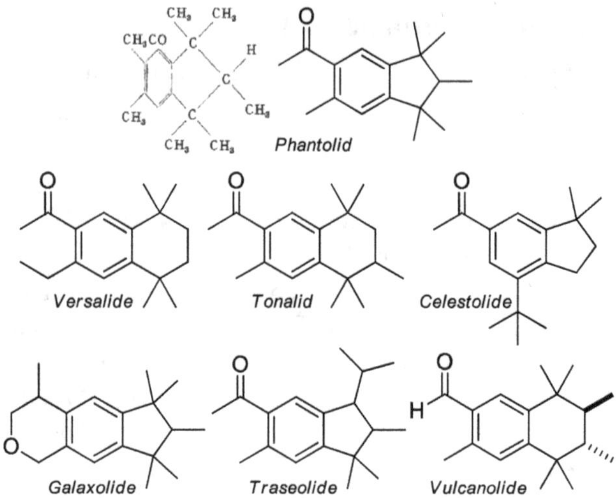

Figure 1.12. Carpenter's depiction of *Phantolid*, all commercially relevant polycyclic musks.

extraordinarily stable in all kinds of media, enabling scenting in any application, fine perfumery, washing powders, softeners, detergents, etc.

In fine perfumery, they enabled perfumers to construct musk complexes by mixing the different families, macrocyclic, polycyclic and nitro-musks, hence avoiding specific anosmia to one or the other type of musk, a very common phenomenon among the population. With other synthetic materials, they shaped the perfumery trends all along the second half of the twentieth century. Their use in soaps, washing powders and softeners actually defined for the general audience the "smell of clean" in the 1960s, which was later transposed in fine perfumery, like in *White Linen* by Sophia Grojsman (*Estée Lauder*, 1978). Their power was perfect to compose perfumes with memorable silage in fashion during the 1980s, like *Paris* (Sophia Grojsman, *Yves Saint Laurent*, 1983), *Obsession* (Jean Guichard, *Calvin Klein*, 1985), *Poison* (Edouard Fléchier, *Dior*, 1985) and *Fahrenheit* (Jean–Louis Sieuzac, *Dior*, 1988). This style culminated in the beginning of the 1990s with *Trésor* by Sophia Grojsman, with the massive use of a quartet of chemicals, methyl ionone, *Hedione*, *Iso E Super* and *Galaxolide*, 20% each, defining what the perfumer called a "hug-me accord" and leaving thus only 20% for the other finishing elements of the composition. But amusingly, the most overdosed perfume compounded with polycyclic musks is the *Kiehl's Original Musk* (1963), containing 92.8% of *Galaxolide*, only a year after its release on the market.

The 1950s were characterised by the formidable impetus towards progress following the end of WWII, R&D activities were sparkling, and numerous new fragrance chemicals flowed out of laboratories, enriching the perfumer's palette. From countless examples, we now examine only three, these rather simple non-natural molecules having opened vast new avenues in perfumery.

1.12 Modernity

The first example is dihydromyrcenol (DHM) discovered by Robert Webb at *The Glidden Co.* in 1956 and produced by the addition of formic acid on dihydromyrcene followed by saponification of the resulting ester. Its fragrance potential was yet recognised but the inefficient process precluded full industrial production. Ten years later, in 1966, Jack Blumenthal at *IFF* patented an improved process, allowing for the mass production of DHM, which could then start a successful career in functional perfumery, bringing fresh and lifting effects to soaps, detergents and cleaners by dints of its chemical stability in these applications. Being olfactorily linked to detergency, perfumers were shy to use it in fine fragrances, but they recognised its potential to freshen perfume top notes, and DHM could thus be found in *Azzaro Pour Homme* by Gerard Anthony, Martin Heiddenreich and Richard Wirtz at 2% (*Azzaro*, 1978). Having experienced the boosting effect of DHM on lavender, perfumers used it in increasing amounts following the huge commercial success of aromatic fougères colognes with male customers: 10% of DHM in *Drakkar Noir* by Pierre Wargnye (*Guy Laroche*, 1982), and then 20% in *Cool Water* by Pierre Bourdon (*Davidoff*, 1988).

The second material we will consider arose from the chemistry of flavours. Intrigued by the taste-enhancing effect of naturally occurring maltol, chemists at Pfizer synthesised derivatives of it and eventually patented ethyl maltol in 1963, which is six times more potent than maltol. It thus gained acclaim among flavourists, but perfumers were slow to exploit its potential for bringing notes of caramel and candyfloss. The first use was for *Vanilia* by Jean–François Laporte (*L'Artisan parfumeur*, 1978), but its full power burst with *Angel* by Olivier Cresp (*Thierry Mugler*, 1992), whose slow but great success inspired a tidal wave of sweet and caramelic perfumes at the turn of the twentieth and twenty-first centuries, culminating with *La Vie Est Belle* by Olivier Polge, Dominique Ropion and Anne Flipo (*Lancôme*, 2012).

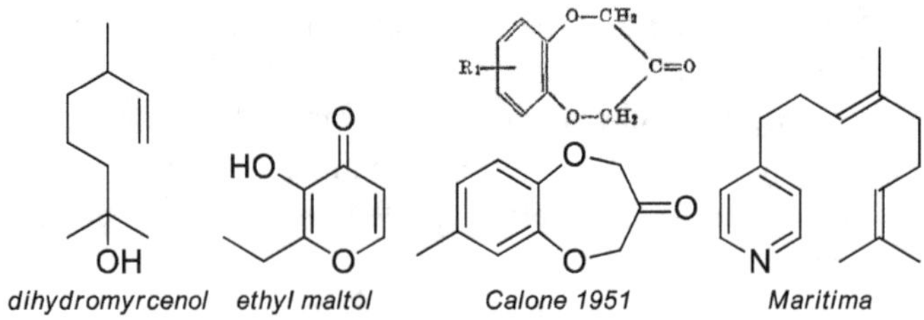

Figure 1.13. Structures of DHM and ethyl maltol, Pfizer's depiction of Calone 1951.

The third molecule to trigger a whole new family of perfumes was *Calone 1951*. While studying oxygenated derivatives of benzodiazepines like *Valium*, *Pfizer* chemists Beerenboom, Cameron and Stephens noted the watermelon-like odour of benzodioxazepines and patented them in 1966. The production was assigned to *Calmilli, Albert & Laloue* (CAL), a division of *Pfizer* specialised in fragrances, and marketed as *Calone 1951*.

This highly unusual marine note rebutted perfumers for 25 years, but Yves Tangy succeeded in taming the compound for composing *New West for Her* (*Aramis*, 1990), marking the birth of the large family of marine perfumes. *Kenzo Pour Homme* composed by Charles Mathieu (*Kenzo*, 1991) and *L'Eau d'Issey* by Jacques Cavallier (*Issey Miyake*, 1992) constituted very successful uses of it. The trend of aquatic and fresh perfumes fostered the development of other marine odorants by structural variation on benzodioxazepines like *Aldolone* (*Firmenich*, 1997) and *Azurone* (*Givaudan*, 2000), or from totally unrelated skeletons like *Maritima* (*IFF*, 1969), *Calypsone* (*Givaudan*, 2005) and *Symmarine* (*Symrise*, 2008) [21] (Figure 1.13).

By the middle of the twentieth century, all major components of the important natural materials used in perfumery were analysed, characterised and industrially reproduced, but advances in chromatographic techniques allowed revealing the presence of trace molecules that, however, have a great impact on the overall olfactory impression given by a flower.

1.13 Hedione

Hence, while 87% of the fragrant compounds of jasmine absolute were known in the 1950s, *Firmenich* founded a doctoral grant to analyse the remaining 13%. Edouard Demole thus started his PhD in 1955 in the laboratory of Edgar Lederer, at the time, the most renowned chemist in analytical science. In October 1957, from five kilos of jasmine absolute, he could uncover the presence of a novel and highly odorant compound present in less than 0.8%, which he named methyl jasmonate. In late 1958, he could synthesise a more accessible derivative, methyl dihydrojasmonate (DHJ), and later, in 1959, methyl jasmonate itself. The unique and highly potent odours of both compounds prompted Roger Firmenich to patent them in 1960. However, only methyl dihydrojasmonate would be brought to the industrial scale, being the most easily preparable, and thus production of the order of multi-kilograms started in 1961. Methyl dihydrojasmonate was synthesised as a racemic mixture of both *cis* and *trans* isomers and used only to manufacture a base called *Jasmin 74D* [22].

The famous perfumer Edmond Roudnitska asked for the pure material instead and found it marvellous; methyl dihydrojasmonate was then sold unmixed as *Hedione* and had its first use in *Eau Sauvage* by him with around 2% (*Dior*, 1966). Analyses performed by the competitor *Roure-Bertrand* revealed that the *cis* isomer of *Hedione* is 70 times more potent than the *trans* but is only present at the level of 10% in the commercial product. A race to the highest content of *cis-Hedione* in the produced mixture thus started. Eventually, *Firmenich* kept the advantage over its competitors (*Zeon*, *IFF*, *Takasago*) and released *Hedione HC* (for high-*cis*) in 1994. During this competition, *Firmenich* realised that one enantiomer of *cis–Hedione* was responsible for the "hedionic" smell, the antipode being very weak, and thus developed academic collaborations with Prof. J.-P. Genêt in Paris to produce enantio- and diastereomerically enriched (+)-*cis*-methyl dihydrojasmonate. This was achieved through an asymmetric hydrogenation process in 1996, the precious material hence obtained taking the name of *Paradisone* (Figure 1.14).

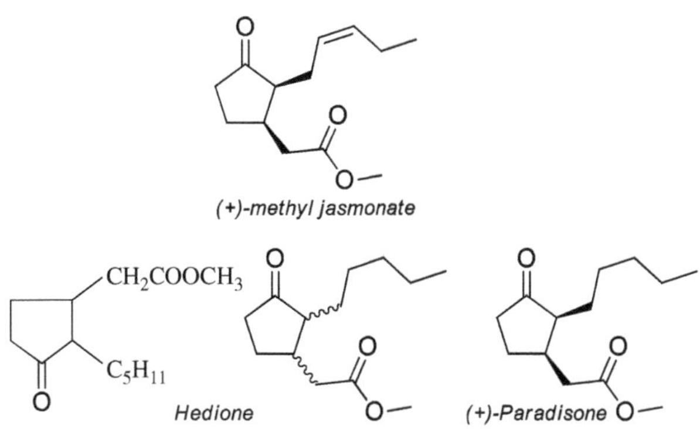

Figure 1.14. Natural methyl jasmonate, Demole's depiction of *Hedione*, *Hedione* and *Paradisone*.

Following the huge success of *Eau Sauvage*, *Hedione* gained a unique place within the perfumer's palette, bringing aeration, light, volume and transparency to perfumes and giving such a joyful touch to floral bouquets that there is hardly a perfume in which it is not present nowadays. Apart from being a precious nuancer, large dosages could bring novel effects to jasminic modules of modern perfumes; thus, Henri Robert used 13% of *Hedione* in *Chanel N°19* (1971), *First* by Jean-Claude Ellena contains 20% (*Van Cleef & Arpels*, 1976), *Samsara* by Jean-Paul Guerlain, over 16% (*Guerlain*, 1990) and 26% in *Cristalle* again by Henri Robert (*Chanel*, 1993). The unique luminous and joyous appeal of *Pleasures* by Alberto Morillas (*Estée Lauder*, 1995) would be inconceivable without the special quality of *Hedione HC*; this holds true for *Perles de Lalique* by Nathalie Lorson (*Lalique*, 2006) with *Paradisone*.

1.14 The Smell of Rose

The same type of investigation conducted on jasmine was performed on rose oil at *Firmenich*. Hence, in 1959, Casimir Seidel and Max Stoll could isolate an unknown oxide of formula $C_{10}H_{18}O$ by careful treatment and fractional distillation of 9 kg of Bulgarian rose oil. Two additional years and the collaboration of Albert Eschenmoser were necessary to ascertain the correct formula of rose oxide possessing a "strong and peculiar" smell, the first step towards the industrial synthesis of this trace element that, however, brings a crucial sharp and almost metallic touch to the smell of rose. Hence, in 1964, *Firmenich* could patent the preparation of synthetic rose oxide, which could then start its career in the perfumer's organ to compose more realistic rose perfumes.

But rose still kept secret some other hidden olfactory jewels. Hence, we find again Edouard Demole and the team of Firmenich chemists now collaborating with Ervin Kováts, a specialist in chromatography at Ecole Polytechnique Fédérale, Lausanne, working to isolate 1.2 mg of a novel enone present at 0.05% in rose oil but exerting a profound olfactory impact on it. This discovery was published in 1970, where the structure damascenone is described, with its synthesis; additionally, the reduced derivative α-damascone was synthesised, which was later identified in natural rose [23] (Figure 1.15).

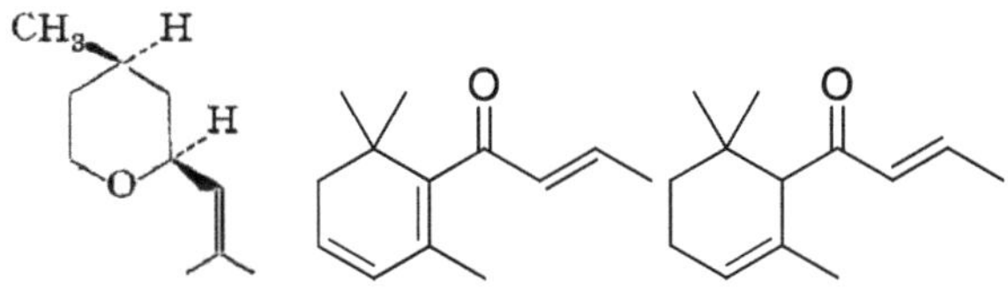

Figure 1.15. Stoll's depiction of rose oxide, Demole's depiction of damascenone, an example of a damascone isomer.

Damascones and their apple-like, fruity, marmalade smells were very important to bring carnal qualities to floral-rosy perfumes, like in *Nahéma* by Jean–Paul Guerlain (*Guerlain*, 1979), to enhance the fruity aspects of the formula already highly rich in natural rose, or like in *Poison*, where Edouard Fléchier combined the damascones with lactones to boost the tuberose theme up to an explosive level (*Dior*, 1985).

The 1970s brought a very special material that will conquer a prominent place in the perfumer's portfolio in much the same way as methyl ionones, musks and *Hedione*, so much so that almost every modern perfume will incorporate at least a small amount and frequently a generous dosage. The discovery of *Iso E Super* is a tortuous story, as we will examine now.

1.15 Iso E Super

The very first studies were conducted by Gunther Ohloff at *Dragoco* around 1956, as he was studying various decalines possessing floral odours and obtained through a Diels–Alder reaction between various α,β-unsaturated ketones with myrcene, followed by acid-catalysed cyclisation. Several cyclemones were patented for *Dragoco* in 1956 and other positional isomers in 1959. Yet another isomer was patented by Garry Kitchens for *Givaudan* in 1959, all exhibiting woody-ambery odours. Then, in 1968, Gunther Ohloff, then working at *Firmenich*, published a chapter on odorants, in which he showed for the first time the formula of what would become *Iso E* and that he described as woody, with notes of olibanum and ambergris, but surprisingly without a related patent on this particular compound.

In fact, John Hall and James Milton Sanders patented the very same compound in 1973 but for *IFF*! The material was then sold as *Isocyclemone*

Ethanone (Iso E) and used for the first time in *Halston Woman* by Bernard Chant (*Halston*, 1975). Later, improved production gave *Iso E Super* with stronger olfactory power. This was rationalised in the 1980s when *Givaudan*'s analysts Neuner–Jehle and Etzweiler performed olfactometry analysis on *Iso E Super* constituents. The material is composed of more than 10 different isomers, but they realised that the major component has a very weak odour, uncharacteristic of *Iso E Super*. Only a minor compound is responsible for the typical smell, with a huge odorant power, since it is present around 5% in the mixture. Modifications in the process could increase this level to nearly 10% in special cases. In 1990, *Givaudan*'s chemists stereoselectively synthesised this interesting isomer and called it *Iso E Super Plus*. Finally, Elias Corey reported in 2005 the enantioselective preparation of both chiral forms of *Iso E Super Plus*, showing that the dextrorotatory antipode was extraordinarily powerful and christened (+)-*arborone*, while the enantiomer was very weak. Sadly, none of these syntheses could be transposed to an industrial scale, while a different process developed in 1995 and starting from homomyrcene could be employed to produce *Georgyone*, a double-bond isomer of arborone that was marketed as *Georgywood* by *Givaudan* (Figure 1.16).

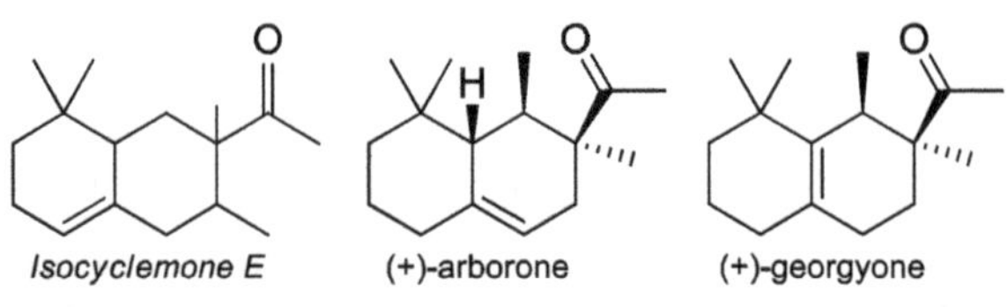

Figure 1.16. Hall's depiction of Isocyclemone E, the main component of *Iso E*, structures of arborone and georgyone, the most odorants isomers of *Iso E Super* and *Georgywood*.

Extremely rapidly, *Iso E Super* became a staple in perfumery, helping the creator in structuring a perfume, like a sturdy architecture on which other notes are placed. It also serves as a nodal point linking the floral notes of ionones with the woody materials like cedar wood, as well as with the ambery notes. Therefore, very significant dosages were progressively used: 25% in *Fahrenheit* (Jean–Louis Sieuzac, *Dior*, 1988), 43% in *Féminité du Bois* (Pierre Bourdon/Christopher Sheldrake, *Shiseido–Serge Lutens*, 1992), 55% in *Terre d'Hermès* (Jean–Claude Ellena, *Hermès*, 2006) and 80% in *Perles de Lalique* (Nathalie Lorson, *Lalique*, 2006). The conceptual brand *Escentric Molecules* is even proposing a creation called *Molecule 01* solely constituted of diluted *Iso E Super*. Like *Hedione*, *Iso E Super* is so well installed in the composing habits of perfumers to give structure, substantivity and diffusivity that nearly all formulas contain these two ingredients.

1.16 Alicyclic Musks

We have seen in the previous paragraphs that nitromusks were the first artificial materials to be employed in fine perfumes, macrocyclic and polycyclic aromatic musks completed the palette; however, the role of musky notes in perfume composition is so important that novel types of musks are highly sought after. Aroma chemists had observed the notable musky facets of certain molecules, like citronellyl ethyl oxalate, or *Rosamusk*, totally unrelated to the known musk families, being neither macrocyclic nor aromatic. Investigations conducted by Karl von Fraunberg and Werner Hoffmann at BASF led to the discovery of *Cyclomusk* in 1975, without subsequent commercialisation, but proving that an alicyclic molecule could possess a strong musk scent. Only 15 years later, in 1990, a compound sharing some similarities with Cyclomusk was patented by Wolfgang Giersch, Karl-Heinrich Schulte-Elte and Cyril Mahaim for *Firmenich* and marketed as *Helvetolide*.

Commercialising a novel type of musk is obviously a huge event in the market of aroma chemistry, and this release triggered vigorous investigation efforts to discover and patent additional members in this musk family. Hence, a diester version was patented in 1998 by *Firmenich* and sold as *Romandolide*, *Givaudan* then patented *Serenolide* in 2001 and *Sylkolide* in 2002, both appended with cyclopropanic esters, while *IFF* protected *Applelide*, now *Edenolide* in 2002 (Figure 1.17).

These novel musks were more important than just additional representatives in the perfumer's organ. Their intrinsic qualities with fruity facets, quite apart from the other families of musks, allowed for the creation of unique blends of musks, bringing new effects in perfumes. Hence, Alberto Morillas could create an unprecedented "cotton and linen" effect by mixing *Helvetolide* with *Habanolide* for *Emporio Armani White for Her* in 2001, which would be reproduced in countless later perfumes and presented as the "white musk" accord.

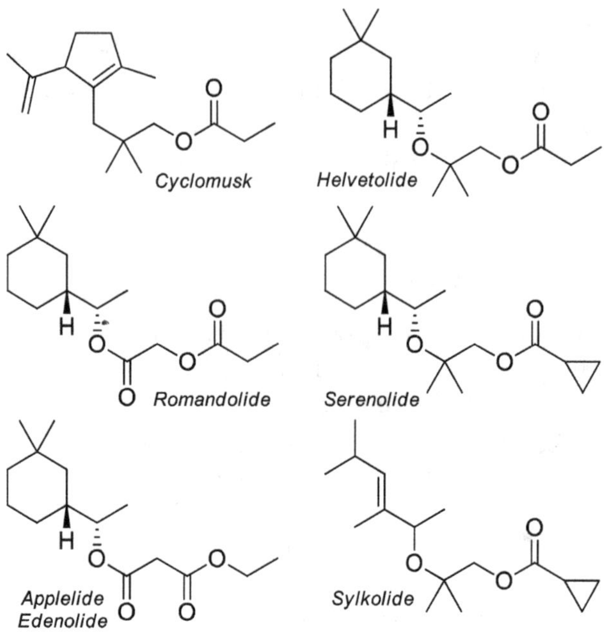

Figure 1.17. Alicyclic musks.

1.17 Present Challenges in Aromachemicals Discovery

Health and environmental regulations are constantly driving fragrance materials to the side; moreover, in a much more extended way than fashion, perfumery also experiences trends with an endless taste for novelty, all of which are fostering a perpetual quest for new odorants. Each year, each major actor in aroma chemistry releases a handful of new molecules. Some have a short lifespan; others are fully adopted by perfumers and eventually can shape the history of perfumery in successful creations. Contemplating the last innovations in the field during the past decade, we can realise that major ingredients have not come out of the chemist's flasks but from fermenters in biotechnology labs [24].

This is exemplified by patchouli-derived ingredients, with usually discarded distillation fractions that are turned into highly valuable materials by bioconversion such as *Clearwood* (*Firmenich*) and *Akigalawood* (*Givaudan*). Chemists still have their role in the R&D process, but limited to the discovery phase, finding novel molecular structures from cheap starting materials. As special emphasis is put on the environmental and health benign ingredients, only biodegradable and non-sensitising candidates are considered nowadays; then, when possible, the production is performed using micro-organisms or enzymes rather than process chemists.

For more prospective considerations, what are the challenges ahead for aroma chemists? As illustrated by the recent launches from all companies, lily-of-the-valley molecules are still an active field, and there never will be too many muguet odorants! Then, if we consider the domain of vetiver odorants, the portfolio is surprisingly narrow; a few artificial compounds are bringing the grapefruit facets of it, but the specific woody character of vetiver is still solely found in the natural material. Other very complex natural ingredients are also waiting for synthetic alternatives, like for costus or oakmoss beyond Evernyl. Another longstanding challenge lies in the modulation of evaporation characteristics without affecting much the olfactory impressions. For instance, can we make citrusy molecules of high molecular weight for retarded evaporation? Floral materials operating in the fond of a perfume? Woody ingredients perceptible in the top notes? Future patents will say.

References

[1] (a) G. Ohloff, W. Pickenhagen, P. Kraft, F. Grau. (2022). *Scent and Chemistry — The Molecular World of Odors, 2nd Edition* (Wiley-VCH, Verlag Helvetica Chimica Acta, Zürich). (b) W. Pickenhagen. (2017). History of Odors and Odorants, in A. Buettner (ed.) *Springer Handbook of Odors* (Springer, Dordrecht), pp. 1–8. (c) E.T. Theimer. (1982). *Fragrance Chemistry: The Science of the Sense of Smell* (Academic Press, New York).

[2] (a) G. Ohloff. (1992). 75 Jahre Riechstoff- und Aroma-Chemie im Spiegel der Helvetica Chimica Acta Teil I, *Helv. Chim. Acta*, 75, 1341–1415. (b) G. Ohloff. (1992). 75 Jahre Riechstoff- und Aroma-Chemie im Spiegel der Helvetica Chimica Acta Teil II, *Helv. Chim. Acta*, 75, 2041–2108.

[3] (a) G. Fritter. (1998). Fragrance Chemistry, *Tetrahedron*, 54, 7633–7703. (b) M. Gautschi. (2001). *Fragrance Chemistry — Milestones and Perspectives*, *Chimia*, 55, 379–387.

[4] P. Kraft, J.A. Bajgrowicz, C. Denis, G. Fráter. (2000). Odds and trends: Recent developments in the chemistry of odorants, *Angew. Chem. Int. Ed.*, 39, 2980–3010.

[5] N. Armanino, J. Charpentier, F. Flachsmann, A. Goeke, M. Liniger, P. Kraft. (2020). What's hot, what's not: The trends of the past 20 years in the chemistry of odorants, *Angew. Chem. Int. Ed.*, 59, 16310–16340.
[6] E. Briot. (2015). *La Fabrique des parfums Naissance d'une industrie de luxe* (Collection Chroniques, Vendémiaire).
[7] P. de Nicolaï. (2008). A smelling trip into the past: The influence of synthetic materials on the history of perfumery, *Chem. Biodiversity,* 5, 1137–1146.
[8] E. Charabot. (1900). *Les Parfums artificiels. Encyclopédie Industrielle* (Libraire J.-B. Baillière et Fils, Paris).
[9] F. Landolph. (1875). *Comptes Rendus Hebdo Acad Sciences*, 81, 97–99.
[10] O.R.P. David. (2017). Artificial nitro-musks, stories of chemists & businessmen, *Eur. J. Org. Chem.*, 1, 4–13.
[11] O.N. Witt. (1901). Ferdinand Tiemann. Ein Lebensbild, *Ber. Dtsch. Chem. Ges.*, 34, 4403–4455.
[12] See Ref. 2b, pp. 2060–2062.
[13] See Ref. 2b, pp. 2062–2069.
[14] P. Kraft, C. Ledard, P. Goutell. (2007). From Rallet N°1 to Chanel N°5 versus Mademoiselle Chanel N°1, *Perfumer & Flavorist*, October, 36–48.
[15] A. Goeke, P. Kraft, D. Lelievre, A.E. Alchenberger. (2018). Discovery of nympheal: The definitive muguet aldehyde, *Perfumer & Flavorist*, March, 24–41.
[16] A.S. Williams. (1999). The synthesis of macrocyclic musks, *Synthesis,* 10, 1707–1723.
[17] See Ref. 2b, pp. 2091–2093.
[18] E. Eichhorn, F. Schroeder. (2023). From ambergris to (−)-ambrox: Chemistry meets biocatalysis for sustainable (−)-ambrox production, *J. Agric. Food Chem.*, 71, 5042–5052.
[19] See Ref. 2b, pp. 2054–2060.
[20] O.R.P. David. (2020). A chemical history of polycyclic musks, *Chem. Eur. J.*, 26, 7537–7555.
[21] See Ref. 4, pp. 16316–16318.
[22] C. Chapuis. (2012). The jubilee of methyl jasmonate and hedione, *Helv. Chim. Acta,* 95, 1479–1685.
[23] See Ref. 2a, pp. 1357–1358.
[24] M. Lecourt, S. Antoniotti. (2020). Chemistry, sustainability and naturality of perfumery biotech ingredients, *ChemSusChem,* 13, 5600–5610.

https://doi.org/10.1142/9781800617780_0002

Chapter 2

Sustainable Extraction of Perfumery Plants

Jean-Baptiste Chagnoleau[*,†,‡], **João A. P. Coutinho**[†,§], **Xavier Fernandez**[*,¶] **and Nicolas Papaiconomou**[*,||]

[*]*Université Côte d'Azur, CNRS, Institut de Chimie de Nice, Nice, France*

[†]*Universidade de Aveiro, CICECO, Campus Universitário de Santiago, Aveiro, Portugal*

[‡]*jean-baptiste.chagnoleau@univ-cotedazur.fr*
[§]*jcoutinho@ua.pt*
[¶]*xavier.fernandez@univ-cotedazur.fr*
[||]*nicolas.papaiconomou@univ-cotedazur.fr*

Abstract

In ancient times, perfumes had hedonic, medicinal and mystical purposes. They were produced by extracting fragrances from natural raw materials. Traditional fragrance extraction techniques such as solvent extraction or hydrodistillation/steam distillation remain in use today. Processes improved throughout the centuries, particularly during the 19th century, and are now well-established industrially. With the development of chemistry that began during the Industrial Revolution, synthetic fragrances are nowadays also used in perfumes. The emergence of a new generation of environmentally concerned consumers has revived the importance of natural extracts, while green chemistry and eco-extraction principles have prompted researchers to revisit conventional extraction processes due to their large energy consumption or use of petrochemical organic solvents. As a response and in order to improve sustainability in the perfume industry, alternative processes have been investigated. Specifically, sustainable solvents and innovative activation techniques – meant to replace petroleum-based solvents and increase extraction and energetic yields, respectively – have been developed in the last decades.

In this chapter, alternative solvents, that is, solvents produced from bioresources (BioSols), deep eutectic solvents (DESs) or ionic liquids (ILs) are introduced. Activation techniques, namely microwave-assisted extraction (MAE), ultrasound-assisted extraction (UAE), pressurised fluid extraction (PFE), enzyme-assisted extraction (EAE) and other intensification techniques, are then presented. Advantages and drawbacks of each technique and solvent are discussed, with respect to their applications in the field of flavours and fragrances. This chapter concludes with a case study dealing with the extraction of fragrances from *Rosa centifolia* using BioSols. This chapter not only describes the state of development of eco-extraction techniques in the fields of flavours and perfumes, as well as the opportunities they offer, but also discusses limitations for their industrialisation.

2.1 Introduction

Since ancient times, combinations of natural ingredients, mainly of plant origin, have been used in order to produce perfumes. Traces of scented products were found in almost all ancient civilisations, including the Egyptian, Persian, Greek, Arab and Roman civilisations [1]. One key step for producing natural perfume ingredients is extraction from plant raw materials. Traditional extraction techniques, such as cold pressing, solvent extraction and hydrodistillation, and more modern techniques, such as supercritical CO_2 extraction, are, as detailed in Chapter 1, currently used in the perfume industry. Some techniques present drawbacks. For example, hydrodistillation results in a large energy consumption due to the high distillation temperature (100°C) and long processing times. Solvent extraction requires large amounts of solvents, generally produced from fossil resources, leading to residual traces of solvents remaining in the extract.

Sustainable-minded consumers became aware of the environmental drawbacks of chemical processes in general and in conventional extraction techniques, which led to distrust of ingredients with safety and ecological issues. Along with concepts of green chemistry and eco-extraction emerging, this led to the research and development of alternative and sustainable extraction techniques. The development of such alternative techniques aimed at proposing eco-conceived new fragrant ingredients is mainly studied along two lines: (i) using alternative solvents that are safer and sustainable and produced from renewable biomass together and (ii) developing innovative extraction intensification techniques.

In this chapter, concepts and principles of green chemistry and eco-extraction are first introduced. Then, innovative sustainable extraction techniques are detailed. Finally, a case study taken from one of our works dealing with the extraction of volatile compounds from *Rosa centifolia* using biobased solvents is discussed.

2.2 Green Chemistry and Eco-Extraction

The amount of chemical products created, distributed and released in the environment has reached 220 billion tonnes in 2017, accounting for 20% of greenhouse emissions [2]. This amount of released compounds, along with their extremely large diversity in structure and properties, is nowadays responsible for serious environmental and health issues [3]. In the 1990s, as an answer to this ever-growing environmental footprint of chemistry, a new concept of so-called "green chemistry" emerged [4]. It was proposed together with 12 principles, paving the way for the development of alternative, sustainable and environment-friendly chemistry processes. Later on, this concept was extended to plant extraction, with the creation of eco-extraction principles [5].

In the fragrance industry, a shift towards more sustainable processes started in the early 1980s when regulatory entities restricted, and even prohibited, petroleum-based solvents in plant extraction processes for health, safety and ecological reasons [6].

2.2.1 *Green chemistry*

Green chemistry is a general concept that was first defined by Anastas and Williamson in 1996 [4]. As stated in their *Green Principles* milestone book, "Green chemistry is a set of techniques and methodology to reduce or eliminate the use or production of raw materials, products, co-products, solvents, reagents ... that are harmful to human health and the environment" [4]. This definition was further detailed in 1998 by Anastas and Warner with the introduction of the 12 principles of green chemistry, as

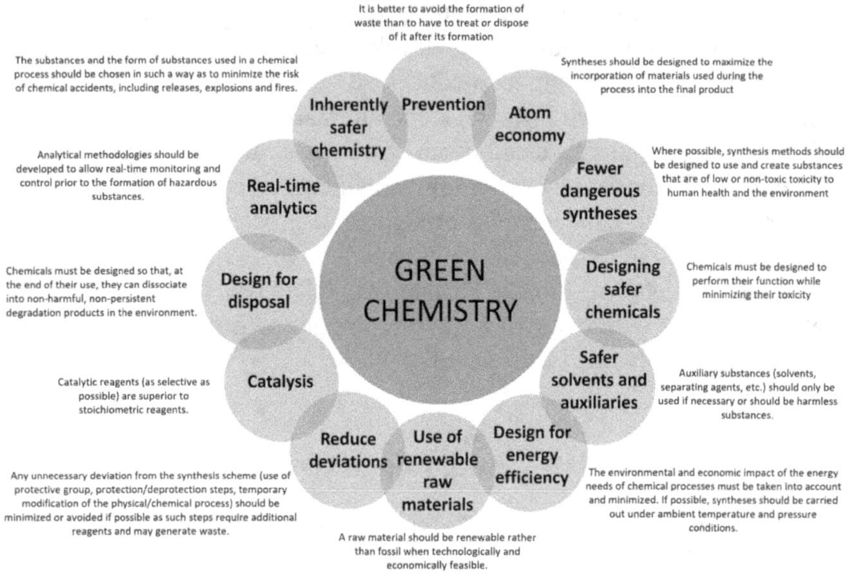

Figure 2.1. The 12 principles of green chemistry.
Source: Adapted from [7].

detailed in Figure 2.1 [7], acting as a general framework for the development of new sustainable chemical processes, taking into account all aspects of a process life cycle.

Green chemistry, or sustainable chemistry, emphasises the reduction of chemistry's environmental impact through the optimisation of the energetic consumption of processes, recycling of both raw materials and by-products created, waste reduction and lowering of health and environmental impacts of such processes, as well as reducing or eliminating the use and generation of hazardous substances [8].

2.2.2 *Eco-extraction*

Soon after the proposition of the new concept of green chemistry, the latter has been applied to a wide range of industrial sectors. Obviously, depending on the chemistry field it is supposed to be applied to, such as plant extraction, some of the 12 principles of green chemistry are not adequate or out of scope. For example, the ninth principle of green chemistry, dealing with catalytic reagents, has been proposed for synthetic chemistry processes [7].

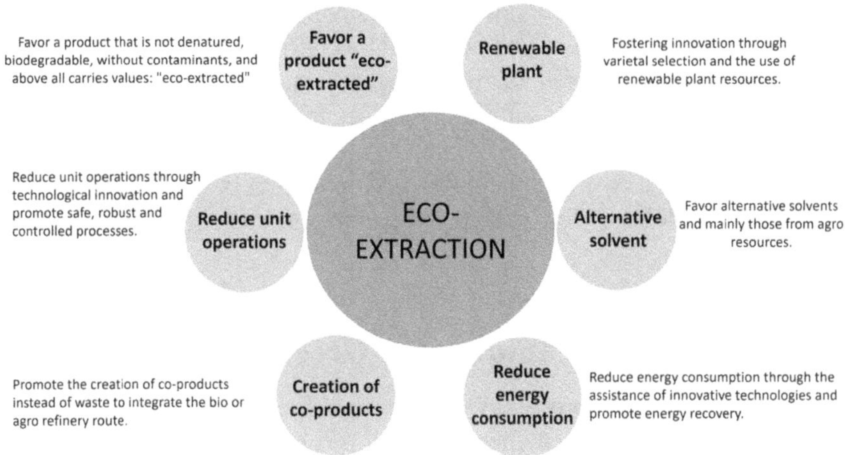

Figure 2.2. The six principles of eco-extraction.
Source: Adapted from [9].

As an extension of green chemistry to plant extraction, the concept and principles of eco-extraction were defined by a group of researchers and professionals who gathered in Grasse in January 2010. A very general definition was adopted: "Eco-extraction is based on the discovery and design of extraction processes in order to reduce energy consumption, but also the use of alternative solvents and renewable plant resources, while guaranteeing a safe and quality product/extract". Six principles of eco-extraction were then proposed (Figure 2.2). These principles have not been identified as rules but rather as examples of innovation to follow [5, 9].

They are listed here as follows:

(1) Encourage a reasonable use of plant renewable resources without over-exploitation to prevent plant extinction because of overutilisation.
(2) Ask researchers and industries to favour the use of solvents produced from renewable sources rather than petroleum-based ones.
(3) Reduce energy consumption, for example by recovering heat released during solvent vapour condensation or by using activation methods such as ultrasound.
(4) Promote the creation of by-products instead of waste during the extraction process in order to integrate further on a bio- or agro-refinery route.
(5) Encourage the use of more robust and controlled processes, considering more compact production units and a reduced number of unit

operations, energy and raw material savings, process safety control and reduction in waste and ecological footprint.

(6) Favour the production of non-denatured and biodegradable extract without contaminants, such as pesticides [5].

The principles of green chemistry and eco-extraction nowadays serve as strong guidelines for the development of sustainable extraction processes.

2.3 Innovative Techniques for Green Extraction

Alternative innovative processes were progressively developed in order to satisfy the criteria of green chemistry and eco-extraction and, in particular, achieve higher efficiency, reduce solvent consumption and lower environmental and health impacts. Applied to traditional raw materials, such innovative processes also represent opportunities to create new products, with their own fragrances and qualities [10].

Several paths of innovation can be considered to improve the sustainability of extraction processes. For instance, the fifth and seventh principles of green chemistry, together with the second principle of eco-extraction, encourage the use of safer solvents produced from renewable resources. In light of the sixth principle of green chemistry for energy efficiency and the third principle of eco-extraction for reduction of energy consumption, new extraction techniques, called intensification techniques, were developed to improve extraction processes.

2.3.1 *New solvents*

The efficiency of solid-liquid extraction is mainly conditioned by the nature of the solvent used: it should be selective, inert regarding the compounds to be extracted and volatile enough to be easily and safely removed at the end of the process [10].

As stated in Chapter 1, volatile organic solvents, in particular hexane and ethanol, are very often used in perfumery plant extraction processes. The use of hexane and, more generally, of non-polar petroleum-based solvents is in contradiction with the seventh principle of green chemistry and the second principle of eco-extraction, as they are produced from fossil resources. Therefore, alternative solvents produced from biomass resources have been developed. Three kinds of alternative solvents would be introduced in this chapter: biobased solvents (BioSols), deep eutectic solvents (DESs) and ionic liquids (ILs). The first kind of alternative solvents, BioSols, are solvents produced from biomass resources, for example through a fermentation process in the case of ethanol and 1-butanol or chemical transformation of biomass derivatives in the case of 2-methyltetrahydrofuran (MeTHF) or cyclopentyl methylether (CPME). DESs are a second family of alternative solvents and are a mixture of at least two compounds exhibiting a significant decrease in the melting point of the mixture compared to the pure substances, which makes their use as an extraction solvent possible. The last family of alternative solvents considered is the one of ILs. ILs are salts composed of an organic cation and an organic or inorganic anion and exhibiting a low melting point, typically below 100°C.

On the one hand, BioSols, DESs and ILs can be considered more sustainable than organic solvents because their production from renewable bioresources complies with the seventh principle of green chemistry and the second principle of eco-extraction. But on the other hand, there are several other criteria that a solvent must meet to be classified as "green" or sustainable. There is no clear definition of a "green" solvent, but several methods were studied previously in order to evaluate the sustainability of a solvent. In 2013, Gu and Jérôme proposed 12 criteria to be met by a solvent to be qualified as "green" [11]. Several environmental, health and safety (EHS) assessments of solvents were previously proposed by pharmaceutical companies Pfizer [12], Sanofi [13] and GlaxoSmithKline (GSK) [14]. A guide, proposed by GSK, is nowadays one of the more important and complete methods for the EHS assessment of a solvent adopting a life-cycle approach [15].

2.3.1.1 *Biobased solvents*

As an alternative to solvents produced from petroleum resources, several BioSols were developed, including ethanol, 1-butanol, MeTHF, CPME, cyrene and γ-valerolactone [16]. However, the constraints of extracting odorous compounds reduce the list of BioSols used in the perfumery field. Solvents generally useful for perfumery are apolar and volatile to be removed after plant extraction [10].

Table 2.1. Names, structure, production method and hydrophobicity of some biobased solvents.

Name	Structure	Octanol–water partition LogP	Production method
Ethanol	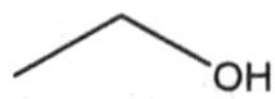	−0.31 [17]	Sugar fermentation
1-Butanol		0.88 [17]	Sugar fermentation
2-Methyltetrahydrofuran (MeTHF)	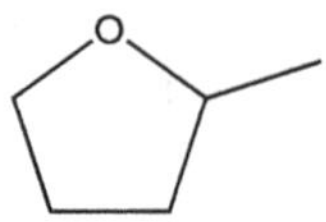	1.01 [18]	Hydrogenation of furfural from lignocellulosic biomass
Cyclopentyl methyl ether (CPME)	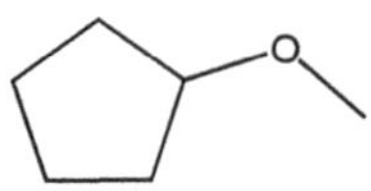	1.65 [19]	Furfural rearrangement

In this section, selected BioSols used in perfumery will be introduced, namely ethanol, 1-butanol, MeTHF and CPME. These solvents are moderately polar to non-polar solvents and exhibit relatively low boiling points, allowing them to be removed by evaporation after extraction. Names, structure, production method and hydrophobicity in terms of octanol–water partition coefficient for these solvents are presented in Table 2.1.

Ethanol

Ethanol is one of the most popular solvents that could be produced from biomass. In 2022, the amount of ethanol produced worldwide reached 107 billion litres [20]. In addition, ethanol can be used to produce apolar solvents such as benzene, toluene and xylene by acid-catalysed oligomerisation [21].

In one process, ethanol is prepared through the acid-catalysed hydration of ethylene, a petrochemical compound [22]. Alternatively, ethanol is produced through biomass fermentation. The resulting ethanol is often called bioethanol. Feedstocks used for this fermentation are either rich in saccharides, such as sugarcane, sugar beets, corn, molasses and potatoes, or rich in starch, such as wheat, barley, oat and rice. Ethanol can also be produced from lignocellulosic materials such as agricultural wastes and woody materials [23]. Pretreatment of biomass prior to fermentation is

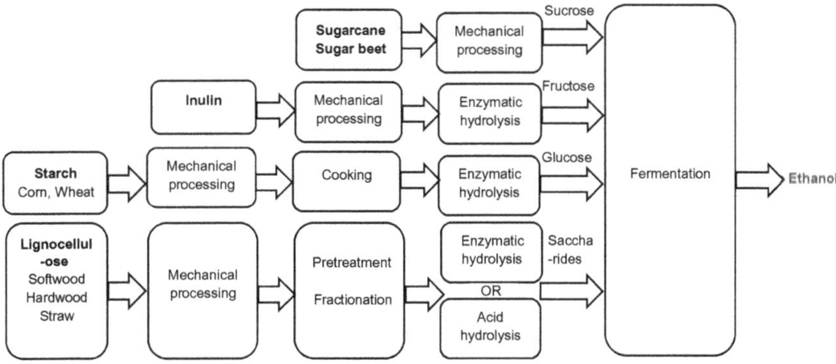

Figure 2.3. Different processes for ethanol according to the biomass used.
Source: Adapted from [23].

crucial and differs from one biomass to the other. Such processes are presented in Figure 2.3.

Production of bioethanol using microorganisms requires fermentable sugars, typically six-carbon sugars, which can be converted into ethanol. Therefore, biomass containing high levels of glucose such as sugarcane or sugar beet are most interesting for ethanol conversion. Sugarcane represents the most widely used resource for bioethanol production. However, using such biomass for ethanol production is generally expensive and competes with the human food chain [24]. It was estimated that the agricultural sources of crops used in order to produce bioethanol could feed up to 200 million people [25].

Starch is another biomass commonly used for ethanol production. Corn, wheat, potato, sweet potato and cassava are starchy materials commonly used around the world for ethanol production. Starch is made up of long chains of glucose molecules, which must be broken before fermentation. For that reason, starch is commonly mixed with water, stirred and heated, followed by the addition of amylase enzymes in order to hydrolyse starch into glucose units [24]. However, as with glucose-rich biomass, such production of ethanol from starch competes with the human food chain. In 2007, 375,000 square kilometres of corn were planted in the US, and one-third of it was used to produce ethanol. This led to increases in corn price of up to 73% at the end of 2010 [26].

Alternatively, lignocellulosic materials, obtained from agricultural and forestry by-products, can be used as an abundant source of carbohydrates.

Lignocellulose is composed of cellulose and hemicellulose, which are polymers of saccharides, and of lignin, a hydrophobic polymer that is highly resistant to chemical or biological degradation. Before fermentation, lignocellulose must be hydrolysed into fermentable saccharides. That is achieved mostly using two methods: acid or enzymatic hydrolysis [23].

After the pretreatment of biomass, baker's yeast, or *S. cerevisiae*, is the most widely used organism for industrial production of ethanol. *S. cerevisiae* presents many advantages: it grows both under aerobic and anaerobic conditions and tolerates a wide range of pH, temperatures up to around 40°C and high concentrations of sugar and ethanol [23].

In perfumery, ethanol has been largely used for a long time for the purification of concretes, leading to so-called absolutes [10]. Ethanol is also used to extract dry raw materials such as balsams, oleo gum resins, oleo resins and resinous products to obtain resinoids. For example, ethanol is used to produce ginger oleoresin, myrrh resinoid, olibanum resinoid or vanilla oleoresin [27]. Ethanol has recently been used in order to extract fresh perfumery plants, such as jasmine (*Jasminum grandiflorum, Jasminum sambac*) or orange blossom (*Citrus aurantium* L. var.amara flowers). But ethanol generally extracts, together with volatile fragrance compounds, some undesirable compounds such as water and sugars that must be removed to obtain a "clearer" fragrance extract. Ethanolic extracts must therefore often be refined. In 2013, Sophie Lavoine–Hanneguelle from Charabot patented a process using supercritical CO_2 in order to obtain flower extracts comparable to traditional absolutes [28,29]. This process remains of limited use today.

2-Methyltetrahydrofuran

MeTHF is a low-polar solvent with very interesting physicochemical properties, such as low miscibility with water, low boiling point and improved stability compared to tetrahydrofuran. MeTHF shows properties similar to those of toluene and dichloromethane and has numerous applications, such as a solvent for synthesis [16].

MeTHF can be produced from furfural. Furfural is prepared from carbohydrates derived from lignocellulose-based biomass, such as polysaccharide hemicellulose, present in many plant materials [30]. Pentose carbohydrates are obtained from pentosan contained in hemicellulose through a hydrolysis reaction. As shown in Figure 2.4, furfural can then be obtained from pentose carbohydrates by a dehydration reaction in acidic conditions using sulphuric or phosphoric acid as catalysts [31].

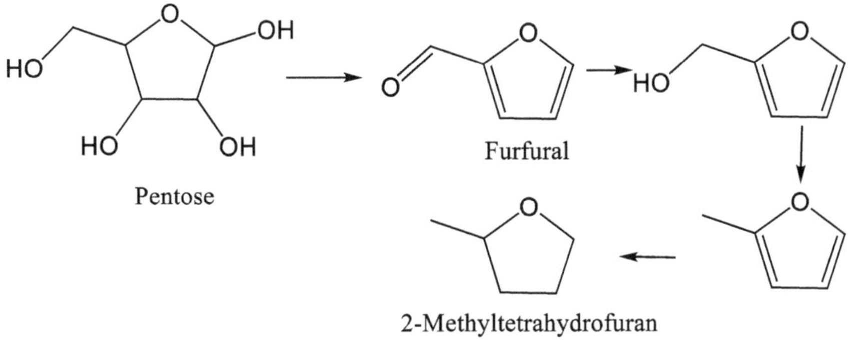

Figure 2.4. Preparation of 2-methyltetrahydrofuran from pentose [30].

From furfural, MeTHF can finally be obtained by successive hydrogenation reactions using Ni-Cu, Fe-Cu, Cu-Zn, or Cu-Cr alloy catalysts [30].

In January 2023, MeTHF has been authorised as an extraction solvent in the food industry. Applications include fractionation of fats, oils or cocoa butter, the preparation of defatted protein products, flours and cereal germs and flavouring with limits with a residual solvent limit varying from 1 to 30 mg/kg of the final product [32].

MeTHF is a promising solvent for the extraction of fragrance compounds. For example, blackcurrant buds (*Ribes nigrum* L.) volatile compounds were extracted using several alternative solvents, namely α-pinene, MeTHF, ethyl acetate, methyl acetate, ethyl lactate, butanol, isopropanol, ethanol and CO_2 supercritical fluid, with the perspective of finding a substitute for hexane traditionally used. MeTHF was found to be the most promising alternative for *n*-hexane substitution with good yields and selectivity of volatile compounds [33].

1-Butanol

1-Butanol is a four-carbon alcohol with relatively low polarity and is used as a solvent and intermediate in syntheses, such as butyl acetate synthesis [16].

1-Butanol could be produced from biological or chemical routes. In 1861, the first route discovered for producing butanol was a fermentation process called Acetone Butanol Ethanol (ABE). The process produced butanol through fermentation of sugars using *Clostridium* species. This process remained the main source of butanol until the 1950s. Later, favoured by cheap oil, a petrochemical process was developed to produce butanol. The fermentation route was then abandoned. More recently, the rise in oil prices has revived the fermentation process. Butanol can thus be produced through fermentation of sugar, glycerol or lignocellulose feedstock in the presence of different micro-organisms from the *Clostridiaceae* family. In 2006, BP launched a production plant capable of producing 420 million litres of butanol per year through fermentation [34].

1-Butanol has been previously used to extract fragrance compounds from peppermint (*Mentha piperita*), lavender and orange peels. 1-Butanol extracts have a comparable odour profile to ethanol extracts but with much higher yields. Compared with hexane and ethanol, 1-Butanol gave higher concentrations of fragrance and flavour molecules. Despite a high boiling point of 117°C, which makes it difficult to eliminate, this demonstrates biobutanol's potential applicability as an effective alternative solvent [35].

CPME

Cyclopentyl methyl ether, or CPME, is an aprotic solvent employed in organic chemistry. Compared to other ethers, CPME is safer with its relatively high boiling point (106°C) and low tendency to generate peroxides [16]. Nevertheless, its high boiling point also represents a strong limit for the extraction of odorants due to the difficulty of eliminating CPME.

Nowadays, CPME is exclusively produced using petrochemical compounds, even if some production processes starting from bioresources are considered. In particular, a synthesis route using furfural, through cyclopentanone or cyclopentanol intermediates, is currently being studied [36].

Furthermore, it has been recently reported that cyclopentanone could be obtained from a rearrangement of furfural, using water as a solvent under a hydrogen atmosphere and a heterogeneous metal catalyst (Figure 2.5) [37]. As furfural could be obtained from carbohydrates derived from lignocelluloses' biomass [30], this reaction allows synthesis of CPME from bioresources.

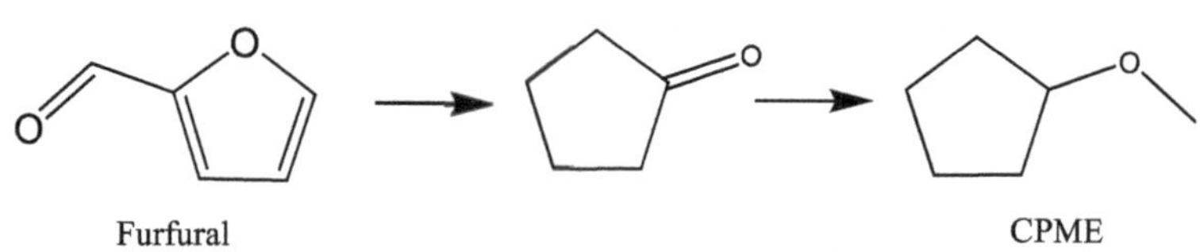

Figure 2.5. Preparation of CPME from furfural [37].

CPME can be a possible solvent for the extraction of volatile compounds. CPME has been compared with a variety of green solvents, in particular with hexane, for the extraction of limonene from orange peels. The extraction yield for limonene obtained using CPME was found to be 80% higher than that obtained using hexane. Using MeTHF, the extraction yield of limonene was found to be 40% higher than that of hexane [38]. Beyond this study on the limonene model compound, it would be necessary to evaluate the impact of the use of CPME on the olfactory profile of the resulting extract. To the best of our knowledge, no study has evaluated the olfactory profiles of extracts produced using CPME. Moreover, the use of CPME in perfumery could be limited by the relatively high boiling point of this solvent, as explained above.

2.3.1.2 *Deep eutectic solvents*

A DESs can be described as a mixture of hydrogen bond acceptor (HBA) and hydrogen bond donor (HBD) compounds, exhibiting a significant decrease in the melting point of the mixture compared to that of each pure substance. Such a mixture is called a eutectic mixture. This is related to the occurrence of unexpected interactions between components within mixture [39]. ADES is "deep" when its melting point is much lower than the ideal one calculated for a mixture according to the colligative properties of a mixture [40].

The concept of eutectic was first observed in 1884 and was limited to metallic mixtures. Mixtures of bismuth with either lead, tin, cadmium or zinc were found to exhibit melting points lower than that of each pure metal or that obtained using an ideal mixing rule [41]. The use of inorganic salts in order to form eutectic solvents was reported by Abbott *et al.* in 2001 [42]. The authors demonstrated the ability of quaternary ammonium salts to form eutectic mixtures when mixed with zinc and/or tin chloride and observed that choline chloride (ChCl) led to mixtures with unexpectedly low melting points [42]. These unexpectedly high decreases in freezing point values for such mixtures led to the definition of "deep eutectic solvents", unlike the ideal eutectic mixtures detailed above.

DESs could be classified into five classes depending on their chemical structure. Abbott *et al.* proposed first in 2014 a classification of DESs into four class [43]. This classification is proposed in Table 2.2. For plant extraction, the main types of DESs used are *type III* and *type V*. In *type III*, the HBA $Cat^{+}X^{-}$ is in principle any ammonium, phosphonium or

Table 2.2. The five types of DESs.

Type	General formula	Examples	References
Type I	$Cat^{+}X^{-}zMCl_x$	$ChCl{:}ZnCl_2$	[44]
Type II	$Cat^{+}X^{-}zMCl_x yH_2O$	$ChCl{:}CuCl_2{\cdot}2H_2O$	[45]
Type III	$Cat^{+}X^{-}zHBD$	ChCl:Urea	[46]
Type IV	$MCl_x (yH_2O)$ zHBD	$FeCl_3$:Acetamide	[47]
Type V	yHBA zHBD	Thymol:Menthol	[40]

sulphonium cation, while X^- is a Lewis' base, generally a halide anion. The more practical and classically studied HBA compound is choline chloride. The HBD is typically an organic molecular component such as an amide, carboxylic acid or polyol. *Type V* DESs is a non-ionic type of DESs, proposed recently by Coutinho *et al.* This latter type of DESs consists of mixtures of non-ionic hydrogen bond acceptor and donor molecules [40].

Some DESs are produced by a mixture of natural compounds. The resulting DESs is then called natural DESs, or NaDES.

For example, NaDESs of *type V* were used to extract vanillin from vanilla pods [48]. DESs were found to exhibit a higher extraction capacity for vanillin than ethanol, while most gave similar or slightly higher yields than methanol. The best extractions yields were obtained with DESs composed of lactic acid and 1,2-propanediol or lactic acid and fructose, respectively. Unfortunately, due to high viscosity and low volatility, these DESs cannot be removed by evaporation after extraction. Nevertheless, as these DESs are composed of chemicals that are safe in high doses and have received the "Generally Recognized As Safe" (GRAS) certification (e.g. FG and FGS), these extracts can therefore be included directly in food products without requiring any removal of DESs [48].

Using a different approach, the non-volatility of DESs may become an advantage. making their use possible as a co-solvent during hydrodistillation. Type III DESs oxalic acid/choline chloride was used in a mixture with 50% of water in a microwave-assisted hydrodistillation of *Litsea cubeba* (Lour.) Pers. Fruits. After 30 minutes of microwave-assisted hydrodistillation in the presence of the DESs, the extraction yield of essential oil was found to be nearly twice that of the yield from hydrodistialltion without any DESs added to water. This result was explained by the capacity of the DESs to destroy the peel and tissue of plant fruits and therefore accelerate the outflow of volatile compounds [49].

Currently, despite their interesting solvent capacities, high viscosity and very low volatility found in most DESs make their removal from an extract at the end of an extraction process tedious if not impossible [10].

2.3.1.3 *Ionic liquids*

ILs are salts exhibiting low melting points. A limit for defining a salt as an IL has been set arbitrarily to a value of 100°C [50]. In addition, most ILs reported in the literature are so-called room-temperature ILs, namely those that remain liquid at or near room temperature.

ILs are considered sustainable solvents considering their numerous advantages, particularly for process safety and the environment. ILs present high thermal, chemical and electrochemical stability, non-flammability and negligible volatility. In addition, ILs can be solubilised into a wide range of compounds and materials, especially those produced naturally [51]. Nevertheless, their non-volatility represents a strong limit for the extraction of odorants due to the impossibility of eliminating ILs by evaporation after extraction. The use of ILs requires the development of alternative processes to recover the volatile compounds extracted.

ILs can be formed using both organic and/or inorganic ions and may contain more than one cation or anion [50]. Most ions are monovalent, and at least one of them, generally the cation, exhibits both an ionic head, typically a saturated or unsaturated heterocycle and one or several side chains. Together with unsymmetrical ion shapes, this leads to steric hindrances and oppositions between short-range and coulombic long-range interactions and a significant decrease in the melting point of the ionic structure [52].

The first IL was observed in 1877 during Friedel–Crafts reactions. Over the course of this reaction, formation of a red liquid phase was observed [53]. The structure of these Friedel–Crafts complexes, given in Figure 2.6, was confirmed in 1952 as heptachlorodialuminate salts [54].

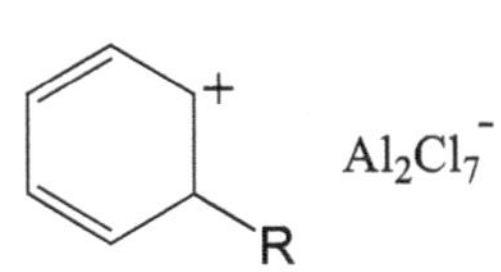

Figure 2.6. Structure of heptachlorodialuminate salts formed during Friedel–Crafts reactions [54].

The first ILs were developed in 1914 by Paul Walden and consisted of alkylammonium nitrate salts being liquid at low temperatures. In particular, ethylammonium nitrate has a melting temperature of 12°C [55]. ILs became then the subject of intense research activities [56–58], and in 1999, commercially available ILs of high quality and at accessible prices appeared [59]. In 2002, Swatloski *et al.* first reported the ability of 1-butyl-3-methylimidazolium chloride IL to dissolve cellulose [60], paving the way for the use of ILs for the treatment of biomass.

One of the many applications of ILs is their use as a solvent for the extraction of essential oil of orange peels [61]. In a first step, orange peels were dissolved in imidazolium-based IL such as 1-methyl-3-ethylimidazolium acetate ($[C_2MIM][AcO]$). In a second step, straightforward distillation of orange peels from $[C_2MIM][AcO]$ was carried out. Due to the extremely low vapour pressure of ILs, a volatile extract without any IL was recovered. Finally, recycling of the IL was achieved by washing with water and charcoal, hence removing biopolymers solubilised within. In this process, the specific properties of ILs allowed a dissolution of the plant matrix, allowing an efficient release of fragrance molecules and a straightforward and mild distillation of these molecules from the IL. This resulted in an increase in yield and improved energy efficiency, but this study does not provide information on the olfactory properties of these extracts [61].

The "greenness" of ILs was questioned due to the poor biodegradability and toxicity of some IL. Furthermore, their high viscosities and the steps required in order to synthesise them have hindered large-scale industrial applications until now [53].

2.3.1.4 *Conclusion*

Greener alternatives to petroleum-based solvents exist, but their sustainability should be carefully considered through a life-cycle assessment approach. However, these solvents have some drawbacks, in particular higher boiling points, which make their elimination difficult and prevent access to all odorous compounds. The ultimate alternative is the development of solvent-free extraction methods. These methods would avoid the use of large volumes of solvent, which is safer and cheaper and gives a resulting extract of higher purity with no traces of residual solvents [10]. This is possible by an intensification technique called solvent-free microwave extraction [62].

2.3.2 *Process intensification technologies*

As stated in Chapter 1, conventional extraction techniques, such as hydrodistillation, are energy-consuming. To minimise this energy consumption, in accordance with the sixth principle of green chemistry and the third principle of eco-extraction, intensification technologies have been developed. These intensification techniques include microwave-assisted extraction (MAE), ultrasound-assisted extraction (UAE), pressurised fluid extraction (PFE), enzyme-assisted extraction (EAE) and other intensification techniques [10]. Particularly, microwaves enable the development of solvent-free processes [62], which are even more advantageous and greener forms of alternative extraction.

2.3.2.1 *Microwave-assisted extraction*

Microwaves are electromagnetic waves with a period typically ranging between 1 m and 1 mm, that is, frequencies ranging between 300 MHz and 300 GHz.

Microwaves interact with molecule dipoles. Submitted to an electric field, a molecule rotates so as to align its dipole with the electric field. Because the latter is oscillating, dipolar molecules irradiated by such microwaves will continuously move upside down. This movement, due to friction with neighbouring molecules, causes a strong and homogeneous heating effect [63].

Therefore, microwave heating is homogeneous throughout the solution and selective to only dielectric materials as well as solvents with permanent dipoles. In the case where a plant cell is submitted to such an electromagnetic field, unlike conventional heating, heat transfer occurs from the inside of the plant cell to the outside (see Figure 2.7).

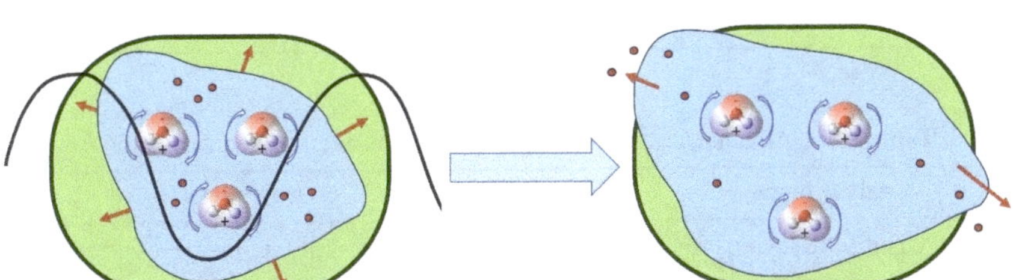

Figure 2.7. Mechanism of microwave extraction of natural compounds (red dots) from a plant cell based on dielectric heating.

The use of microwaves for extracting natural compounds was first proposed in 1986 [64] for the extraction of crude fats, antinutritives (i.e. chemical compounds that prevent the absorption of nutrients) and organophosphate pesticides from soya bean, cottonseed, walnut and fava beans. MAE appeared to be more effective than conventional methods with savings in energy and time [64]. Afterwards, MAE was used to extract various compounds from plant materials, including fats and oils [65], antioxidants and food colours [66], and essential oils and aromas [67].

For essential oils and aroma extraction, several MAE methods were developed [67]. These techniques are summarised in Table 2.3.

Advantages and drawbacks of MAE are presented in Table 2.4.

Table 2.3. Microwave-assisted extraction methods for extraction of essential oils and aromas with examples of biomass extracted and characteristics of the extracts obtained using these techniques.

Technique	Example of biomass extracted	Extract characteristics	References
Compressed air microwave distillation	*Lippia sidoides* Cham	Composition similar to conventional extract with reduced extraction time	[68]
Microwave hydrodistillation	Boldo leaves	Olfactive profile similar, slightly improved compared to conventional extract	[69]
Solvent-free microwave extraction	*Rosmarinus officinalis* L.	More odoriferous than conventional extract with reduced extraction time	[70]
Microwave hydrodiffusion and gravity	*Rosmarinus officinalis* L.	Superior olfactive properties better reproducing the smell of rosemary leaves compared to conventional extract	[71]

Table 2.4. Advantages and drawbacks of microwave-assisted extraction.

Advantages	Drawbacks
• Energy and time savings [64] • Possibility to perform solvent-free extraction [72]	• Higher cost of microwave-assisted extractor [73] • Promote certain chemical reactions [74]

Compressed air microwave distillation

Compressed air microwave distillation (CAMD) is recognised as the first microwave-assisted method for extraction of essential oils. It was developed in 1989 by Craveiro *et al.* [68]. This technique is similar to steam distillation except that compressed air is injected instead of steam. A scheme for the experimental setup is shown in Figure 2.8. A flask containing the plant material is heated in a microwave oven while a compressor injects a flow of compressed air within the flask. The steam produced from the microwave heating of the raw material becomes saturated in volatile molecules and is then driven to a recovery vessel immersed in a refrigeration system situated outside the microwave oven. This technique has been applied to the extraction of essential oils from *Lippia sidoides* Cham. The essential oil obtained using CAMD has a composition similar to the one obtained by conventional steam distillation, while CAMD significantly reduces the extraction time [68]. Unfortunately, the authors did not give information on the olfactory properties of these extracts.

Microwave hydrodistillation

Microwave-assisted hydrodistillation (MWHD) was developed by Stashenko *et al.* in 2004 [75]. This method is based on a conventional hydrodistillation apparatus, except the heat is provided by a microwave oven. A scheme of the MWHD apparatus is presented in Figure 2.9. This technique was applied to the extraction of Colombian *X. aromatica* fruit essential oils [75]. Hydrodistillation and MWHD gave essential oils with similar compositions in which the compounds responsible for the different

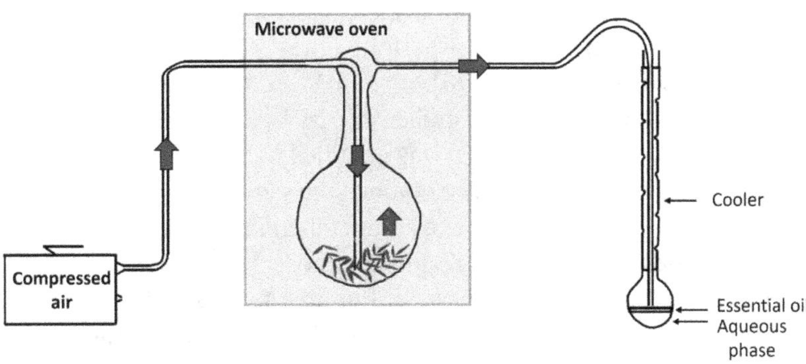

Figure 2.8. Compressed air microwave distillation apparatus.

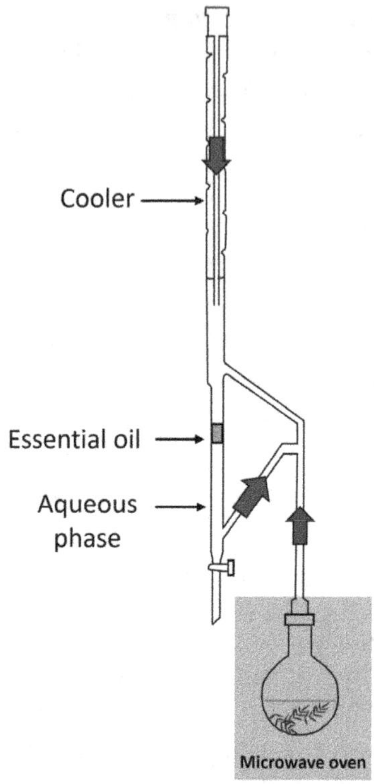

Figure 2.9. Microwave hydrodistillation apparatus.

olfactory notes of these fruits are found without providing an olfactory analysis of the extracts. For obtaining the same amount of essential oils, the time needed using MWHD was four times shorter than when using conventional hydrodistillation [75].

In 2014, Petigny *et al.* [69] studied the MWHD of Boldo leaves. The authors observed that the essential oils obtained by MWHD or the conventional hydrodistillation technique contained the same major compounds but in different relative proportions. The authors then investigated the sensorial properties of these extracts. The results are presented in Figure 2.10.

It can be observed that the two different essential oils have similar olfactive profiles with a slightly better overall perception for the MWHD. The use of MWDH reduced the boiled smell and improved the quality of the essential oil [69].

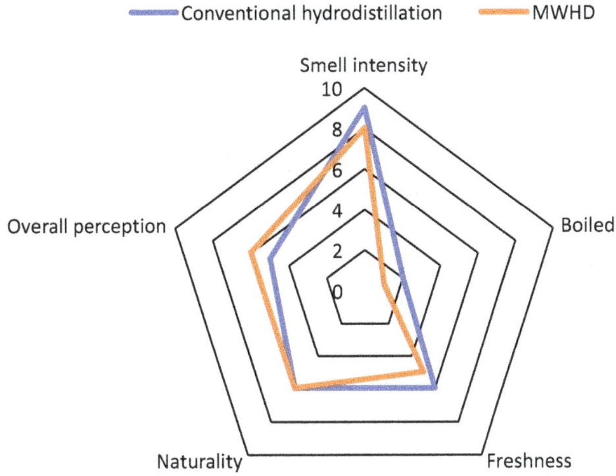

Figure 2.10. Sensorial analysis of essential oils produced by conventional hydrodistillation or MWHD.

Source: Adapted from [69].

A modification of this technique was proposed in 2007 by Flamini *et al.* by introducing *in situ* microwave heating through an insulated microwave coaxial antenna placed inside the glass flask [76]. Another improvement was proposed by Sahraoui *et al.* in 2008. This innovation is based on the principle of steam distillation, with a steam generator placed outside the extraction reactor which is irradiated with microwaves [77].

Solvent-free microwave extraction

Solvent-free microwave-assisted extraction (SFME) was developed and patented in 2004 by Chemat *et al.* [72]. In this process, also called microwave-assisted dry distillation, a fresh plant matrix is placed in a microwave reactor without any water or organic solvent added. Heating of the water constituting the raw material causes the breaking of cell walls containing the essential oil, thereby releasing the oil. The latter distils with the steam produced from matrix water [72]. This method has been applied to several fresh and dry plants, such as spices (ajowan, cumin and star anise), aromatic herbs (basil, mint and thyme) and citrus fruits. For extraction of essential oils, SFME is a more efficient and quicker method than conventional hydrodistillation. Also, essential oils produced using SFME were found to contain more oxygenated compounds, which are more odoriferous than monoterpene hydrocarbons [70].

Microwave hydrodiffusion and gravity

Microwave hydrodiffusion and gravity (MHG) extraction was patented by Chemat *et al.* in 2008 [78]. As in the previous method, microwaves induce warming of the water contained in a plant matrix, causing the destruction of cells containing essential oils. A cooling system is placed outside the microwave below the plant for condensing the distillate [71]. This technique combines microwaves for the hydrodiffusion of volatile compounds from the inside to the exterior of biological material and Earth's gravity to improve the collection and separation of the extract. It is important to note that the extract obtained cannot be called "essential oil" because the oil produced by this technique does not correspond to the definition of essential oil described in the ISO 9235 standard [79]. In addition, non-volatile compounds can be solubilised in the extracted mixture. A scheme for the experimental setup is shown in Figure 2.11. This technique was applied to spearmint (*Mentha spicata* L.) and pennyroyal (*Mentha pulegium* L.). MHG for 15 min allowed the extraction of extracts quantitatively (yield) and qualitatively similar to those extracted from conventional hydrodistillation in 90 min. Therefore, the MHG method is advantageous in terms of time and energy [80].

In 2009, Bousbia *et al.* [71] compared hydrodistillation and MHG for the extraction of volatile compounds from rosemary leaves. The use of MHG improved the composition of the extract with a higher amount of oxygenated compounds. In addition, the extract produced with MHG presented superior olfactive properties, better reproducing the smell of rosemary leaves [71].

2.3.2.2 *Ultrasound-assisted extraction (UAE)*

Low-frequency ultrasounds are sonic waves with frequencies ranging from 16 kHz to 100 kHz and are used for extraction and processing of biomass. Ultrasounds irradiating a liquid phase lead to a cavitation phenomenon. This corresponds to the formation in the solvent of microscopic air bubbles/cavities, whose size and shape oscillate in response to the successive compression/relaxation cycles according to the wavelength of the sonic wave. These cavities always reach an unstable size and implode in such a way that pressure and temperature during the implosion can reach

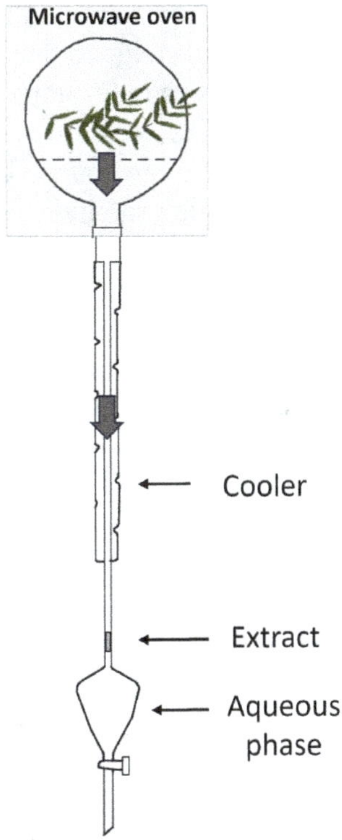

Figure 2.11. Microwave hydrodiffusion and gravity apparatus.

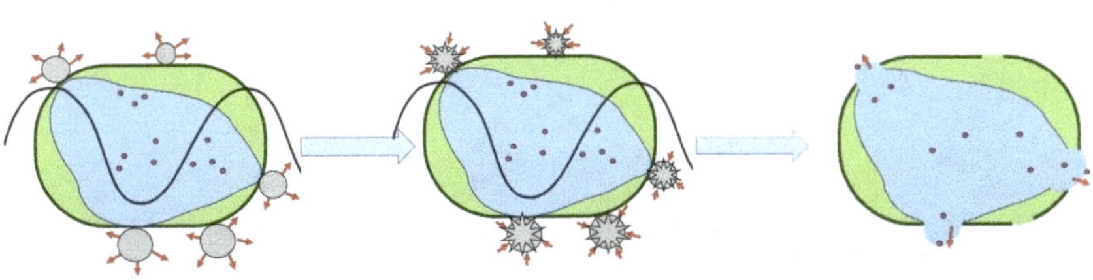

Figure 2.12. Mechanism of ultrasound extraction of natural compounds (red dots) from a plant cell based on cavitation (grey circles).

locally, where the cavitation bubble implodes, several hundreds of atmospheres and thousands of kelvins, respectively [81]. This results in a disruption of the biological cell membranes [82]. A schematic representation of the effect of ultrasounds is proposed in Figure 2.12.

UAE is mainly used at the industrial scale in the aroma sector, as this technique is known to cause lesser damage to thermal-sensitive molecules than conventional techniques. For example, fresh garlic (*Allium sativum*) cloves were extracted using UAE and conventional hydrodistillation [83]. Variations among sulphur-containing molecules, which are thermo-sensitive compounds, were observed. Polysulphide concentration decreased, whereas that of vinyl dithiin isomers increased when the isolation techniques were changed from conventional hydrodistillation to UAE. Therefore, UAE appears to be a promising approach in order to preserve and extract the compounds responsible for the characteristic odour and taste of freshly chopped garlic [83].

UAE was also used in combination with other techniques. UAE was used, coupled with supercritical fluid, to extract oil from adlay seeds. Ultrasounds led to an increase in the extraction yield and a significant decrease in the extraction time [84]. Ultrasounds were also used combined with hydrodistillation in order to enhance the extraction of essential oil of celery seeds (*Apium graveolens* L.) [85]. Ground celery seeds were sonicated prior to the hydrodistillation process. Sonication accelerates water absorption, leading to a faster swelling of plant material and an increase in pore size on the cell walls, thus facilitating recovery of essential oil during the hydrodistillation process. With the help of ultrasounds, extraction yield increased by 48% compared to conventional hydrodistillation, while chemical composition was similar using both methods [85]. Nevertheless, even though there are several applications of UAE in the aroma industry, there is little use in the fragrance industry [10]. Advantages and drawbacks of UAE are presented in Table 2.5.

Table 2.5. Advantages and drawbacks of ultrasound-assisted extraction (UAE).

Advantages	Drawbacks
• Cause less damage to thermal-sensitive molecules [83] • Energy and time savings [84] • Compared with other activation techniques, ultrasound equipment is economical [86]	• In an ultrasound bath, waves vary depending on position, ultrasound probes are more efficient [86] • Scale-up difficult [87]

2.3.2.3 *Pressurised fluid extraction (PFE)*

Pressure is another interesting parameter in order to intensify extraction processes. Pressurised fluid extraction methods can be divided into four classes:

(i) Supercritical fluid extraction (SFE), already presented in Chapter 1, for supercritical CO_2.
(ii) Pressurised solvent extraction (PSE), also known as accelerated solvent extraction, pressurised liquid extraction, or subcritical solvent extraction (SSE), accordingly.
(iii) Pressurised low-polarity water extraction (PLPW), also known as subcritical water extraction, pressurised hot water extraction or superheated water extraction [10].
(iv) Pressurised gas extraction (PGE) [88].

In PSE and PLPW, elevated temperatures increase the extraction efficiency, whereas elevated pressure keeps the solvent in a liquid state above its boiling point. High pressures also improve the diffusion of solvents to regions of the plant matrix that could not be reached at atmospheric pressure. However, one must keep in mind that high temperatures are incompatible with the extraction of fragile raw material and heat-sensitive compounds [10]. In 2008, Giray *et al.* [89] studied the extraction of *Lavandula stoechas* flowers using PLPW. The authors observed the appearance of thermal degradation with burning smell and the blackening of extracts when the temperature was increased in particular at 150°C [89].

PSE was investigated for the extraction of volatile compounds of *Thymus vulgaris* L. [90]. Maximum extraction efficiency was obtained using hexane as an extraction solvent, an extraction time of 10 min, an extraction pressure of 30 bar and a temperature of 100°C. PSE was compared with steam distillation, Soxhlet extraction, supercritical CO_2 extraction and solid-phase microextraction. PSE was the most efficient and the most appropriate sample preparation method in determining volatile compounds from thyme herb. Even though the authors did not provide information on the olfactive profile of their extract, they highlighted several advantages of PSE, such as high yields and short extraction time, but this method presents the disadvantage of co-extracting non-volatile component together with volatile compounds [90]. In addition, PSE equipment expensive since materials constituting this equipment must withstand high pressures and temperatures [91].

PLPW extraction, using pressurised hot water as a solvent, can be considered an even more sustainable version of PSE, as no organic solvents are used in this method. Water is a polar solvent, but when heated at 200°C, its dielectric constant drops from 79 to 35, displaying a polarity similar to that of acetonitrile [92]. PLPW was applied to extraction of fragrance compounds from *Rosa canina*. The author showed that the best extraction conditions are 100°C with 2 ml/min water flow rate and 50 bar pressure for 2 h. This method was compared with Soxhlet extraction using hexane for 12 h. Under these conditions, PLPW and Soxhlet extraction gave similar yields of benzyl alcohol and phenylethyl alcohol, while tetrahydroional, eicosane and 2,6,11-trimethyl dodecane are much better extracted with PLPW. In addition, some fragrance compounds, namely benzaldehyde, α-pinene and limonene, were found only in the PLPW extract. The PLPW method is therefore quicker, more environmentally friendly and more efficient than Soxhlet, but no information on the olfactory profile of these extracts was provided [93]. In 2013, Miyashita and Hideo [94] showed that PLPW improves the aroma profile of green-tea extracts. Green tea leaves were extracted by PLPW at 30 bar and 130°C for 1 min. The resulting extract had higher catechin extraction yields, together with less bitterness and astringency than the conventional sample extracted at 90°C for 10 min. The authors explain this result by the improved capacity of PLPW in extracting arginine and water-soluble pectin, which masks the bitterness and astringency of green tea [94]. Advantages and drawbacks of PFE are presented in Table 2.6.

In PGE, a gas is used under pressure but below its critical point to extract natural compounds from plants. The use of a gaseous solvent allows the production of extracts at low temperatures to preserve heat-sensitive compounds and moderate extraction pressures. The use of gaseous solvents simplifies the elimination of the extraction solvent, which

Table 2.6. Advantages and drawbacks of pressurised fluid extraction (PFE).

Advantages	Drawbacks
• Energy and time savings [95] • Extraction of apolar compounds with water by decreasing water's dielectric constant [92]	• High temperatures are incompatible with extraction of heat-sensitive compounds [10] • High equipment cost [91]

can be removed by simple depressurisation to obtain an extract without solvent traces [88]. PGE has been used for the extraction of volatile compounds from several plant species, particularly using a halogenated gas, such as 1,1,1,2-tetrafluoroethane (freon R134a) [96], or a hydrocarbon gas, such as *n*-butane [97]. In 2011, Nenov *et al.* [98] studied the low-temperature extraction of cinnamon barks (*Cinnamomum zeylanicum* Nees) with 1,1,1,2-tetrafluoroethane. The authors observed that extracts produced with this technique have odours, tastes and physico-chemical properties comparable with those of cinnamon essential oil [98]. In 2017, Rapinel *et al.* [97] studied the extraction of lipophilic compounds from sunflower seeds, carrot chips and caraway seeds using pressurised *n*-butane. The authors observed that the chemical compositions of extracts were unchanged when the reference solvent, hexane, was substituted with *n*-butane, but the extraction yields were reduced. The authors attributed the reduced extraction yields to a slow mass transfer inside the solvent due to the lack of stirring [97]. Unfortunately, the authors did not provide information on the sensory properties of their extracts. In 2021, Dobreva *et al.* [96] used pressurised 1,1,1,2-tetrafluoroethane extraction to compare volatile compounds present in four rose species, namely, *Rosa gallica* L., *Rosa damascena* Mill., *Rosa alba* L. and *Rosa centifolia* L. The chemical compositions obtained were close to those of the traditional rose oils and absolutes, but no information on olfactive profiles was provided in this study [96]. Advantages and drawbacks of PGE are presented in Table 2.7.

Table 2.7. Advantages and drawbacks of pressurised gas extraction (PGE).

Advantages	Drawbacks
• Moderate extraction pressures [88] • Solvent removed by simple depressurisation [88] • Extracts without solvent traces [88]	• Environmental concerns around halogenated gas [96] • Reduction of extraction yield due to slow mass transfer [97]

2.3.2.4 *Enzyme-assisted extraction* (*EAE*)

In a plant matrix, compounds of interest are contained within plant cells. The presence of large amounts of various polysaccharides such as hemicelluloses, starch, pectin and of lipid bilayers composing cell walls reduces extraction efficiency using conventional extraction methods. Enzymes,

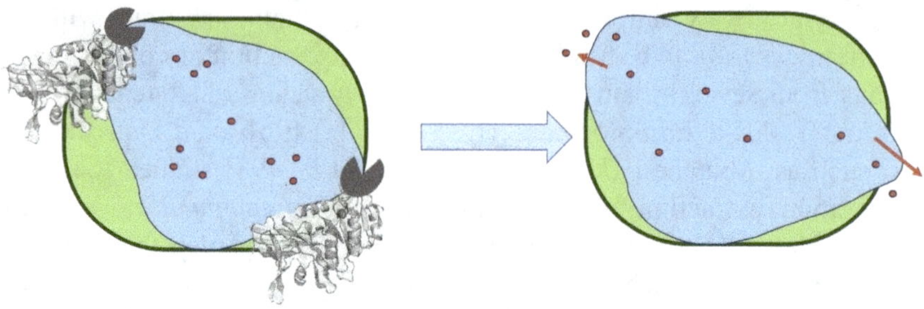

Figure 2.13. Mechanism of enzyme-assisted extraction of natural compounds (red dots) from a plant cell based on enzyme hydrolysis (grey).

proteins with specific hydrolytic properties, are able to degrade the plant matrix and therefore gain access to active components. A schematic representation of the effect of enzymes is proposed in Figure 2.13. Enzymes are classified according to the target molecule they hydrolyse [99]. Cellulase enzymes (such as endoglucanases, cellobiohydrolases and β-glucosidases) hydrolyse celluloses, hemicellulases (such as endoxylanases and β-xylosidases), hemicelluloses and pectinases (such as pectintranseliminase, polygalacturonase and pectinesterase), and pectin. Very often, cooperative effects of multiple enzymes are necessary to increase the degradation of cell walls, resulting in higher extraction yields. Therefore, commercially available multiple enzyme blends were developed [100].

EAE was used in order to extract essential oil from black pepper and cardamom [101]. A mixture of enzymes, called Lumicellulae and composed of cellulase, β-glucanase, pectinase and xylanase, was used for the pre-treatment of black pepper and cardamom. After enzymatic pre-treatment, essential oils were recovered by hydrodistillation. Compared with conventional hydrodistillation, EAE increased the extraction yields from 0.9 to 1.8% in black pepper and 1.9 to 2.5% in cardamom. Also, compared to conventional hydrodistillation, the mass fraction of major active compounds in spices, such as β-caryophyllene in black pepper and α-terpenyl acetate in cardamom, increased from 15.03 to 25.58% and 38.91 to 48.6%, respectively. EAE was therefore found to improve extraction yield and quality of essential oils by increasing fractions of major active compounds [101]. Unfortunately, the authors did not provide a description of the olfactive profiles of their extracts. Enzymes are already used in various domains, including the food industry for texturing, lactose removal, or flavouring [102]. But today, to the best of our knowledge,

Table 2.8. Advantages and drawbacks of enzyme-assisted extraction (EAE).

Advantages	Drawbacks
• Energy and time savings [83] • Higher extraction yields [101]	• Enzymes are pH and temperature sensitive [83] • Enzymes' incorporation into products is regulated [103]

enzymes have no industrial applications for aroma or perfume extraction. Advantages and drawbacks of EAE are presented in Table 2.8.

2.3.2.5 *Other intensification techniques*

Other intensification techniques were developed at a laboratory scale for intensifying plant extraction processes, with limited industrial applications.

Instant controlled pressure drop

Instant controlled pressure drop (ICPD) is a technique used at an industrial scale for drying, decontamination and texturing of material applications. ICPD is a two-step technique. First, the raw material is submitted to a rapid (usually between 5 and 60 s) thermal treatment with high-pressure saturated steam, typically between 0.1 and 0.7 MPa. This stage is followed by an instant pressure drop towards a vacuum at about 5 kPa. This provokes an autovaporisation of a part of the water in the product, an instantaneous cooling of the products together and a mechanical effect breaking the cell walls.

This technique has been applied to the extraction of essential oils from Algerian myrtle leaves [104]. ICDP improved the essential oil extraction yield by 10% compared to conventional hydrodistillation, in a shorter extraction time of 2 min compared to 180 min [104].

Pulsed electric field (PEF)

Under the effect of PEF at low electric fields, a biological membrane is electrically pierced and becomes permeable. This phenomenon is called electroporation. PEF has been first used for killing microorganisms [105]. Later, PEF was found to be useful for extraction of compounds from plants. For example, PEF was applied to the extraction of functional compounds from cinnamon. Plant powder, mixed with ethanol as a solvent, was placed between electrodes. In order to perform extraction, a pulse generator was used with a voltage from 2 to 6 kV/cm and 40–60 pulses. The PEF-assisted extraction was applied to the extraction of aroma compounds from aromatic plants and food by-products. The use of PEF improved the extraction yield of vanillin from vanilla pods by 14%, theobromine and caffeine from cocoa bean shells by, respectively, 25 and 34%, linalool from vermouth mixture by 114%, and limonene from orange peels by 33%, as compared with untreated samples [106].

2.3.2.6 *Conclusion*

Intensification techniques present several advantages, such as a reduction of time and energy consumption compared with conventional methods. Furthermore, innovative technologies are expected to yield ingredients with properties and olfactory profiles different from those obtained from conventional methods. Such original ingredients can be very valuable for specific sectors, such as niche perfumery. Nevertheless, some limitations, such as regulation or higher extraction equipment costs, hinder the widespread use of these techniques.

2.4 A Case Study: Sustainable Extraction of Fragrances from *Rosa centifolia*

Rosa centifolia L. belongs to the genus *Rosa* in the Rosaceae family. Roses will always remain one of the most important groups of ornamental and perfumery plants. Nowadays, about 1,000 genotypes of roses are known [107]. *Rosa centifolia* L. is an ancient rose formed by a complex hybridisation of *R. gallica, R. moschata, R. canina* and *R. damascena* [108].

In the perfumery field, the fragrance of *R. centifolia* L. cannot be extracted by hydrodistillation because essential oil yields are too low. Therefore, the extraction of *R. centifolia* L. is obtained by solvent extraction, producing a concrete and an absolute. [107]

Rose concrete is currently obtained using hexane, a petroleum-based solvent. MeTHF and CPME, two alternative sustainable solvents produced from bioresources, were evaluated as hexane substitutes for obtaining rose concretes and absolutes. These two solvents were selected as low-polar bio-based solvents. Also, these solvents exhibit relatively low boiling points and thus can be removed by evaporation after extraction.

Rosa centifolia flowers (Figure 2.14) were collected from the garden of the International Perfume Museum in Grasse (France) and extracted by solid-liquid extraction using hexane, MeTHF or CPME for 2 h at room temperature under constant magnetic stirring. The compositions of the resulting extracts were determined using gas chromatography flame ionisation detector coupled with mass spectroscopy (GC-FID/MS). For comparison, an additional extraction using hexane, the solvent industrially used for rose extraction, was carried out. Extraction yields were evaluated using GC-FID with the predicted response factor [109] by summing the amount of each compound identified and reported in Figure 2.15.

With hexane, a yield of 0.20% was obtained. Using CPME and MeTHF as alternative solvents led to extraction yield improvements to

Figure 2.14. Pictures of *Rosa centifolia* L. taken in the garden of the International Perfume Museum in Grasse (France).

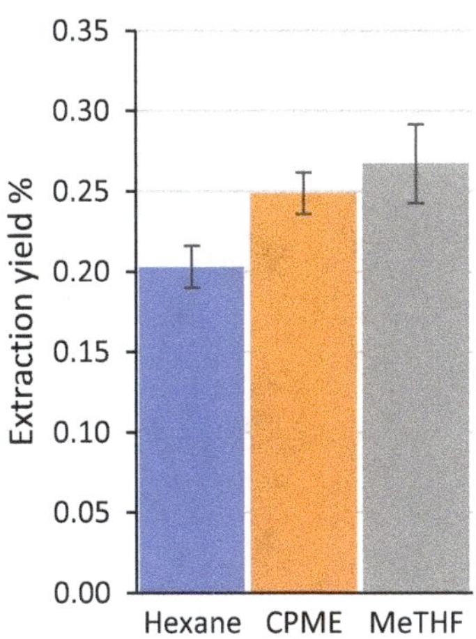

Figure 2.15. Rose petals extraction yields as function of different solvents used during maceration.

0.25% and 0.27%, respectively. This corresponds to increases of 25% and 35%, respectively.

Then, chemical compositions of these extracts were investigated. In all extracts, major compounds identified using GC-MS were 2-phenylethanol, nonadecane, geraniol and citronellol. Compound structures are presented in Figure 2.16. These compounds correspond to those previously reported in *Rosa centifolia* extracted through hexanoic solvent extraction [110]. But in rose extracts, some minor compounds present at trace levels also contribute to the rose smell. These compounds include citronellol acetate, rose oxides, nerol oxides, farnesol and damascenone [111]. Traces of rose oxides were found only in MeTHF extracts, which should have some influence on the scent of this extract. The other compounds were not detected in these three extracts. These compounds are either absent in the extracts or present at concentrations below the limit of detection of GC-MS.

The major compounds extracted were quantified in GC-FID using external calibration curves. Results are presented in Figure 2.17.

Compared to hexane, CPME and MeTHF improved extraction of 2-phenylethanol from 124 mg/g of fresh weight (FW) of flower to 166 and 176 mg/g FW, respectively. On the contrary, there was no significant difference in extraction yield for citronellol or geraniol. Finally, the use of CPME and MeTHF decreased the extraction of nonadecane from 16 mg/g FW to 8 and 9 mg/g FW, respectively. Differences in polarity between hexane and CPME or MeTHF explain our results. Hexane is the least polar of all three solvents, exhibiting a value for its octanol-water partition

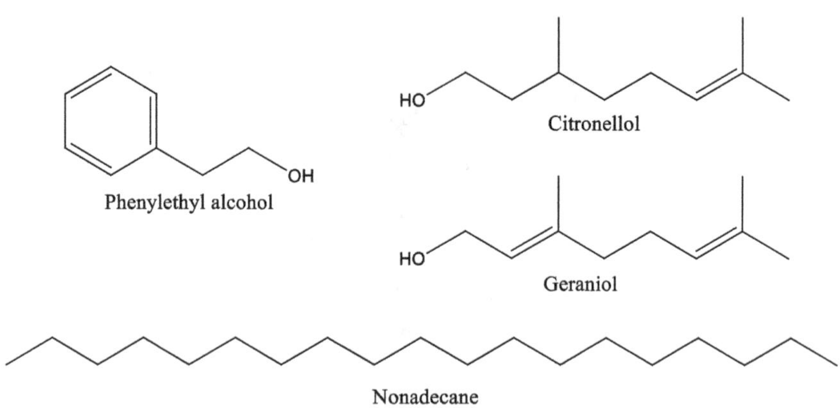

Figure 2.16. Structure of major compounds identified in rose petal extracts.

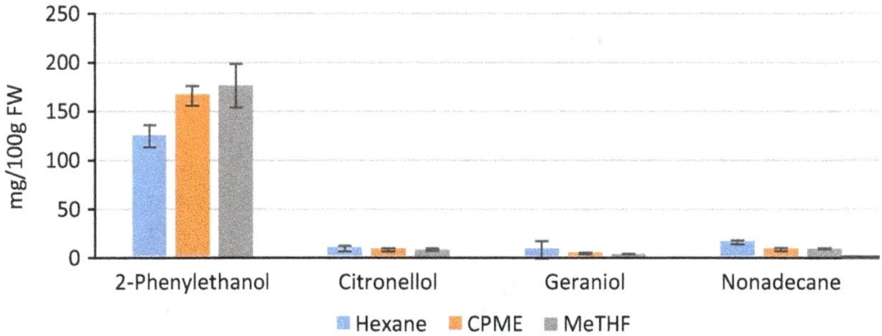

Figure 2.17. Composition of rose extracts obtained using different solvents. Blue: Hexane. Orange: CPME. Grey: MeTHF. Only major compounds are shown.

coefficient (logP) of 3.90 [17], compared to 1.01 [18] and 1.65 [19] for MeTHF and CPME, respectively. 2-phenylethanol, exhibiting a logP value of 1.36 [17], close to that of MeTHF and CPME, is therefore expected to be better extracted using these two solvents than hexane. On the contrary, nonadecane exhibiting a logP value of 10.39 [112] is better extracted in hexane.

CPME and MeTHF therefore appear to be promising alternatives to hexane for the extraction of volatile compounds from *Rosa centifolia*. Nevertheless, use of such solvents significantly changes the composition of an extract. Impact of these changes on olfactive profiles was therefore evaluated with the help of a perfumist. Because CPME and MeTHF are volatile, volatile compounds were first separated from the solvent by rotary evaporation. The resulting concrete was then mixed with warm ethanol. Upon cooling down the ethanolic solution in a freezer, non-volatile compounds including waxes and fatty acids precipitate. An absolute, shown in Figure 2.18, is then obtained after filtration and evaporation of ethanol [10].

This procedure was performed for extracts obtained with MeTHF and hexane. CPME was excluded because of its high boiling point (106°C). Evaporating CPME leads to the loss of a non-negligible fraction of volatile compounds. Olfactive notes from absolutes obtained from hexane or MeTHF extraction were determined by a perfumist[1] and rated on a scale of 0–5, where 0 corresponds to an absence of this note in the extracts and 5 a strong presence of this note. For clarity reasons, olfactive notes are presented on a radar chart in Figure 2.19.

---

[1]Virginie Armand fragrance artist, InnovaGrasse, Grasse, France.

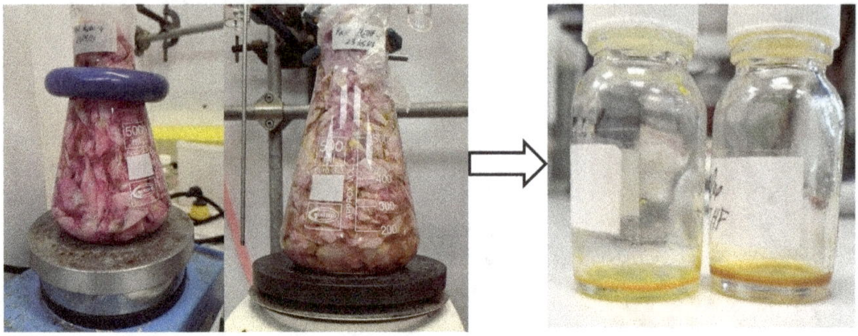

Figure 2.18. Picture of *Rosa centifolia* petals maceration and absolute produced with hexane (left) and MeTHF (right).

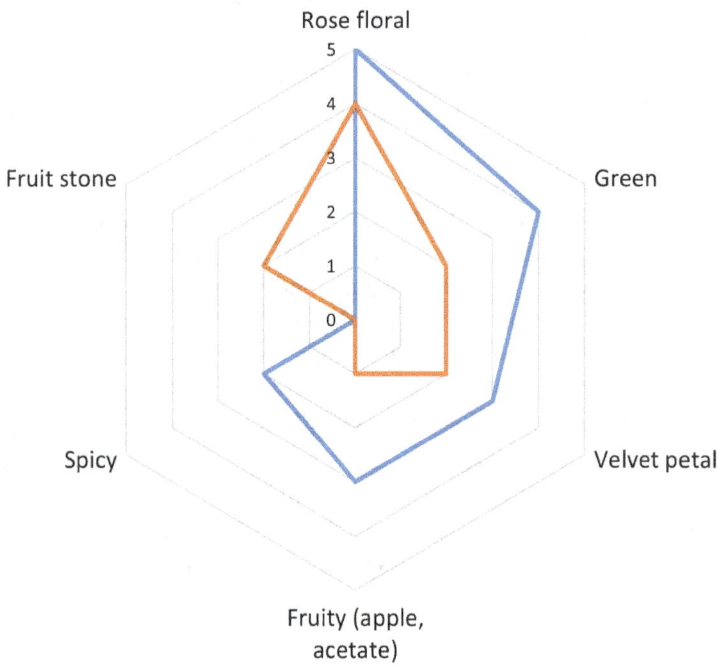

Figure 2.19. Odour description graphs for *Rosa centifolia* petals absolutes. Blue: concrete extracted with hexane, orange: concrete extracted with MeTHF.

Olfactive profiles for hexane and MeTHF absolutes are significantly different. Both absolutes share a common characteristic rose floral note. But the one obtained from MeTHF is less spicy and green. Also, a new fruit stone note appeared. MeTHF is therefore able to extract the characteristic rose floral note from *Rosa centifolia* and complete it with additional notes not present in the conventional absolute obtained from hexane. The use of MeTHF allows us to produce a new type of rose extract that will be evaluated in a perfume composition to confirm the applicability of this new rose absolute.

2.5 Conclusion

This chapter shows that numerous innovative technologies are currently being developed in order to create more sustainable natural ingredients in agreement with the principles of green chemistry and eco-extraction as well as today's consumer concerns. Nevertheless, proposing sustainable alternatives to conventional ingredients and integrating these new ingredients into the perfumer's palette will require much effort. Today, the vast majority of the flavour and fragrance ingredients are manufactured by traditional methods presented in Chapter 1. There are a certain number of obstacles to the industrialisation of processes using alternative solvents and/or modern activation techniques. In particular, the high cost or the absence of large-scale equipment on the market makes many of the concepts economically unrealistic. There are also several regulatory limits that prevent further developments and widespread use of these techniques on an industrial scale. Finally, because any modification in the extraction process leads to a modification of the extract composition and its sensorial profile, as well as for all the reasons mentioned above, it is unlikely to see in the near future such modern technologies and green solvents replacing conventional methods currently used in the fragrance sector.

But because niche perfumery looks for new and innovative fragrance extracts and because volumes of extracts required in this niche field remain quite restricted, such alternative techniques might have good opportunities to invest in such a sector.

Finally, considering a regulatory environment constantly evolving towards more environmentally friendly processes and safer ingredients, current non-sustainable ingredients or extraction techniques will eventually be prohibited or restricted. Eco-extracts obtained from sustainable processes are in line with the general movement of chemistry towards safer and greener processes and products and represent a valuable alternative to current ingredients.

Acknowledgements

This work was partly developed within the scope of the project CICECO-Aveiro Institute of Materials, UIDB/50011/2020, UIDP/50011/2020 and LA/P/0006/2020, financed by national funds through the Portuguese Foundation for Science and Technology (FCT)/MCTES. J-B C. is grateful to the EUR Spectrum — Graduate School of Formal, Physical and Engineering Sciences for financing his PhD. The authors thank the International Perfume Museum Garden and, in particular, Christophe Mège, head gardener, for allowing them to graciously collect the rose flowers necessary for this study. The authors also thank Virginie Armand, independent perfumist, for having graciously carried out the olfactory analysis of our rose absolutes.

References

[1] J.-P. Brun, X. Fernandez. (2015). *Parfums antiques de l'archéologue au chimiste* (Silvana Editoriale, Milano).

[2] J. Cribb. (2017). *Surviving the 21st Century-Humanity's Ten Great Challenges and How We Can Overcome Them* (Springer International Publishing).

[3] R. Naidu, B. Biswas, I.R. Willett, J. Cribb, B. Kumar Singh, C. Paul Nathanail, F. Coulon, K.T. Semple, K.C. Jones, A. Barclay, R. John Aitken. (2021). Chemical pollution: A growing peril and potential catastrophic risk to humanity, *Environ. Int.*, 156, 106616.

[4] P.T. Anastas, T.C. Williamson. (1996). Green chemistry: An overview, *ACS Symp. Ser.*, 626.

[5] F. Chemat, M.A. Vian, G. Cravotto. (2012). Green extraction of natural products: Concept and principles, *Int. J. Mol. Sci.*, 13, 8615–8627.

[6] M. Pagliaro, R. Delisi, R. Ciriminna. (2016). Green fragrances: A critically important technology for the new cosmetic industry, *Househ. Pers. Care Today*, 11, 67–71.

[7] P.T. Anastas, N. Eghbali. (2010). Green chemistry: Principles and practice, *Chem. Soc. Rev.*, 39, 301–312.

[8] P. Anastas, J. Warner (1998) Green chemistry: theory and practice, (Oxford University Press, Oxford).

[9] F. Chemat. (2011). *Eco-extraction du végétal: Procédés innovants et solvants alternatifs*.

[10] P. Burger, H. Plainfossé, X. Brochet, F. Chemat, X. Fernandez. (2019). Extraction of natural fragrance ingredients: History overview and future trends, *Chem. Biodiv.*, 16, 10, e1900424.

[11] Y. Gu, F. Jérôme. (2013). Bio-based solvents: An emerging generation of fluids for the design of eco-efficient processes in catalysis and organic chemistry, *Chem. Soc. Rev.*, 42, 9550–9570.

[12] K. Alfonsi, J. Colberg, P.J. Dunn, T. Fevig, S. Jennings, T.A. Johnson, H.P. Kleine, C. Knight, M.A. Nagy, D.A. Perry, M. Stefaniak. (2008). Green chemistry tools to influence a medicinal chemistry and research chemistry based organisation, *Green Chem.*, 10, 31–36.

[13] D. Prat, A. Wells, J. Hayler, H. Sneddon, C.R. McElroy, S. Abou-Shehada, P.J. Dunn. (2015). CHEM21 selection guide of classical- and less classical-solvents, *Green Chem.*, 18, 288–296.

[14] C.M. Alder, J.D. Hayler, R.K. Henderson, A.M. Redman, L. Shukla, L.E. Shuster, H.F. Sneddon. (2016). Updating and further expanding GSK's solvent sustainability guide, *Green Chem.*, 18, 3879–3890.

[15] L. Soh, M.J. Eckelman. (2016). Green solvents in biomass processing, *ACS Sustain. Chem. Eng.*, 4, 5821–5837.

[16] F.G. Calvo-Flores, M.J. Monteagudo-Arrebola, J.A. Dobado, J. Isac-García. (2018). Green and bio-based solvents, *Top. Curr. Chem.*, 376, 1–40.

[17] C. Hansch, A. Leo, D. Hoekman. (1995). *Exploring QSAR – Hydrophobic, Electronic, and Steric Constants*, 2nd edn. (American Chemical Society, Washington, DC).

[18] The Good Scents Company Information System. (n.d.). 2-methyl tetrahydrofuran. https://www.thegoodscentscompany.com/data/rw1179261.html (accessed August 24, 2023).

[19] T. Oshima, Y. Iwakiri, A. Inada. (2023). Development of solvents for Au(III) extraction in hydrochloric acid media based on ion solvation with reference to the Hansen solubility parameters: Success of dicyclohexylketone, *Hydrometallurgy*, 220, 106106.

[20] Statista. (2023). Volume of ethanol production in the world in 2022, by country (in billion liters). https://fr.statista.com/statistiques/753650/production-d-ethanol-par-pays-monde/ (accessed August 25, 2023).

[21] C.H. Christensen, J. Rass-Hansen, C.C. Marsden, E. Taarning, K. Egeblad. (2008). The renewable chemicals industry, *ChemSusChem*, 1, 283–289.

[22] N.S. Hidzir, A. Som, Z. Abdullah. (2014). Ethanol production via direct hydration of ethylene: A review, *Int. Conf. Glob. Sustain. Chem. Eng.*, 8–13.

[23] A. Rudolf, K. Karhumaa, B. Hahn-Hägerdal. (2009). Ethanol Production from Traditional and Emerging Raw Materials, in T. Satyanarayana, G. Kunze (eds.) *Yeast Biotechnol. Divers. Appl.* (Springer), pp. 489–513.

[24] C. Pasha, L.V. Rao. (2009). Thermotolerant yeasts for bioethanol production using lignocellulosic substrates, in *Yeast Biotechnol. Divers. Appl.*, pp. 551–584.
[25] M.C. Rulli, D. Bellomi, A. Cazzoli, G. De Carolis, P. D'Odorico. (2016). The water-land-food nexus of first-generation biofuels, *Sci. Rep.*, 6, 1–10.
[26] D. Graham-Rowe. (2011). Beyond food versus fuel, *Nature*, 474, S6–S8.
[27] H. Surburg, J. Panten. (2006). Natural Raw Materials in the Flavor and Fragrance Industry, in *Common Fragr. Flavor Mater.* (Wiley), pp. 177–238.
[28] S. Lavoine-Hanneguelle, C. Périchet, N. Schnaebele, M. Humbert. (2014). Development of new natural extracts, *Chem. Biodivers.*, 11, 1798–1820.
[29] S. Lavoine-Hanneguelle. (2013). Process for obtaining a scented extract of fresh flowers and/or leaves using natural solvents, US 2013/0338241.
[30] A.-G. Sicaire, M.A. Vian, A. Filly, Y. Li, A. Bily, F. Chemat. (2014). 2-Methyltetrahydrofuran: Main Properties, Production Processes, and Application in Extraction of Natural Products, in *Altern. Solvents Nat. Prod. Extr. Green Chem. Sustain. Technol.* (Springer), pp. 253–268.
[31] R. Adams, V. Voorhees. (2003). Furfural, in *Org. Synth.* (Wiley).
[32] European Commission. (2023). COMMISSION DIRECTIVE (EU) 2023/175 of 26 January 2023 amending Directive 2009/32/EC of the European Parliament and of the Council as regards 2-methyloxolane, *Off. J. Eur. Union.*
[33] A. Filly, A.S. Fabiano-Tixier, Y. Lemasson, C. Roy, X. Fernandez, F. Chemat. (2014). Extraction of aroma compounds in blackcurrant buds by alternative solvents: Theoretical and experimental solubility study, *C. R. Chim.*, 17, 1268–1275.
[34] B. Ndaba, I. Chiyanzu, S. Marx. (2015). n-Butanol derived from biochemical and chemical routes: A review, *Biotechnol. Rep.*, 8, 1–9.
[35] F. Gilman, R. Syrenne, C. Keyari, R. Kropp. (2019). Bio-Based Butanol as a Solvent for Essential Oil Extractions, *Perfum. Flavorist.* https://www.perfumerflavorist.com/flavor/regulatory-research/article/21856934/biobased-butanol-as-a-solvent-for-essential-oil-extractions.
[36] C. Aranda, G. De Gonzalo. (2020). Biocatalyzed redox processes employing green reaction media, *Molecules*, 25, 1–26.
[37] M. Hronec, K. Fulajtarová. (2012). Selective transformation of furfural to cyclopentanone, *Catal. Commun.*, 24, 100–104.
[38] B. Ozturk, J. Winterburn, M. Gonzalez-Miquel. (2019). Orange peel waste valorisation through limonene extraction using bio-based solvents, *Biochem. Eng. J.*, 151, 107298.

[39] D. Rodríguez-Llorente, A. Cañada-Barcala, S. Álvarez-Torrellas, V.I. Águeda, J. García, M. Larriba. (2020). A review of the use of eutectic solvents, terpenes and terpenoids in liquid-liquid extraction processes, *Processes*, 1–54.

[40] D.O. Abranches, M.A.R. Martins, L.P. Silva, N. Schaeffer, S.P. Pinho, J.A.P. Coutinho. (2019). Phenolic hydrogen bond donors in the formation of non-ionic deep eutectic solvents: The quest for type V DESs, *Chem. Commun*, 55, 10253–10256.

[41] F. Guthrie. (1884). On eutexia, *Proc. Phys. Soc. London*, 6, 124–146.

[42] A.P. Abbott, G. Capper, D.L. Davies, H.L. Munro, R.K. Rasheed. (2001). Preparation of novel, moisture-stable, Lewis-acidic ionic liquids containing quaternary ammonium salts with functional side chains, 2010–2011.

[43] E.L. Smith, A.P. Abbott, K.S. Ryder. (2014). Deep eutectic solvents (DESs) and their applications, *Chem. Rev.*, 114, 11060–11082.

[44] A.D. Potchamyou Ngatcha, A. Zhao, S. Zhang, W. Xiong, M. Sarker, J. Xu, M.A. Alam. (2023). Determination of active sites on the synthesis of novel Lewis acidic deep eutectic solvent catalysts and kinetic studies in microalgal biodiesel production, *RSC Adv.*, 13, 10110–10122.

[45] X. Liu, Y. Zhai, Z. Xu, Y. Zhu, Y. Zhou, Z. Wang, L. Liu, W. Ren, Y. Xie, C. Li, M. Xu. (2023). The novel application of type II deep eutectic solvents (DESs) for sludge dewatering, *Sep. Purif. Technol.*, 306, 122714.

[46] A.P. Abbott, G. Capper, D.L. Davies, R.K. Rasheed, V. Tambyrajah. (2003). Novel solvent properties of choline chloride/urea mixtures, *Chem. Commun.*, 70–71.

[47] S. Higashino, A.P. Abbott, M. Miyake, T. Hirato. (2020). Iron(III) chloride and acetamide eutectic for the electrodeposition of iron and iron based alloys, *Electrochim. Acta*, 351, 136414.

[48] C.G. González, N.R. Mustafa, E.G. Wilson, R. Verpoorte, Y.H. Choi. (2018). Application of natural deep eutectic solvents for the "green"extraction of vanillin from vanilla pods, *Flavour Fragr. J.*, 33, 91–96.

[49] Y. Guo, Y. Li, Z. Li, L. Jiang, X. Cao, W. Gao, J. Wang, D. Luo, F. Chen. (2021). Deep eutectic solvent-homogenate based microwave-assisted hydrodistillation of essential oil from Litsea cubeba (Lour.) Pers. fruits and its chemical composition and biological activity, *J. Chromatogr. A*, 1646.

[50] J.L. Shamshina, O. Zavgorodnya, R.D. Rogers. (2019). Ionic Liquids, in *Encycl. Anal. Sci.*, Third Edit (Elsevier), pp. 218–225.

[51] H. Ullah, C.D. Wilfred, M.S. Shaharun. (2019). Ionic liquid-based extraction and separation trends of bioactive compounds from plant biomass, *Sep. Sci. Technol.*, 54(4), 559–579.

[52] M.F.C. Gomes, J.N.C. Lopes, A.A.H. Padua. (2009). Thermodynamics and Micro Heterogeneity of Ionic Liquids, in *Ion. Liq. Top. Curr. Chem.* (Springer), pp. 161–183.

[53] C. Friedel , J. Crafts. (1877). Sur une méthode générale nouvelle de synthése d'hydrocarbures, d'acetone, ect, *Comptes Rendus*, 84, 74–77.

[54] H.C. Brown, H.W. Pearsall. (1952). The action of the catalyst couple aluminum chloride-hydrogen chloride on toluene at law temperatures; the nature of Friedel-Crafts cmplexes, *J. Am. Chem. Soc.*, 74, 191–193.

[55] P. Walden. (1914). Molecular weights and electrical conductivity of several fused salts, *Bull. l'Académie Impériale Des Sci. St.-Pétersbg.*, 1800, 405–422.

[56] J.S. Wilkes. (2002). A short history of ionic liquids — From molten salts to neoteric solvents, *Green Chem.*, 4, 73–80.

[57] K.R. Seddon. (2003). Ionic liquids: A taste of the future, *Nat. Mater.*, 2, 363–365.

[58] R.D. Rogers, K.R. Seddon. (2003). *Ionic Liquids as Green Solvents: Progress and Prospects* (American Chemical Society).

[59] T. Welton. (2018). Ionic liquids: A brief history, *Biophys. Rev.*, 10, 691–706.

[60] R.P. Swatloski, S.K. Spear, J.D. Holbrey, R.D. Rogers. (2002). Dissolution of cellose with ionic liquids, *J. Am. Chem. Soc.*, 124, 4974–4975.

[61] K. Bica, P. Gaertner, R.D. Rogers. (2011). Ionic liquids and fragrances — direct isolation of orange essential oil, *Green Chem.*, 13, 1997–1999.

[62] M.E. Lucchesi, F. Chemat, J. Smadja. (2004). Solvent-free microwave extraction of essential oil from aromatic herbs: Comparison with conventional hydro-distillation, *J. Chromatogr. A*, 1043, 323–327.

[63] P. Tatke, Y. Jaiswal. (2011). An overview of microwave assisted extraction and its applications in herbal drug research, *Res. J. Med. Plant.*, 5, 21–31.

[64] K. Ganzler, A. Salgo, K. Valko. (1986). Microwave extraction: A novel sample preparation method for chromatography, *J. Chromatogr.*, 371, 299–306.

[65] M.D. Luque de Castro, M.A. Fernández-Peralbo, B. Linares-Zea, J. Linares. (2012). The Role of Microwaves in the Extraction of Fats and Oils, in *Microwave-Assisted Extr. Bioact. Compd.* (Springer), pp. 69–102.

[66] Y. Li, A.-S. Fabiano-Tixier, M. Abert-Vian, F. Chemat. (2012). Microwave-Assisted Extraction of Antioxidants and Food Colors, in *Microwave-Assisted Extr. Bioact. Compd.* (Springer), pp. 103–126.

[67] F. Chemat, M. Abert-Vian, X. Fernandez. (2012). Microwave-Assisted Extraction of Essential Oils and Aromas, in *Microwave-Assisted Extr. Bioact. Compd.* (Springer), pp. 53–68.

[68] A.A. Craveiro, F.J.A. Matos, J.W. Alencar, M.M. Plumel. (1989). Microwave oven extraction of an essential oil, *Flavour Fragr. J.*, 4, 43–44.

[69] L. Petigny, S. Périno, M. Minuti, F. Visinoni, J. Wajsman, F. Chemat. (2014). Simultaneous microwave extraction and separation of volatile and non-volatile organic compounds of Boldo leaves. from lab to industrial scale, *Int. J. Mol. Sci.*, 15, 7183–7198.

[70] A. Filly, X. Fernandez, M. Minuti, F. Visinoni, G. Cravotto, F. Chemat. (2014). Solvent-free microwave extraction of essential oil from aromatic herbs: From laboratory to pilot and industrial scale, *Food Chem.*, 150, 193–198.

[71] N. Bousbia, M. Abert Vian, M.A. Ferhat, E. Petitcolas, B.Y. Meklati, F. Chemat. (2009). Comparison of two isolation methods for essential oil from rosemary leaves: Hydrodistillation and microwave hydrodiffusion and gravity, *Food Chem.*, 114, 355–362.

[72] F. Chemat, E. Lucchesi, M, J. Smadia. (2004). Solvent-free microwave extraction of volatile natural substances, US 2004/0187340 A1.

[73] R. Ciriminna, D. Carnaroglio, R. Delisi, S. Arvati, A. Tamburino, M. Pagliaro. (2016). Industrial feasibility of natural products extraction with microwave technology, *ChemistrySelect*, 1, 549–555.

[74] E. Destandau, T. Michel, and C. Elfakir. (2022). Chapter 4: Microwave-assisted Extraction in Natural Product Extraction: Principles and Applications, in ed. M. A. Rostagno and J. M. Prado, *The Royal Society of Chemistry*, 2013, pp. 113–156.

[75] E.E. Stashenko, B.E. Jaramillo, J.R. Martínez. (2004). Analysis of volatile secondary metabolites from Colombian *Xylopia aromatica* (Lamarck) by different extraction and headspace methods and gas chromatography, *J. Chromatogr. A*, 1025, 105–113.

[76] G. Flamini, M. Tebano, P.L. Cioni, L. Ceccarini, A.S. Ricci, I. Longo. (2007). Comparison between the conventional method of extraction of essential oil of *Laurus nobilis* L. and a novel method which uses microwaves applied in situ, without resorting to an oven, *J. Chromatogr. A*, 1143, 36–40.

[77] N. Sahraoui, M.A. Vian, I. Bornard, C. Boutekedjiret, F. Chemat. (2008). Improved microwave steam distillation apparatus for isolation of essential oils. Comparison with conventional steam distillation, *J. Chromatogr. A*, 1210, 229–233.

[78] F. Chemat, M. Vian, F. Visinoni. (2008). Microwave hydrodiffusion for isolation of natural products, EP 1 955 749 A1.

[79] Norme ISO. (2021). ISO 9235 Matières premières aromatiques naturelles – Vocabulaire. https://standards.iteh.ai/catalog/standards/sist/599cef4e-41ea-48e7-8dd1-2039e0ec6fbf/iso-9235-2021.

[80] M.A. Vian, X. Fernandez, F. Visinoni, F. Chemat. (2008). Microwave hydrodiffusion and gravity, a new technique for extraction of essential oils, *J. Chromatogr. A*, 1190, 14–17.

[81] K.S. Suslick. (1989). The chemical effect of ultrasound, *Sci. Am.*, 260, 80–86.

[82] F. Chemat, Zill-E-Huma, M.K. Khan. (2011). Applications of ultrasound in food technology: Processing, preservation and extraction, *Ultrason. Sonochem.*, 18, 813–835.

[83] A.C. Kimbaris, N.G. Siatis, D.J. Daferera, P.A. Tarantilis, C.S. Pappas, M.G. Polissiou. (2006). Comparison of distillation and ultrasound-assisted extraction methods for the isolation of sensitive aroma compounds from garlic (*Allium sativum*), *Ultrason. Sonochem.*, 13, 54–60.

[84] A. jun Hu, S. Zhao, H. Liang, T. qiu Qiu, G. Chen. (2007). Ultrasound assisted supercritical fluid extraction of oil and coixenolide from adlay seed, *Ultrason. Sonochem.*, 14, 219–224.

[85] J. Zorga, A. Kunicka-Styczyńska, R. Gruska, K. Śmigielski. (2020). Ultrasound-assisted hydrodistillation of essential oil from celery seeds (*Apium graveolens* L.) and its biological and aroma profiles, *Molecules*, 25, 5322.

[86] A. Carreira-Casais, P. Otero, P. Garcia-Perez, P. Garcia-Oliveira, A.G. Pereira, M. Carpena, A. Soria-Lopez, J. Simal-Gandara, M.A. Prieto. (2021). Benefits and drawbacks of ultrasound-assisted extraction for the recovery of bioactive compounds from marine algae, *Int. J. Environ. Res. Public Health*, 18, 9153–9177.

[87] J.M. Prado, P.C. Veggi, M.A.A. Meirele. (2017). Chapter 20 Scale-Up Issues and Cost of Manufacturing Bioactive Compounds by Supercritical Fluid Extraction and Ultrasound Assisted Extraction, in *Glob. Food Secur. Wellness* (Springer), pp. 377–433.

[88] V. Rapinel, N. Rombaut, N. Rakotomanomana, A. Vallageas, F. Chemat. (2021). Gaz liquéfiés comme solvants alternatifs appliqués à l'éco-extraction du végétal, *Tech. l'Ingenieur.*

[89] E.S. Giray, S. Kirici, D.A. Kaya, M. Türk, Ö. Sönmez, M. Inan. (2008). Comparing the effect of sub-critical water extraction with conventional extraction methods on the chemical composition of *Lavandula stoechas*, *Talanta*, 74, 930–935.

[90] A.L. Dawidowicz, E. Rado, D. Wianowska, M. Mardarowicz, J. Gawdzik. (2008). Application of PLE for the determination of essential oil components from Thymus vulgaris L., *Talanta*, 76, 878–884.

[91] A. Perez-Vazquez, M. Carpena, P. Barciela, L. Cassani, J. Simal-Gandara, M.A. Prieto. (2023). Pressurized liquid extraction for the recovery of bioactive compounds from seaweeds for food industry application: A review, *Antioxidants*, 12, 612–638.

[92] D. Luong, M.A. Sephton, J.S. Watson. (2015). Subcritical water extraction of organic matter from sedimentary rocks, *Anal. Chim. Acta*, 879, 48–57.

[93] M.Z. Özel, A.A. Clifford. (2004). Superheated water extraction of fragrance compounds from Rosa canina, *Flavour Fragr. J.*, 19, 354–359.

[94] T. Miyashita, H. Etoh. (2013). Improvement of the bitterness and astringency of green tea by sub-critical water extraction, *Food Sci. Technol. Res.*, 19, 471–478.

[95] P. Mottahedin, A. Haghighi Asl, M. Khajenoori. (2017). Extraction of curcumin and essential oil from *Curcuma longa L.* by subcritical water via response surface methodology, *J. Food Process. Preserv.*, 41, 1–9.

[96] A. Dobreva, D. Nedeltcheva-Antonova, N. Nenov, K. Getchovska, L. Antonov. (2021). Subcritical extracts from major species of oil-bearing roses–a comparative chemical profiling, *Molecules*, 26, 1–15.

[97] V. Rapinel, N. Rombaut, N. Rakotomanomana, A. Vallageas, G. Cravotto, F. Chemat. (2017). An original approach for lipophilic natural products extraction: Use of liquefied *n*-butane as alternative solvent to *n*-hexane, *LWT – Food Sci. Technol.*, 85, 524–533.

[98] N. Nenov, V. Gochev, T. Girova, I. Stoilova, T. Atanasova, V. Stanchev, A. Stoyanova. (2011). Low temperature extraction of essential oil bearing plants by liquefied gases. 6. barks from cinnamon (*Cinnamomum zeylanicum nees*), *J. Essent. Oil-Bearing Plants*, 14, 67–75.

[99] W.N. Arnold. (1987). Hydrolytic Enzymes, in D.R. Berry, I. Russell, G.G. Stewart (eds.) *Yeast Biotechnol.* (Springer, Dordrecht), pp. 369–400.

[100] S.S. Nadar, P. Rao, V.K. Rathod. (2018). Enzyme assisted extraction of biomolecules as an approach to novel extraction technology: A review, *Food Res. Int.*, 108, 309–330.

[101] J. Chandran, K.P.P. Amma, N. Menon, J. Purushothaman, P. Nisha. (2012). Effect of enzyme assisted extraction on quality and yield of volatile oil from black pepper and cardamom, *Food Sci. Biotechnol.*, 21, 1611–1617.

[102] O. Kirk, T.V. Borchert, C.C. Fuglsang. (2002). Industrial enzyme applications, *Curr. Opin. Biotechnol.*, 13, 345–351.

[103] P. Streimikyte, P. Viskelis, J. Viskelis. (2022). Enzymes-assisted extraction of plants for sustainable and functional applications, *Int. J. Mol. Sci.*, 23, 2359–2373.

[104] B. Berka-Zougali, A. Hassani, C. Besombes, K. Allaf. (2010). Extraction of essential oils from Algerian myrtle leaves using instant controlled pressure drop technology, *J. Chromatogr. A*, 1217, 6134–6142.

[105] W.A. Hamilton, A.J.H. Sale. (1967). Effects of high electric fields on microorganisms. II. Mechanism of action of the lethal effect, *BBA — Gen. Subj.*, 148, 789–800.

[106] S. Carpentieri, A. Režek Jambrak, G. Ferrari, G. Pataro. (2022). Pulsed electric field-assisted extraction of aroma and bioactive compounds from aromatic plants and food by-products, *Front. Nutr.*, 8, 1–12.

[107] M. Mileva, Y. Ilieva, G. Jovtchev, S. Gateva, M.M. Zaharieva, A. Georgieva, L. Dimitrova, A. Dobreva, T. Angelova, N. Vilhelmova-Ilieva, V. Valcheva, H. Najdenski. (2021). Rose flowers — A delicate perfume or a natural healer?, *Biomolecules*, 11, 1–32.

[108] H. Iwata, T. Kato, S. Ohno. (2000). Triparental origin of Damask roses, *Gene*, 259, 53–59.

[109] T. Cachet, H. Brevard, A. Chaintreau, J. Demyttenaere, L. French, K. Gassenmeier, D. Joulain, T. Koenig, H. Leijs, P. Liddle, G. Loesing, M. Marchant, P.H. Merle, K. Saito, C. Schippa, F. Sekiya, T. Smith. (2016). IOFI recommended practice for the use of predicted relative-response factors for the rapid quantification of volatile flavouring compounds by GC-FID, *Flavour Fragr. J.*, 31, 191–194.

[110] A. Younis, M. Aslam, K. Asif, A. Khan, A. Riaz, M.A. Pervez. (2007). Effect of different extraction methods on yield and quality of essential oil from four rosa species, *Floric. Ornam. Biotechnol.*, 1, 73–76.

[111] E. sz. Kováts. (1987). Composition of essential oils. Part 7. Bulgarian oil of rose (*Rosa damascena* Mill.), *J. Chromatogr.*, 406, 185–222.

[112] The Good Scents Company Information System. (n.d.). Nonadecane. https://www.thegoodscentscompany.com/data/rw1069421.html (accessed December 6, 2023).

https://doi.org/10.1142/9781800617780_0003

Chapter 3

Modern or Renewed Techniques and Approaches for the Detection of Adulteration of Essential Oils

Marissa Pierson[*,†,§] **and Sylvain Antoniotti**[*,‡,¶]

[*]*Université Cote d'Azur, CNRS, Institut de Chimie de Nice, Nice, France*

[†]*European Research Institute on Natural Ingredients (ERINI), Grasse, France*

[‡]*Université Cote d'Azur, Institute for Innovation and Partnerships in Flavor, Fragrance, Cosmetics, Grasse Biotech, Grasse, France*

[§]*marissa.ann.pierson@gmail.com*
[¶]*sylvain.antoniotti@univ-cotedazur.fr*

Abstract

Essential oils are natural complex substances obtained from selected parts of plants by distillation techniques, or, in some cases of citrus plants, by cold expression. They are widely used materials in perfumery, the food and drink industry, cosmetics and well-being applications and aromatherapy. Essential oils of high demand and/or of limited offer are difficult to find, and their value can reach several thousands of euros per kilogram. This has resulted in economically motivated adulteration (EMA), a fraudulent practice consisting of adding cheap material to an expensive one to increase profit. Adulteration of essential oils is typically observed in four different forms, namely dilution with a solvent featuring similar physico-chemical properties, addition of a cheaper essential oil, addition of individual compounds, either natural or synthetic, or substitution with another essential oil. Analytical chemistry is thus required to characterise adulteration both qualitatively and quantitatively. Adulteration techniques have evolved continuously along with analytical countermeasures to reach a high level of sophistication. In this chapter, we analyse the literature on the topic that appeared between 2015 and 2021.

3.1 Introduction

Essential oils are complex natural products used in applications ranging from cosmetics to well-being products and pharmaceuticals. They have seen a rise in popularity over the past 20 years as consumer goods for at-home use. As with other agricultural products like honey, milk, olive oil and wine, essential oils are susceptible to adulteration. All of these products are at risk of economically motivated adulteration, that is, a producer or seller of the product will dilute or modify the product in order to sell at a higher profit margin. For example, cheaper vegetable oils can be added to olive oil and sold at the same price as quality olive oil, increasing the profit in two ways: first, selling more olive oil that they have produced, and second, selling cheap oil for a higher price. Similar unethical techniques are often used to adulterate essential oils.

Essential oils are steam distilled or cold pressed from the flowers, fruits, leaves, barks, stems and roots of wild or cultivated plants. The yields are small percentages of the bulk weight of the plant material. The plants can be exotic, best grown in tropical countries. Some are wild harvested, limiting the ability to increase supplies even as demand rises. These peculiarities of essential oils increase their value, and with increasing value, the temptation and pay-off for adulteration increase as well.

In order to police unscrupulous tactics during the process from raw materials to consumer goods, scientific techniques are applied to detect adulteration. Some techniques, like chiral gas chromatography, look for specific modifications of the natural product. Others use a chemical fingerprinting approach to determine if a sample is significantly different than the expected composition such as near-infrared (NIR) spectroscopy. In all cases, the analysis is to identify if a product has been altered, resulting in the rejection of the raw material or commercial product before its distribution on the market.

This chapter reviews the techniques used to analyse essential oils, starting with well-established techniques that have seen improvements in the past six years. Previous review articles and book chapters might be of interest to the reader dealing generally with the theme of authentication or with specific aspects [1–4].

3.2 Current State of Established Approaches

Essential oils have long been analysed by gas chromatography using a range of columns and detectors. But as advances come for these techniques in general, we also see advances in other techniques applied to essential oil analysis.

3.2.1 *Chirality*

Essential oils are metabolomic products, and their biosynthesis is guided by enzymes; they are often characteristically enantiomerically enriched in one or more chiral compounds. In contrast, cheaply produced synthetic products are often racemic mixtures. Identifying the expected chiral ratio within an essential oil can rule out the addition of a synthetic or foreign compound.

Enantiomeric compounds have the same atomic composition and relative spatial distribution and connectivity but are not superimposable. Because of the nearly identical chemical structure and, therefore, boiling points and chemical affinities to standard stationary phases, chiral separations can be a challenge. The use of cyclodextrins in chromatographic stationary phases is the standard approach. The cyclic molecules are made of 5–7 chiral sugar molecules that preferentially retain one enantiomer over its chiral partner. Chiral stationary phases generally separate the chiral compounds and are often used as the second dimension in GCxGC applications. Hong *et al.* paired an RTX-5MS column with a Supelco β-DEXTM 225 column in a two-gas chromatograph multi-dimensional (MDGC) system to determine the volatile flavour components in citrus fruits [5]. Optimising MDGC is still a challenging undertaking, as Hong *et al.* did not succeed in separating the α-pinene enantiomers and did not baseline separate the sabinene enantiomers. As the determination of chiral ratios becomes easier, so does the identification of an adulterated natural product.

Chiral columns are regularly employed to separate enantiomers both for gas and liquid chromatographic systems. But chiral compounds were first analysed using their optical rotation power. A chiral compound will physically turn plane-polarised light. This unique characteristic of chiral compounds has been exploited by Lafhal *et al.* in using a polarimetric HPLC detector to create chiroptical fingerprints of lavender and lavandin essential oil [6]. After separation on a chiral column, the resulting chromatogram contains positive and negative peaks corresponding to the relative rotational power of the compound. Non-chiral compounds are not detected by the polarimetric detector, and compounds with a low power of optical rotation are difficult to detect. The authors describe the resulting chromatogram as a chiroptical fingerprint because it is characteristic of the sample but does not represent a complete chromatogram of the sample. The authors subject the chiroptical fingerprints to statistical methods of discrimination and were able to discriminate lavender from lavandin species using the first two principal components: camphor and linalool content.

Vibrational circular dichroism (VCD) is an instrumental spectroscopy technique that capitalises on the unique characteristics of chiral compounds. Like IR spectroscopy, VCD measures the absorbance of infrared light due to bond vibrations, but in this instance, the chiral asymmetry is also accounted for in the spectra, allowing for the differentiation of enantiomers and verified by theoretical predictions. Bulk IR-VCD analysis of white wormwood (*Artemisia herba-alba*) essential oil was compared, as a means of validation, to GC-MS, chiral HPLC and VCD analysis of purified chiral standards. This essential oil has three major chiral compounds, (−)-α-thujone, (+)-β-thujone and (−)-camphor, which dominate the VCD spectrum, and their concentration and enantiomeric ratio affect the resulting spectrum in a way that the authors plan to exploit to identify adulteration [7].

3.2.2 *Mass spectrometric techniques*

Gas chromatography was a natural choice to couple with mass spectrometry as a complementary detection approach for several reasons. Identification of compounds is of particular interest while analysing complex and natural products and the direct linkage of gas-phase analytes into electron impact ionisation and quadrupole mass detectors allows for direct identification of molecules without major complexing products. GC-MS generates nearly pure mass spectra that are reproducible across mass spectrometric instruments, thus allowing the application of spectral libraries for reliable peak identification.

In the past five years, along with the increasing application of statistical profiling to be covered in more detail later in the chapter, some researchers have bypassed the need for chromatographic separation before MS analysis. Taverna *et al.* report the use of paper spray mass spectrometry (PS-MS), to directly analyse a drop of essential oil. The resulting spectra contain mass fragment information of all of the components in the oils [8]. The profile is then subjected to multivariate chemometric analysis to discriminate between the patterns. Linear discriminant analysis (LDA) and soft independent modelling of class analogy (SIMCA) pattern recognition procedures were able to distinguish bergamot essential oil samples and other citrus essential oils that could be used for bergamot oil reconstitution, as well as admixtures of bergamot oil with other oils.

As the scientific community turns increasingly to computer modelling, it is expected that the same trend will be applied to nearly all sectors, including essential oil analysis. Beale *et al.* note that the current ISO and AFNOR norms for identifying essential oils are generally too broad to use to determine the market quality of an essential oil [9]. In the case of lavender essential oil, a wide range of oils that fit the required ranges of the seven compounds mandated in the ISO norm may demand significantly different values on the market due to more subtle variations. The researchers used GC-MS data and, again, a SIMCA modelling method to denote 15 peaks that are significantly different in high- and low-quality lavender essential oils.

3.2.3 *Isotopic methods*

First published in 1976, isotopic ratio mass spectrometry (IRMS) is a precision instrument that allows for highly precise measurement of isotopes of carbon, hydrogen and nitrogen [10]. A typical instrument includes a high-resolution capillary gas chromatograph, reference gas in-line, combustion reactor, water removal and mass-spectrometer with a multicollector. Samples are separated via GC, and the effluent peaks are combusted at over 800°C to convert molecules into CO_2, NO_x and H_2O. In carbon isotope ratio analysis, H_2O is carefully removed to avoid HCO_2^+ interference. A reference gas of the same target element is introduced prior to analysis by the mass spectrometer. An array of Faraday cups simultaneously measures the relative abundance of predetermined masses; for carbon analysis, these are 44, 45 and 46 *m/z*, corresponding to $^{12}C^{16}O_2$, $^{13}C^{16}O_2$ and $^{12}C^{18}O^{16}O$, respectively [11]. The ratio of each isotopomer is compared to the reference gas, and the difference from the reference is reported.

This information is used to assess the origins of the material being analysed. Plants integrate isotopes depending on their photosynthetic and respiratory processes. The resulting naturally occurring isotopic ratio in plants is characteristic of their cycle type. Isotopic ratios also vary between natural and synthetic (petroleum-based) chemicals. IRMS is used to identify hormonal doping in athletes as well as synthetic adulteration in essential oils. Finally, with robust libraries of data, geographic origins of samples can be identified. IRMS, especially compared to other techniques discussed in this summary, is an expensive instrument which requires highly trained users and high-quality reference materials and databases. Despite the cost, IRMS is also considered integral in some food authentication practices.

3.2.3.1 $^{13}C/^{12}C$

The natural abundance of ^{13}C is 1.109%. Geographic location and biological processes can affect the incorporation of ^{13}C into the metabolism of a plant (or animal). ^{13}C abundance is not homogenous across the globe; therefore, plants growing in an environment with a higher abundance of the isotope will also have a higher $^{13}C/^{12}C$ ratio. In the case of metabolism-driven variations, the kinetics of ^{13}C *on* and *off* of reaction sites is slightly slower than ^{12}C due to the primary isotopic kinetic effect and can lead to preferential incorporation of ^{13}C in C_4 and CAM plants. Preferential incorporation of isotopes can be measured by IRMS in order to identify natural products that have been adulterated by synthetic compounds and natural compounds with different origins than the target product.

While IRMS is considered integral to the analysis of some essential oils, other oils' isotopic data are still being added to the available published literature. Nagarmotha (*Cyperus scariosus*) essential oil and rhizomes extract were analysed by IRMS with the isotopic range being −15.78‰ to −11.15‰ for the oil compared to −16.17‰ to −11.53‰ in the corresponding rhizome raw material [12]. White truffles from Italy, a high-value unprocessed foodstuff, were found to have $\delta^{13}C$ values between −42.6‰ and −33.9‰; additionally, knowing the compound with the signature aromatic properties, in this case bis(methylthio)methane, gives an extra point of quality evaluation during analysis [13].

The Sciarrone truffle analysis also makes use of multidimensional gas chromatography (MDGC) to achieve better chromatographic resolution before IRMS analysis. 35 saffron samples were similarly evaluated by Rocchi *et al.* in 2019; Italian saffron was shown to have δ13C between –27.37 to –31.52 but non-Italian saffron fell between –25.19 and –27.10 [14]. Studies like these are strengthened by having several authentic samples to determine the acceptable range, and then the authors, or other laboratories, can apply the ranges to real samples to assess the authenticity of the ingredients.

Schipilliti *et al.* are among the most prolific authors on the application of IRMS for essential oil analysis, with several articles applying IRMS to authenticate essential oils and other natural products, like vanilla and orange flavour [15]. C-IRMS is used to add a layer of data to the authentication practices of essential oil, and thus a base of data must be available to compare new samples to the literature. In 2018, the group analysed citrus petit grain oils in order to provide a second set of $\delta^{13}C_{VBDP}$ data for mandarin and lemon petit grain oils, and a first set of published values for bitter orange petit grain oil [16]. In the same analysis, a deviation in MNM-anthranilate was attributed to the addition of MNM-anthranilate to correct the oil, demonstrating the use of IRMS analysis for the authentication of the oils. The Dugo research group has plans to multiplex multidimensional GC with IRMS in order to increase the resolution of peaks prior to combustion, a noted requirement for optimal $\delta^{2}H$ measurements. As of 2017, the application of GCxGC-IRMS to essential oils had not been published, according to Tranchida *et al.* [17].

For *Helichrysum italicum*, the choice of two internal standards allowed Schipilliti *et al.* to differentiate between genotypes [18]. Because C-IRMS effectively measures the way in which a plant uptakes carbon from the environment and how that carbon gets passed through the plant's metabolism, using an internal standard, a chemical found in the natural product that is linked to the compounds of interest metabolically, can correct for the environmental changes and show, instead, the variation caused by the plants' metabolism, which is dictated by the genetics of the plant.

3.2.3.2 *^{14}C*

^{14}C radiocarbon is an unstable carbon isotope that is found in living materials and begins to decay at a half-life of 5,730 years upon the death of the organism. This is the basis of the science of carbon dating. When a synthetic chemical is produced using petrochemical starting materials, it effectively has no remaining ^{14}C radioisotope. Measuring the amount of ^{14}C in a natural product can be used to identify the addition of petrochemically sourced synthetic material.

According to Cuchet *et al.*, naturally occurring ^{14}C in biologically sourced chemicals will have a measurable 13.7 ± 0.1 dpm/gC in 2017 (disintegrations per minute per gram of carbon) measured by liquid scintillation counting (LSC) or by accelerator mass spectrometry (AMS) [19]. Cuchet *et al.* used this physical property and the LSC method to identify wintergreen essential oil that had been adulterated with synthetic methyl salicylate, which showed a lower activity. The research also showed, shockingly, a sample adulterated with ^{14}C-labelled synthetic methyl salicylate.

Garlic essential oil has also been used as a test case for ^{14}C analysis for the determination of biological source. Five garlic essential oils were tested, and all five passed the traditional GC-MS quality controls, but only one had 13.78 dpm/g; one had 9.21 dpm/g and was determined to be only 67% biosourced [1].

Where an essential oil has a profile that can be easily mimicked or corrected with a cheaper synthetic alternative, the analysis using ^{14}C can quite directly detect the adulteration as the target, 13.7 dpm/gC is uniform across all samples.

3.3 Bidimensional Chromatography

3.3.1 *Comprehensive GCxGC*

The chemical complexity of essential oils makes them an interesting chemical question to researchers, with some oils containing hundreds of compounds. The quest to effectively resolve all of the naturally occurring peaks has pushed the technology of separation science [20]. Standard essential oil practices generally include a multistep analysis on a range of stationary phases in GC: polar, non-polar and chiral columns in a variety of chemistries. But even multiple analyses often leave unresolved peaks. The advent of multi-dimensional gas chromatography (MDGC) and its optimisation results in a much more comprehensive chromatogram. MDGC can come in a variety of formats and technological components. Lebanov *et al.* note that while the analysis of essential oils benefits from the MDGC systems, essential oils are often prime samples for the optimisation of their configuration as well.

MDGC can be as complex as a lab can imagine and build. Pantò *et al.* complexed a HPLC and three GCs to create two preparatory chromatographic systems [21]. Each system will depend on the resources of the lab and the separation desired. The simplest format is a single GC instrument with two columns in line. The first column is long, and the second is much shorter, allowing for a typical separation on the first dimension with a fast separation of some coelution in the second dimension. Beyond this format, the complexity comes from a variety of modifications, including the use of one or multiple GCs for unique oven ramps, heart cutting or comprehensive sampling, dean switches or cryogenic refocusing, modulation method and time, column lengths and stationary phase chemistries and, finally, the detection system. Systems can also be split into dual columns in the same dimension or split into a variety of detectors. Each modification generally aims to increase the practical peak capacity of the chromatographic system so as to more completely separate the essential oils. Additionally, a fully resolved peak results in more accurate quantification.

By including an enantioselective column as the second-dimension column, the enantiomeric fraction of chiral compounds can be determined. Enantiomeric fractions are one of many criteria that can signal the presence of foreign ingredients in an essential oil. Wong *et al.* use a heart-cutting approach to separate tea tree oil in about 25 minutes, where the second dimension separates three chiral pairs for comparison to authentic enantiomeric fractions [22,23].

The choice of column for each dimension is an ongoing challenge in this field, as each combination will provide better separation for some components while reintroducing coelutions at other points in the chromatography. Balancing the variables of this type of system is not simple. One option is to continue to add dimensions and splits in the system to analyse the samples from several angles, but this results in data that become increasingly more difficult to interpret.

3.3.2 *LC-GCxGC coupling*

A new take on an old approach seen in a 2015 article describes pre-separation of citrus essential oils to obtain the sesquiterpene hydrocarbon (SH) fraction using a silica LC column. Citrus essential oils were separated using an LC system, and the SH fractions were collected after PDA detection. Fractions were concentrated under nitrogen before being injected on a GCxGC-qMS system. This triple separation set-up resulted in detailed identification of the SH fraction of several citrus oils. However, the SH fraction, being chemically similar in polarity, did not appear to benefit greatly from 2D GC separation on D^1 SLB-5ms and D^2 Supelcowax-10 (100% PEG) and may leave room for further study for optimal 2D separation parameters [24].

3.4 Analytical Techniques Combined with Statistical Methods: Metabolomics

The analytical field, as a whole, has pulled increasingly from the —omics approach to using datasets of ever-increasing size and applying statistical methods to data analysis. Chemical and statistical fingerprinting allow for non-specific data analysis. That is to say, statistical methods can view the data as a whole and sort the data set by statistically significant similarities and differences. Complex data can be modelled and classified according to the most relevant data in ways that are not always obvious to the naked eye.

Metabolomics is the study of all the small molecules generated and consumed in the metabolism of an organism. Because of the sheer enormity of this undertaking, large dataset handling techniques have been developed, and their utility is applicable in a variety of fields. Essential oils are often classified as secondary metabolites; thus, it follows that metabolomic methods are logically applied. The small molecules found in essential oils are formed during biosynthetic processes in plants and are subject to variations based on genetics and plant environment. The complexity of essential oils and the natural variation inherent in natural products require data handling that can compare major and minor variations between samples.

The following sections cover recent publications that have combined analytical techniques with statistical analysis. Traditionally, analytical techniques are employed specifically, meaning signals are tied to a particular chemical or chemical characteristics and responses are standardised in order to quantify the chemical or characteristic. While these techniques can still result in specific information, the general approach is to regard the data as a whole, often referred to as a fingerprint, and less so the individual chemical species. Therefore, data acquisition must be more standardised, so each fingerprint can be compared. In addition, successful models need many samples to generate the model and even more to test the model before they can be used on unknown samples.

The most commonly used data analysis techniques are partial least squares discriminant analysis (PLS-DA) and principal component analysis (PCA). Both of these techniques are mathematical operations applied to data to reduce their dimensionality. That is, the many data points measured are transformed into components, and the components that are the most different (greatest variance) are deemed the most important for classification. The points are projected onto a new set of axes, resulting in a statistical model which clusters similar data sets and separates dissimilar data sets.

3.5 NMR

Nuclear magnetic resonance (NMR) spectroscopy is classically employed for the identification and structural verification of organic molecules. In general, the molecules are purified and analysed individually for structural elucidation. But with the increase in complex data modelling methods, NMR spectra are increasingly used as spectral fingerprints, with data handling flowing from metabolomics and machine learning algorithms [25,26].

NMR can be used to identify and quantify specific chemical compounds with the help of internal standards. In fennel essential oil, 400 MHz qNMR has been used to quantify the *trans*-anethole content, while simultaneously showing vegetable oil adulteration in affected samples [27]. In minimally complex oils, like 60–80% *trans*-anethole containing fennel, qNMR is a reasonable approach to simple and rapid quality control, but in oils with more varied compositions, the ^{1}H-NMR spectra can become too complicated to isolate a signal for a single compound. Adulteration can also affect the capacity of ^{1}H-NMR to quantify components, as the adulterant can obscure the peak used for quantification, as with the case of DPG contamination of *Cymbopogon* [28].

With the increasing ability to interpret and analyse complex data comes increasingly complex data to be analysed. If the modelling method can manage multiplexed chromatography data or NMR data of natural products, then the next logical test could be to use both sets of data in one model. Mehl *et al.* compared the resulting models of GC-FID, UHPLC-TOF-MS+/– and ^{1}H-NMR to better model the differences among several extraction methods using only 64 samples [29]. This type of modelling has also been demonstrated on a range of Brazilian edible oils, providing a method of oil identification based on NMR spectra [30].

High-field NMR (200, 400 and 500 MHz) produces well-resolved spectra but can be prohibitively expensive for quality control environments, often requiring a dedicated, highly trained operator as well as a constant supply of liquid helium and liquid nitrogen even during non-use. Low-field NMR (60–100 MHz) trades resolution for much lower operating costs. The resulting spectra are often treated more like a fingerprint and considered as a whole rather than analysing each peak for its chemical information. Patchouli essential oil was adulterated by 20% using 17 adulterants and analysed, undiluted, by LF-NMR. Nine of the 17 adulterants produced extraneous signals and were easy to identify by eye when the spectra were overlaid with a reference spectrum, while others produced empty regions in the spectra. The authors note the technique is not as informative as GC-MS, but the analysis time is merely minutes compared to the often-hour-long separation times used in GC protocols [31].

3.6 Other Types of Spectroscopy

Spectroscopic methods of analysis measure the way light interacts with chemical compounds. Incident light can be absorbed, reflected, refracted and rotated depending on the sample and can be used to identify and quantify chemicals. Advantageously, most spectroscopic methods are non-destructive, so expensive samples are preserved. Spectroscopic methods can be applied specifically to pure compounds or as bulk analysis of a mixture. Bulk analysis results in complicated spectra that are analysed as fingerprints using statistical methods of comparison. Again, as data handling methods become more easily managed by computer software, spectroscopic bulk analysis becomes more popular, especially in quality control environments.

For these reasons, for non-destructive bulk analysis and statistical comparisons, essential oils are an obvious use case. When essential oils can sell for over $100/mL, there is always an incentive to reduce or eliminate the sampling volume. What's more, an analysis that does not require a lengthy separation step like in HPLC or GC and instead reduces analysis times to a few seconds will increase testing throughput.

3.6.1 *Raman*

Raman spectroscopy measures inelastically scattered photons depending on the polarisability of the sample molecules. The resulting spectra contain characteristic bands, measured in wave numbers, which can be assigned to chemical structures and their movements such as bends, wags, rocks and stretches. Raman can be used to characterise a single molecule's structure or fingerprint a mixture. The acquisition of data is fast — taking only a few seconds — and non-destructive. Data can be interpreted by hand to identify bands which correspond to particular compounds in the oil, or they can be processed with PCA analysis or other statistical or data analysis methods to partition a set of samples.

These techniques have long been applied to pure compounds and more recently applied to complex mixtures with the help of multivariate analysis programs. As is expected, essential oils are also being treated with Raman spectroscopy for the detection of adulteration. Clove essential oil was found to be adulterated with benzyl alcohol and vegetable oil using Raman spectroscopy [32]. The research used PCA to group samples using two components as well as a comparison to a benzyl alcohol spectrum to validate the source of the adulteration. A notable upgrade to the instrumentation itself is the miniaturisation of the spectrometer into a handheld device. Portable instrumentation streamlines testing processes in industrial settings, thereby reducing or eliminating sampling time, transport time and reporting time for quality control.

3.6.2 *Fluorescence*

Fluorescence spectroscopy is a non-destructive analytical technique which measures the excitation and emission responses of a sample. With the excitation spectrum being the absorption spectra, the emission spectrum results from the fluorescent response of the sample. Each compound will have an excitation, and some will have an emission spectrum which represents the compound but is not necessarily unique. With the popularity of statistical modelling techniques for handling otherwise uninterpretable data, fluorescence spectroscopy can be applied as a fingerprinting technique. Essential oils, being a complex mixture of chemical compounds, will result in a complex fluorescence signature, due primarily to the presence of benzene rings or Π-conjugation-containing compounds in the sample.

Riza *et al.* have applied PCA to two fluorescence techniques to determine the geographical origins of patchouli oil from Indonesia [33,34]. Using about 50 patchouli essential oil samples from seven regions of Indonesia, the first three principal components were used to predict the regions of test samples. The authors did not investigate the sources of the fluorescence response in the samples, nor did they test the system with samples that are outside of the geographic region or of non-patchouli samples. The method as it stands is lacking robustness validation and the chemical explanation of the results.

Excitation-emission matrix fluorescence spectra of essential oil samples have been used as a non-specific technique combined with a type of artificial neural network (ANN), called a multilayer perceptron (MLP) [35]. MLP analysis was able to classify samples into the varying oils as well as further distinguish the pure oils from the adulterated oils. What is more, the pure oils were adulterated with vegetable oil, ethanol and another essential oil. The MLP further categorised the oils by adulterant. The limitation of this technique is the identification of adulteration at the 5% adulteration level. This study used only five essential oils and sunflower vegetable oil, along with the lab-prepared dilutions, totalling 96 samples. While this is one of the smallest sample sets reviewed, there is no testing dataset outside of the samples used to create the network.

The technology of neural networks is still being tested and understood; as the technology becomes easier to deploy, we will see its use expand. The current field appears to be lacking good analytical chemistry guidance as far as experimental design and validation are concerned.

3.6.3 *Infra-red*

Infra-red (IR) spectroscopy is a technique used to measure the rotational and vibrational absorbances of chemical compounds. In a pure sample, functional groups like C=O or –OH can be identified by characteristic peaks. IR spectroscopy comes in several formats, including near (NIR), middle (MIR), Fourier transform (FTIR) and attenuated total reflection (ATR). Much like other techniques discussed in this review, the chemical complexity of samples results in a highly complicated spectral signature that is not easily interpreted by eye, but rather analysed as a whole by a statistical method. IR techniques are popular because they are non-destructive, require little to no sample preparation and can be used on solid, liquid or gaseous samples; they can even be used in-line for production monitoring or in portable formats for on-site analysis.

Recently ATR-FTIR was proposed as a lower-cost and efficient method for the identification of vegetable oil, mineral oil or Vaseline used as diluents in commercial essential oils [36]. The method identifies several characteristic bands for alkane and aliphatic functional groups, and their presence in the sample would signal an adulteration. Gallart-Mateu *et al.* combined MIR and PLS-DA chemometric modelling to classify tea tree essential oils as authentic or non-authentic with a 98% certainty using a set of 267 samples of known quality [37]. The results are split into only authentic and non-authentic groups and do not give further information about the reason a sample is rejected.

The essential oils of lavender (*L. angustafolia* Miller) and lavandin (*L. angustafolia* x *L. latifolia* Linnaeus fils Medikus) are difficult to differentiate by typical chromatographic methods alone, as the majority of the compounds appear in both oils and at similar levels. By applying PLS-DA regression to NIR data, Lafhal *et al.* were able to 100% discriminate the species from its hybrid, as well as the sub-varieties within the two species, with 96–100% correct rates [38]. This is not a singular case; NIR has been widely employed as a raw material identification technique across industries and other essential oils [39].

Infra-red spectroscopy combined with statistical methods has been thoroughly demonstrated as a low-cost, non-destructive, widely applicable method of essential oil quality control. The technique distinguishes essential oils of varying identities, qualities and geographic varieties. It can also be applied to other raw materials, making it an asset in a quality control lab. A potential drawback is that this technique is not necessarily capable of describing what is chemically different amongst samples to give a deeper understanding or a reason that a sample is grouped a certain way and therefore rejected or not.

3.6.4 *Electronic nose*

Science has been attempting to capture the sense of smell in a repeatable and measurable way for many decades. The ability to mimic our sense of smell is elusive compared to the ways we can measure sight with film and digital cameras, touch with pressure-sensing piezoelectric measurements or sound with microphones. The other human senses can be directly converted into electronic signals to be measured and converted back into a form we can sense. This allows for measurement and verification by a scientist that the measurement corresponds to our human senses. This has proven to be much more complicated for chemical senses, i.e. olfaction and gustation [40–42].

Currently, the main approach to measuring odour besides olfactometry is through the use of electronic noses (e-nose). E-noses detect an odour as a whole, contrary to chromatographic approaches that separate each molecule to be measured individually. An e-nose is usually created using an array of different sensor types and can include miniature mass spectrometers, metal–oxide semiconductors, conducting polymers and more. These sensors generate electronic signals in response to the presence of different molecules and, as an array, generate a packet of responses that will be unique for a unique mixture of molecules. The approach is considered a fingerprinting method as the data set is used as a whole and compared using statistical models or artificial intelligence.

A tin-oxide seven-sensor array was used to classify rose essential oils into three quality classes [43]. To classify cumin samples and identify adulteration from coriander, Tahri *et al.* used a combination of an e-nose, a VE-tongue and SPME-GC-MS [44]. The VE-tongue uses noble and non-noble metallic electrodes rather than metal–oxide sensors and operates on liquid rather than gaseous samples. All of the data are processed using chemometric techniques, including PCA, hierarchical cluster analysis (HCA) and support vector machines (SVMs), to ultimately classify pure cumin by region and distinguish pure cumin from adulterated samples.

3.7 Non-Volatile Fraction Analysis by HPLC

Essential oil chromatographic analysis is primarily accomplished using GC. The composition of essential oils being small volatile molecules makes gas chromatography the logical choice. However, there are some cases in which HPLC can provide useful additional information about the authenticity or provenance of an essential oil sample. In the case of cold-pressed citrus essential oils, there is a fraction of the oil that is non-volatile. Likewise, in natural extracts such as rose or violet absolutes, part of the product will be non-volatile. In these cases, HPLC is a useful method of authentication.

Citrus oils are a particular class of essential oils because they are not distilled but cold pressed. Cold pressing is a mechanical process in the fruits themselves, thus resulting in the waxes and natural pigments that are soluble in the oil to remain in the final product. This contrasts with a distilled oil, where waxes and pigments are not volatile enough to carry over during the distillation process. The compounds most notable in the citrus oils' non-volatile fraction are oxygenated heterocyclic compounds such as coumarins, furocoumarins, psoralens and flavonoids. The content and quantity of the non-volatile fraction vary by species, much like the volatile fraction of essential oils, and therefore constitute another means of authentication [45].

While the analysis of citrus essential oils by HPLC is not a new approach, [46] technological updates in the field of LC are applied to essential oil research in turn. Russo revisited the citrus non-volatile analysis in 2015 with fused-core technology UPLC columns and significantly reduced analysis times by producing a 10-minute program capable of separating and quantifying 38 compounds. Better separation of compounds can help identify markers of adulteration. Russo specifically mentions the potential adulteration of bergamot with lime oil, which, according to their study, should result in some 19 extraneous peaks using this method [47].

In much the same way that many of the analytical methods previously discussed have applied metabolomics data handling, HPLC-MS can be applied in a non-targeted fingerprinting approach. Using PLS-DA, citrus species could be reliably discriminated from each other, while oils adulterated 10% with cheaper citrus oils were at least somewhat de-grouped [48]. However, these results need larger datasets for better models.

3.8 Biological Assays

3.8.1 *Protein post-translational modification profiling*

Advances in proteomics and genomics lead to the development of protein post-translational modification profiling. Exposing a cell to a new environment results in the proteins in the cell being dynamically regulated through normal cell metabolic processes. The phosphorylation or dephosphorylation of a protein can be observed throughout the cell proteome and is directly linked to the environment to which the cell has been exposed.

Protein post-translational modification profiling is used to study allergenicity and cell response to a number of challenges such as infection and pollution [49]. While the technique is incredibly insightful to the study of cellular biology, it can also be applied to study the environment rather than the cell. As the cell is responding to the environment, where we would normally study the cell, we can also interpret the cell response as an output of the environment itself. Exposing the HepG2 human liver cancer cell line to different essential oils resulted in the change in phosphorylation levels of several protein isoforms measured by immunofluorescent assays. The PPTM analysis of adulterated essential oils resulted in outputs observably different than those of quality oils of the same species, opening a new biological approach to essential oil quality assessment that can be achieved in a nanofluidic format [50].

3.9 Machine Learning

3.9.1 *Non-targeted*

A large portion of this update has focused on the non-targeted analytical advancements as applied to essential oils. Towards the goal of a universal quality assessment of essential oils, a non-specific comparative method that can successfully differentiate between quality materials and low-quality or compromised materials is an ongoing goal. Machine learning is a hot-topic technology, being widely applied to many sectors. Machine learning and deep learning are computer modelling methods that require a learning data set from which the algorithm identifies patterns and uses those patterns to make decisions. The algorithm can continue to incorporate new data points as they are given, and these updates continuously improve the algorithm. It is therefore important to ensure the quality of the metadata, i.e. the classification of sample quality, in the learning dataset. There are two schools of machine training: guided and unguided. Guided training involves giving the algorithm areas of interest; for example, in essential oils, the major compounds or a particular composition are considered ideal and the rest are non-conformant. Unguided algorithms are given data as a whole and left to discern the important details from the rest. There are trade-offs between these approaches, one major factor being the number of samples necessary to train and validate the model.

There have been many examples of PCA and partial least squares analysis as statistical models already discussed above where data are treated as a fingerprint of the chemical composition. Recent reviews have covered extensively how spectroscopic techniques have been combined with chemometric data analysis to identify, characterise and qualify plant extracts and beyond [51–53].

Beyond PCA and PLS-DA statistical models, other chemometric approaches have been applied to essential oils. GC-MS data modelled using PCA and PLS-DA identified 15 compounds to differentiate between two quality classes of lavender oil [9]. Yang *et al.* added similarity analysis (SA) and hierarchical clustering analysis (HCA) to quantitative GC-FID of patchouli extracts to further aid in the quality assessment [54].

3.9.2 *Classification*

Cluster analysis methods aim to identify similar samples based on their data. Similar samples clustered together can then be labelled with respect to known samples within the clustering model. Other clustering methods make use of other pieces of metadata, like geographic origin, to see if the clustering model will group these samples together, resulting in a model that can verify the metadata of an unknown sample. Classification can help clarify where the important features are contained in the data. For example, the quality of lavender oils can be largely determined by the linalool and linalyl acetate content of the oil as well as the total peak area of a gas chromatograph, and this is validated by HCA [55].

There are many factors which affect the essential oil of a given plant, generally referred to as natural variation. Essential oils are complex mixtures; moreover, their already complex compositions are not fixed, so the make-up of the oil varies by season, geographic location, genetics and stressors like drought, disease and pests. It is likely that these factors affect the composition in a specific and, therefore, detectable manner, but as these conditions are difficult to control and can interact simultaneously, the information contained in the small variations of oil composition is difficult to tease out with traditional data analysis. Intersecting computer science with essential oil analysis is advancing the state of the art in this complex field. ANNs use computer models to mimic the way we envision the human brain to work by grouping data into nodes and linking those nodes together. Once an ANN has been trained, it can classify new data inputs. In one initial study, bitter orange samples were classified into ripening stage subgroups [56]. The future of this approach will hopefully combine the analytical expertise of chemists with the technical expertise of engineers to maximise the data interpretation contained in these complex natural products.

3.10 Conclusion

3.10.1 *Looking at sectors facing similar issues*

Distillers purchasing raw plant materials can benefit from applying a variety of these techniques to the raw plant materials as well [26, 30, 44, 57–61]. Honey, beer, olive oil and spices suffer from many of the same challenges with adulteration that are seen in essential oils. While each product will have unique challenges, the solutions can help improve the control of quality in all sectors. McGrath has outlined a range of samples analysed using spectroscopic methods from alcohols and dairy products to oils and vinegar to detect adulteration, differentiation or geographic origin [62].

3.10.2 *DNA barcoding*

Correct classification of raw materials is a concurrent challenge in the essential oil industry. Unfortunately, distilled essential oils will not contain DNA beyond very small fragments due to the distillation process. While adulteration of an essential oil occurs after the distillation or cold pressing of the oil, it is possible, especially if the distiller is purchasing the raw material from another party, that the raw material is adulterated or contains mis-identified plant material. Genetic testing has become more accessible for quality testing of raw materials. DNA barcoding uses short regions of DNA which are used as genetic markers for a specific species. These markers are identified through plastid genome sequencing first and then used to differentiate between plant (or other organism) species [63]. One type of barcode, Inter Simple Sequence Repeat (ISSR), was used with hierarchical cluster analysis (HCA) to differentiate the leaves of the medicinal plant *Saraca asoca* from its adulterants [64]. Bishoyi *et al.* have begun the process of identifying PCR primers for key loci in *Cymbopogon* (lemongrass) species, as this is a key early step in genetic validation [65].

3.10.3 *Future of essential oil analysis*

As technology advances, we see its features applied to all areas of science and daily life. Essential oil analysis will continue to benefit from the newest analytical instrumentation and the advancement of computing power and statistical modelling. There has been a long trend towards PCA and PLS-DA interpretation of data. The non-specific data analysis approach will continue to improve the speed and cost of high-throughput quality control laboratories that need fast and reliable raw material qualification. However, often a simple pass/fail judgement of a sample is just the answer to the first question. Eventually, a scientific justification of the quality judgement will be requested by a customer, or even a court of law. So, the advancement of both non-targeted and targeted analytical technologies will serve the essential oil industry in the long term. In that regard, authorities should also increase the surveillance of online platforms where ephemeral companies can sell directly to end-users without control and regulations, exposing the consumers to fraud and safety breach, despite the heavy regulations applying to the conventional flavour, fragrance and cosmetics industry [66].

References

[1] H. Gershon, A. Lykkeberg, F. Goren, S. Mason. (2019). Identifying Fraudulent Natural Products: A perspective on the application of carbon-14 analysis, *J. Agric. Food Chem.*, 67, 13393–13399.

[2] T.K.T. Do, F. Hadji-Minaglou, S. Antoniotti, X. Fernandez. (2015). Authenticity of essential oils, *Trends Anal. Chem.*, 66, 146–157.

[3] E. Schmidt, J. Wanner. (2020). Adulteration of Essential Oils, in K.H.C. Başer & G. Buchbauer (eds.) *Handbook of Essential Oils: Science, Technology, and Applications* 3rd Edition (CRC Press, Boca Raton), p. 38.

[4] P. Satyal, W.N. Setzer. (2019). *Adulteration Analysis in Essential Oils* (Springer International Publishing, Cham), pp. 261–273.

[5] J.H. Hong, N. Khan, N. Jamila, Y.S. Hong, E.Y. Nho, J.Y. Choi, C.M. Lee, K.S. Kim. (2017). Determination of volatile flavour profiles of citrus spp. fruits by SDE-GC-MS and enantiomeric composition of chiral compounds by MDGC-MS, *Phytochem. Anal.*, 28, 392–403.

[6] S. Lafhal, I. Bombarda, N. Dupuy, M. Jean, K. Ruiz, P. Vanloot, N. Vanthuyne. (2020). Chiroptical fingerprints to characterize lavender and lavandin essential oils, *J. Chromatogr. A*, 1610, 460568.

[7] M.E.-A. Said, P. Vanloot, I. Bombarda, J.-V. Naubron, E.M. Dahmane, A. Aamouche, M. Jean, N. Vanthuyne, N. Dupuy, C. Roussel. (2016). Analysis of the major chiral compounds of Artemisia herba-alba essential oils (EOs) using reconstructed vibrational circular dichroism (VCD) spectra: En route to a VCD chiral signature of EOs, 903, 121–130.

[8] D. Taverna, L. Di Donna, F. Mazzotti, A. Tagarelli, A. Napoli, E. Furia, G. Sindona. (2016). Rapid discrimination of bergamot essential oil by

paper spray mass spectrometry and chemometric analysis, *J. Mass Spectrom.*, 51, 761–767.
[9] D.J. Beale, P.D. Morrison, A.V. Karpe, M.S. Dunn. (2017). Chemometric analysis of lavender essential oils using targeted and untargeted GC-MS acquired data for the rapid identification and characterization of oil quality, *Molecules*, 22, 1339–1350.
[10] M. Sano, Y. Yotsui, H. Abe, S. Sasaki. (1976). A new technique for the detection of metabolites labelled by the isotope 13C using mass fragmentography, *Biomed. Mass Spectrom.*, 3, 1–3.
[11] W. Meier-Augenstein. (1999). Applied gas chromatography coupled to isotope ratio mass spectrometry, *J. Chromatogr. A*, 842, 351–371.
[12] A. Kumar, A. Niranjan, A. Lehri, R.K. Srivastava, S.K. Tewari. (2016). Effect of geographical climatic conditions on yield, chemical composition and carbon isotope composition of nagarmotha (*Cyperus scariosus R. Br.*) essential oil, *J. Essent. Oil-Bear. Plants*, 19, 368–373.
[13] D. Sciarrone, A. Schepis, M. Zoccali, P. Donato, F. Vita, D. Creti, A. Alpi, L. Mondello. (2018). Multidimensional gas chromatography coupled to combustion-isotope ratio mass spectrometry/quadrupole MS with a low-bleed ionic liquid secondary column for the authentication of truffles and products containing truffle, *Anal. Chem.*, 90, 6610–6617.
[14] R. Rocchi, M. Mascini, A. Faberi, M. Sergi, D. Compagnone, V. Di Martino, S. Carradori, P. Pittia. (2019). Comparison of IRMS, GC-MS and E-Nose data for the discrimination of saffron samples with different origin, process and age, 106, 106736.
[15] L. Schipilliti, I.L. Bonaccorsi, L. Mondello. (2017). Characterization of natural vanilla flavour in foodstuff by HS-SPME and GC-C-IRMS, *Flavour Fragr. J.*, 32, 85–91.
[16] L. Schipilliti, I.L. Bonaccorsi, L. Mondello. (2019). Evaluation of the carbon isotope ratios of selected volatiles determined in several citrus authentic petitgrain oils. Bigarade (*C. aurantium*) petitgrain oil's first case report, *J. Essent. Oil Res.*, 31, 99–110.
[17] P.Q. Tranchida, F.A. Franchina, L. Mondello. (2017). Analysis of essential oils through comprehensive two-dimensional gas chromatography: General utility, *Flavour Fragr. J.*, 32, 218–227.
[18] L. Schipilliti, I.L. Bonaccorsi, S. Ragusa, A. Cotroneo, P. Dugo. (2016). Helichrysum italicum (Roth) G. Don fil. subsp. italicum oil analysis by gas chromatography — Carbon isotope ratio mass spectrometry (GC-C-IRMS): A rapid method of genotype differentiation?, *J. Essent. Oil Res.*, 28, 193–201.
[19] A. Cuchet, P. Jame, A. Anchisi, F. Schiets, C. Oberlin, J.-C. Lefèvre, E. Carénini, H. Casabianca. (2019). Authentication of the naturalness of

wintergreen (Gaultheria genus) essential oils by gas chromatography, isotope ratio mass spectrometry and radiocarbon assessment, 142, 111873.

[20] L. Lebanov, L. Tedone, M. Kaykhaii, M.R. Linford, B. Paull. (2018). Multidimensional gas chromatography in essential oil analysis. Part 1: Technical developments, *Chromatographia*, 82, 377–398.

[21] S. Panto, D. Sciarrone, M. Maimone, C. Ragonese, S. Giofre, P. Donato, S. Farnetti, L. Mondello. (2015). Performance evaluation of a versatile multidimensional chromatographic preparative system based on three-dimensional gas chromatography and liquid chromatography-two-dimensional gas chromatography for the collection of volatile constituents, *J. Chromatogr. A*, 1417, 96–103.

[22] Y.F. Wong, N.W. Davies, S.-T. Chin, T. Larkman, P.J. Marriott. (2015). Enantiomeric distribution of selected terpenes for authenticity assessment of Australian Melaleuca alternifolia oil, *Ind. Crops Prod.*, 67, 475–483.

[23] Y.F. Wong, R.N. West, S.T. Chin, P.J. Marriott. (2015). Evaluation of fast enantioselective multidimensional gas chromatography methods for monoterpenic compounds: Authenticity control of Australian tea tree oil, *J. Chromatogr. A*, 1406, 307–315.

[24] M. Zoccali, I.L. Bonaccorsi, P.Q. Tranchida, P. Dugo, L. Mondello, G. Dugo. (2015). Analysis of the sesquiterpene fraction of citrus essential oils by using the off-line combination of high performance liquid chromatography and gas chromatography-based methods: A comparative study, *Flavour Fragr. J.*, 30, 411–422.

[25] J.L. Wolfender, G. Marti, A. Thomas, S. Bertrand. (2015). Current approaches and challenges for the metabolite profiling of complex natural extracts, *J. Chromatogr. A*, 1382, 136–164.

[26] T. Kuballa, T.S. Brunner, T. Thongpanchang, S.G. Walch, D.W. Lachenmeier. (2018). Application of NMR for authentication of honey, beer and spices, *Current Opinion in Food Science*, 19, 57–62.

[27] S. AbouZid. (2016). Use of nuclear magnetic resonance spectroscopy in analysis of fennel essential oil, *Natural Product Sciences*, 22, 30–34.

[28] C.I. Cerceau, L.C.A. Barbosa, E.S. Alvarenga, C.R.A. Maltha, F.M.D. Ismail. (2020). 1H-NMR and GC for detection of adulteration in commercial essential oils of Cymbopogon ssp, *Phytochem. Anal.*, 31, 88–97.

[29] F. Mehl, G. Marti, P. Merle, E. Delort, L. Baroux, H. Sommer, J.-L. Wolfender, S. Rudaz, J. Boccard. (2015). Integrating metabolomic data from multiple analytical platforms for a comprehensive characterisation of lemon essential oils, *Flavour Fragr. J.*, 30, 131–138.

[30] J.S. Santos, G.B. Escher, J.M. da Silva Pereira, M.T. Marinho, L.d. Prado-Silva, A.S. Sant'Ana, L.M. Dutra, A. Barison, D. Granato. (2018). 1H NMR combined with chemometrics tools for rapid characterization of edible oils and their biological properties, *Ind. Crops Prod.*, 116, 191–200.

[31] A. Krause, Y. Wu, R. Tian, T.A. van Beek. (2018). Is low-field NMR a complementary tool to GC-MS in quality control of essential oils? A case study: Patchouli essential oil, *Planta Med.*, 84, 953–963.

[32] P. Vargas Jentzsch, F. Gualpa, L.A. Ramos, V. Ciobotă. (2018). Adulteration of clove essential oil: Detection using a handheld Raman spectrometer, *Flavour Fragr. J.*, 33, 184–190.

[33] D.F. Al Riza, S. Widodo, Y.A. Purwanto, N. Kondo. (2019). Authentication of the geographical origin of patchouli oil using front-face fluorescence spectroscopy and chemometric analysis, *Flavour Fragr. J.*, 34, 15–20.

[34] D.F. Al Riza, S. Widodo, Y.A. Purwanto, N. Kondo. (2019). Combined fluorescence-transmittance imaging system for geographical authentication of patchouli oil, *Spectrochim. Acta A Mol. Biomol. Spectrosc.*, 218, 155–160.

[35] W. Mbogning Feudjio, H. Ghalila, M. Nsangou, Y. Majdi, Y. Mbesse Kongbonga, N. Jaïdane. (2017). Fluorescence spectroscopy combined with chemometrics for the investigation of the adulteration of essential oils, *Food Anal. Methods*, 10, 2539–2548.

[36] K. Bounaas, N. Bouzidi, Y. Daghbouche, S. Garrigues, M. de la Guardia, M. El Hattab. (2018). Essential oil counterfeit identification through middle infrared spectroscopy, *Microchem. J.*, 139, 347–356.

[37] D. Gallart-Mateu, C.D. Largo-Arango, T. Larkman, S. Garrigues, M. de la Guardia. (2018). Fast authentication of tea tree oil through spectroscopy, *Talanta*, 189, 404–410.

[38] S. Lafhal, P. Vanloot, I. Bombarda, J. Kister, N. Dupuy. (2016). Chemometric analysis of French lavender and lavandin essential oils by near infrared spectroscopy, *Ind. Crops Prod.*, 80, 156–164.

[39] S. Kuriakose, I.H. Joe. (2013). Feasibility of using near infrared spectroscopy to detect and quantify an adulterant in high quality sandalwood oil, *Spectrochim. Acta A Mol. Biomol. Spectrosc.*, 115, 568–573.

[40] J. Golebiowski, L. Charlier, J. Topin, S. Fiorucci, S. Antonczak, V. Ferreira, R. Lopez. (2014). Chapter 96 — Molecular Features Underlying the Chemoreception of Odorant Binding Proteins and Olfactory Receptors. Insights from Molecular Modeling and Biophysical Data, in Academic Press (San Diego), pp. 519–523.

[41] C. Spence. (2019). Perceptual learning in the chemical senses: A review, 123, 746–761.

[42] J.N. Lundström, S. Boesveldt, J. Albrecht. (2011). Central Processing of the Chemical Senses: An Overview, *ACS Chem. Neurosci.*, 2, 5–16.

[43] A. Gorji-Chakespari, A.M. Nikbakht, F. Sefidkon, M. Ghasemi-Varnamkhasti, E.L. Valero. (2017). Classification of essential oil composition in *Rosa damascena* Mill. genotypes using an electronic nose, *Journal of Applied Research on Medicinal and Aromatic Plants*, 4, 27–34.

[44] K. Tahri, C. Tiebe, N. El Bari, T. Hübert, B. Bouchikhi. (2016). Geographical provenience differentiation and adulteration detection of cumin by means of electronic sensing systems and SPME-GC-MS in combination with different chemometric approaches, *Anal. Methods*, 8, 7638–7649.

[45] H. Fan, Q. Wu, J.E. Simon, S.N. Lou, C.T. Ho. (2015). Authenticity analysis of citrus essential oils by HPLC-UV-MS on oxygenated heterocyclic components, *J. Food Drug Anal.*, 23, 30–39.

[46] G. Dugo, I.S. d'Alcontres, A. Cotroneo, P. Dugo. (1992). On the genuineness of citrus essential oils. Part XXXV. detection of added reconstituted mandarin oil in genuine cold-pressed mandarin essential oil by high resolution gas chromatography with chiral capillary columns, *J. Essent. Oil Res.*, 4, 589–594.

[47] M. Russo, I. Bonaccorsi, R. Costa, A. Trozzi, P. Dugo, L. Mondello. (2015). Reduced time HPLC analyses for fast quality control of citrus essential oils, *J. Essent. Oil Res.*, 27, 307–315.

[48] J. Masson, E. Liberto, J.C. Beelor, H. Brevard, C. Bicchi, P. Rubiolo. (2016). Oxygenated heterocyclic compounds to differentiate Citrus spp. essential oils through metabolomic strategies, *Food Chem.*, 206, 223–233.

[49] K. Smiljanic, I. Prodic, D. Apostolovic, A. Cvetkovic, D. Veljovic. (2019). In-depth quantitative profiling of post-translational modifications of Timothy grass pollen allergome in relation to environmental oxidative stress, *Environ Int.*, 126, 644–658.

[50] Y. Urasaki, T.T. Le. (2019). Differentiation of essential oils using nanofluidic protein post-translational modification profiling, *Molecules*, 24, 2383–2397.

[51] M. Kharbach, I. Marmouzi, M. El Jemli, A. Bouklouze, Y. Vander Heyden. (2020). Recent advances in untargeted and targeted approaches applied in herbal-extracts and essential-oils fingerprinting — A review, *J. Pharm. Biomed. Anal.*, 177, 112849.

[52] K. Kucharska-Ambrożej, J. Karpinska. (2020). The application of spectroscopic techniques in combination with chemometrics for detection adulteration of some herbs and spices, *Microchem. J.*, 153, 104278.

[53] Y.B.R. Monakhova, Douglas N. (2020). Independent components analysis (ICA) at the "cocktail-party" in analytical chemistry, *Talanta*, 208, 120451.

[54] Y. Yang, W. Kong, H. Feng, X. Dou, L. Zhao, Q. Xiao, M. Yang. (2016). Quantitative and fingerprinting analysis of Pogostemon cablin based on GC-FID combined with chemometrics, *J. Pharm. Biomed. Anal.*, 121, 84–90.

[55] O. Marincaş, I. Feher. (2018). A new cost-effective approach for lavender essential oils quality assessment, *Ind. Crops Prod.*, 125, 241–247.

[56] S. Taghadomi-Saberi, S. Mas Garcia, A. Allah Masoumi, M. Sadeghi, S. Marco. (2018). Classification of bitter orange essential oils according to fruit ripening stage by untargeted chemical profiling and machine learning, *Sensors*, 18, 1922–1935.

[57] P. Galvin-King, S.A. Haughey, H. Montgomery, C.T. Elliott. (2019). The rapid detection of sage adulteration using Fourier transform infra-red (FTIR) spectroscopy and chemometrics, *J. AOAC Int.*, 102, 354–362.

[58] S.M. van Ruth, I.C.J. Silvis, M. Alewijn, N. Liu, M. Jansen, P.A. Luning. (2019). No more nutmegging with nutmeg: Analytical fingerprints for distinction of quality from low-grade nutmeg products, *Food Control*, 98, 439–448.

[59] S. Kiani, S.M. van Ruth, L.W.D. van Raamsdonk, S. Minaei. (2019). Hyperspectral imaging as a novel system for the authentication of spices: A nutmeg case study, *LWT*, 104, 61–69.

[60] P. Mishra, A. Kumar, A. Nagireddy, D.N. Mani, A.K. Shukla, R. Tiwari, V. Sundaresan. (2016). DNA barcoding: An efficient tool to overcome authentication challenges in the herbal market, *Plant Biotechnol. J.*, 14, 8–21.

[61] B. Gouilleux, B. Charrier, S. Akoka, F.-X. Felpin, M. Rodriguez-Zubiri, P. Giraudeau. (2016). Ultrafast 2D NMR on a benchtop spectrometer: Applications and perspectives, *TrAC*, *Trends Anal. Chem.*, 83, 65–75.

[62] T.F. McGrath, S.A. Haughey, J. Patterson, C. Fauhl-Hassek, J. Donarski, M. Alewijn, S. van Ruth, C.T. Elliott. (2018). What are the scientific challenges in moving from targeted to non-targeted methods for food fraud testing and how can they be addressed? — Spectroscopy case study, *Trends Food Sci. Technol.*, 76, 38–55.

[63] S. Marakli. (2018). A brief review of molecular markers to analyse medicinally important plants, *International Journal of Life Sciences and Biotechnology*, 1, 29–36.

[64] S. Hegde, A. Saini, H.V. Hegde, S.D. Kholkute, S. Roy. (2018). Molecular identification of *Saraca asoca* from its substituents and adulterants, *3 Biotech*, 8, 161.

[65] A.K. Bishoyi. (2017). A report on identification of sequence polymorphism in barcode region of six commercially important *Cymbopogon* species, *Molecular Biology Reports*, *44*, 19–24.

[66] M. Pierson, X. Fernandez, S. Antoniotti. (2021). Type and magnitude of non-compliance and adulteration in neroli, mandarin and bergamot essential oils purchased on-line: Potential consumer vulnerability, *Sci. Rep.*, 11, 11096.

https://doi.org/10.1142/9781800617780_0004

Chapter 4

Full Complementarity of Three Aroma Extraction Methods and Two GC Detection Techniques for the Analysis of Natural Odorants in Fermentation Media: Focus on Terpenoids, Phenols, Esters and Polyfunctional Thiols

Cécile Chenot[*] and Sonia Collin[†]

Unité de Brasserie et des Industries Alimentaires,
LIBST Institute, Université catholique de Louvain,
Louvain-la-Neuve, Belgium

[*]*cecile.chenot@uclouvain.be*
[†]*sonia.collin@uclouvain.be*

Abstract

The analysis of odorants in complex matrices poses significant challenges due to the trace-level presence of these compounds (typically less than 1%) amidst a wide variety of heavy compounds at much higher concentrations. Consequently, the initial separation of odorants from non-volatiles is a crucial step that requires careful optimisation. This selective extraction, which aims for high recovery, can exploit various physicochemical properties including solubility, volatility, electric charge or specific interactions, such as the affinity of thiols for metallic atoms. Concentration of the clean extract is then often necessary to meet the sensitivity limitation of the analytical tool. Gas chromatography (GC) remains the primary technique for separating and quantifying individual volatile compounds. It can be coupled with a variety of detectors, ranging from universal to highly specific, enabling precise quantitation of targeted compounds.

Beer, as a fermented matrix, exemplifies a complex matrix containing a diverse spectrum of oxygen-, sulphur- and nitrogen-containing volatiles. These compounds, with sensory thresholds sometimes as low as nanograms per litre, exhibit a significant organoleptic impact on the final product and are therefore worth studying. To investigate the influence of yeast on fusel alcohols, esters and terpenoids, solvent-assisted flavour evaporation (SAFE) followed by GC-mass spectrometry appeared to be an effective method. In contrast, polyfunctional thiols, present at even lower concentrations, require more specific extraction techniques. Solid-phase extractions using mercury- or silver-based phases have proven highly efficient and complementary for different thiol recovery. Additionally, the pulsed flame photometric detector (PFPD) offers excellent sensitivity for detecting and quantifying thiols in beer.

4.1 Introduction

An odorant is an organic volatile which interacts with olfactory receptors, either by the orthonasal (odour) or the retronasal (aroma) route.

To characterise an odour-active compound in a complex matrix, sensory descriptors (with or without hedonic tones) and quantitative data are needed. Sensory quantifications can be achieved through GC-olfactometry or by using a number of flavour units (concentration/sensory threshold ratio) of each odorant in the matrix.

The aim of this chapter is to give a rapid overview of the analytical techniques available for investigating traces of odorants in a complex matrix containing a wide variety of high-molecular-weight compounds. Results obtained by means of three extraction procedures and two detection techniques are further discussed as a tool for screening different yeasts in a fermentation medium.

4.2 Analysis of Aroma Compounds

Analysing a complex matrix for volatile compounds always requires the following steps: rough extraction allowing high recovery of the compounds of interest while leaving the rest behind; concentration (especially if the compounds are present at trace levels in the original sample); fine separation (by gas chromatography); identification (by mass spectrometry, retention index, etc.); and quantitation (by standard addition with or without an internal standard, isotopic dilution, etc.).

Aromas usually constitute less than 1% of the matrix (food, beverage, or plant). Therefore, the initial separation from non-volatiles is a key step to be optimised.

4.2.1 *Extraction techniques*

The rough extraction must be selective (for cleaning and concentration purposes), with a good recovery rate. It must also be repeatable, easy to implement, fast, and cheap. Many physico-chemical properties can be exploited for this purpose.

4.2.1.1 *Solubility*

Most volatile compounds have limited solubility in water. Therefore, they can be preferentially extracted from an aqueous phase or a non-fat-containing matrix (2) with an apolar organic solvent (1), such as dichloromethane or diethyl ether. For very apolar compounds, pentane or supercritical CO_2 can also be used.

Each compound (x) has its thermodynamic distribution coefficient (K_d) for a two-phase (1 & 2) system:

$$K_d = \frac{\text{Weight}_1^x}{\text{Weight}_2^x} \times \frac{\text{Volume 2}}{\text{Volume 1}}. \tag{4.1}$$

This equation shows the necessity of using a sufficient volume of solvent (2). It is also recommended to apply successive extractions with fractions of the total volume rather than only one extraction with the whole volume.

When applied to solid matrices, solvent extraction can be conducted in a Soxhlet apparatus (extensively used for delipidation) [1].

Before applying any other extraction procedure at a temperature above 30°C, it is also advisable to apply a liquid/liquid extraction with diethyl ether or dichloromethane to eliminate all reducing sugars (Maillard reaction reagents).

Unfortunately, liquid/liquid extraction cannot be used in all cases, especially for more polar aromas (for which it is difficult to find an efficient non-miscible organic solvent) and for protein-rich systems (where emulsions complicate separation of the two phases). To avoid emulsion issues and to reduce the solvent volume used, other solubility-based techniques can be applied. Among them, solid-phase extraction (SPE) is particularly appreciated. According to the nature of the sorbent (C18, XAD-2, etc.) and the solvent used for elution, the compounds are either washed, bound and/or eluted. Desorption requires a solvent for which the compounds have a stronger affinity than for the sorbent [2–5].

In its immersion version, solid-phase microextraction (SPME) is based on similar principles. A silica fibre coated with an adsorbent is immersed in the aqueous sample. Compounds of interest will be partitioned between the aqueous phase and the fibre. Once bound to the fibre, they can be thermally desorbed directly into a GC apparatus for analysis. Unfortunately, although this method is easy to implement, its reproducibility (from fibre to fibre) and its loading capacity are often low. An alternative to SPME is the stir bar sorptive extraction (SBSE, e.g. Twister®), where the immersed fibre is replaced with a stir bar, offering a greater adsorption surface. Both techniques have proved effective for aroma extraction from matrices such as fermentation media. Despite their prices, being fast and solvent-free makes these techniques really attractive [6–8].

4.2.1.2 *Volatility*

The high volatility of aromas often allows separating them from major compounds in the matrix. A wide variety of specific techniques, including static headspace technology, dynamic cold trap sorptive extraction, vacuum distillation and solvent-assisted flavour evaporation (SAFE), are commercially available.

The static headspace method exploits the equilibrium between a liquid and a gas phase in a sealed environment. Most of the volatile compounds tend to move into the gas phase (headspace), while the non-volatile or less volatile ones remain in the liquid phase. The more volatile the compound, the more concentrated it will be in the headspace. After simply waiting for an equilibrium to be reached, a headspace sample is taken and injected into the gas chromatograph (the volume is much greater, usually 500 μL of headspace, than the 1 μL of liquid injected when a chemical solvent is present). Standard addition in the presence of an internal standard (see section hereunder) is required to take individual recovery factors into account. The static headspace procedure is much appreciated (due to its automatic system, no solvent requirement, etc.) but is limited to the most volatile fraction (for example, fruity esters in a fermentation medium).

With the dynamic headspace method, the diversity and quantity of volatile compounds that can be analysed are greater because the method includes a step aiming to push the volatiles out (e.g. flushing the gas or liquid phase). The intermediate trap is either a cold trap (at −80°C, for example, in the purge and trap system) or a chemical phase (e.g. Tenax™, Figure 4.1) that does not retain ethanol and is thus very suitable for alcoholic beverage analysis [6, 9, 10].

The SPME method discussed above can also be used in headspace mode (where the fibre is placed above the liquid medium).

For distillation, the matrix (in the liquid state or dissolved in water) can be heated only slightly to avoid thermal degradation. Therefore, the whole system needs to be under reduced pressure to decrease the boiling points of the aromas. One, two or even more cold traps then re-condense the aromas in flasks for further solvent extraction, as discussed above. In this case, no lipids are present to create emulsions during liquid/liquid extractions or in liquid samples. The vessel also usually allows starting with a big volume, so that compounds of interest can be highly concentrated before

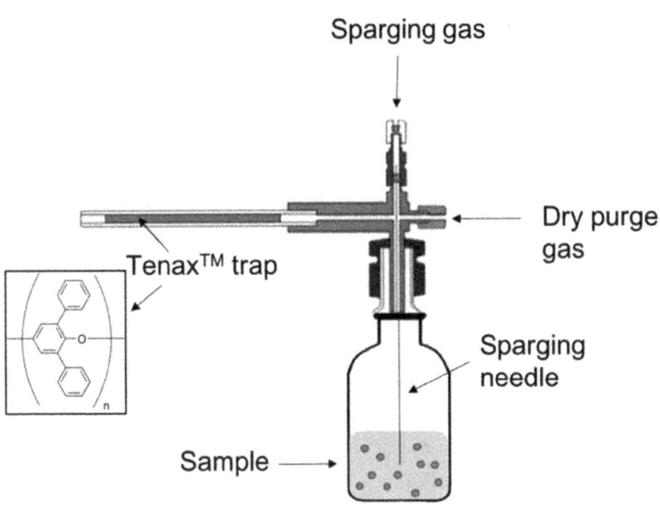

Figure 4.1. Dynamic headspace with Tenax™ trap.

GC injection (which is very suitable for trace analyses in fermentation media, an example being the cardboard defect brought by *trans*-2-nonenal in aged beers [11]). The sole limitation is that this method is time-consuming and is not automated [12].

SAFE (illustrated in Figure 4.2) is a variant of vacuum distillation allowing faster and careful isolation (recovery factor often above 75%) of most volatiles (complete extraction method illustrated in Figure 4.3) from either solvent-pre-extracted aqueous or lipidic media (A_1) (e.g. beer,

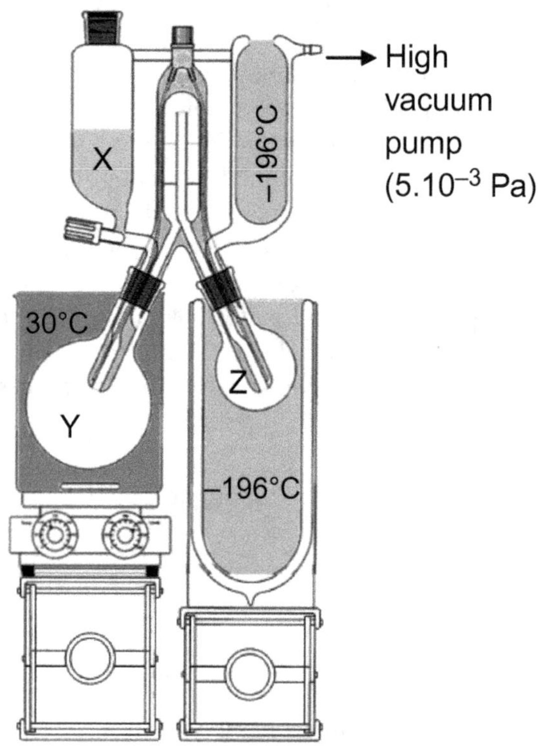

Figure 4.2. SAFE apparatus (modified from Engel *et al.* 1999 [13]): X: dropping funnel through which liquid extracts are introduced; Y: distillation vessel immersed in a 30°C bath, where liquid extracts first arrive at low flow (in the case of solids, a solvent is directly used in Y, with no use of X). Aromas leave vessel Y in the vapour state through channels (thermostated with water) leading to vessel Z. Z: distillation vessel immersed in a liquid nitrogen bath (−196°C) to instantly condense the aroma vapours. An additional cooling trap filled with liquid nitrogen prevents aroma losses. The whole apparatus is linked to a high-vacuum pump.

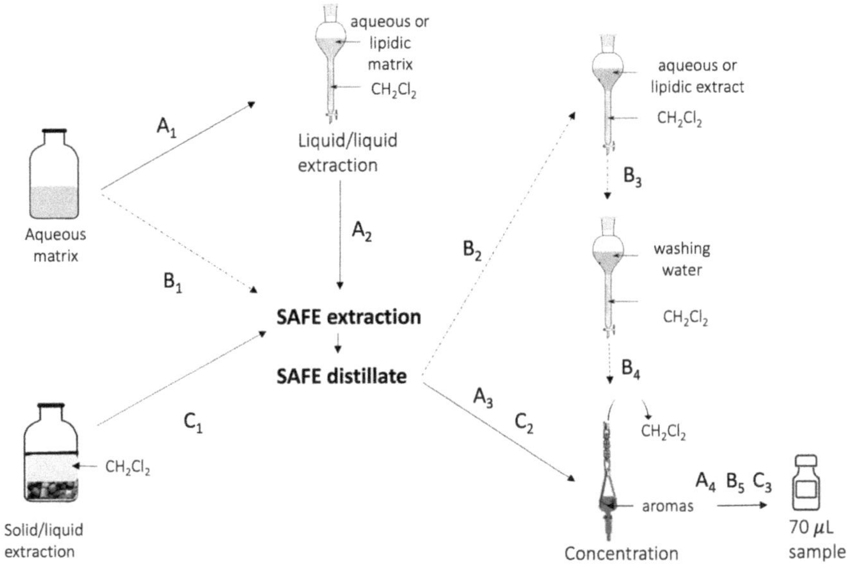

Figure 4.3. SAFE extraction *modus operandi.*

butter, or fruit pulp) or solid matrices (C_1) (e.g. cocoa). In the former case, the dichloromethane or diethyl ether sample is injected dropwise into the system for optimal aroma evaporation (A_2). A frozen distillate is obtained (A_3 or C_2) and further concentrated (A_4 or C_3) (as detailed in a following section). Original aqueous samples can also be directly inserted into the SAFE apparatus (B_1), but the resulting distilled extract requires further solvent extraction and washing with water (B_2, B_3, B_4) before concentration (B_5). Using this compact distillation unit not only allows efficient aroma isolation but also saves time [6, 13–15].

The Lickens–Nikerson apparatus (Figure 4.4) takes advantage of both the volatility and lipophilicity of aromas to extract them from a complex matrix (e.g. hop, sugar-free honey extracts, fermented olives, and odorant plastics). Only the most volatile compounds are able to leave flask A, which is heated to around 140°C (not possible for a sugar matrix, which would create artefacts through Maillard reactions). Only lipophilic compounds are moved into flask B after liquid/liquid extraction in the intermediate funnel. Very nice, representative extracts can be obtained by this procedure when protection from oxygen and optimised extraction times are applied [16, 17].

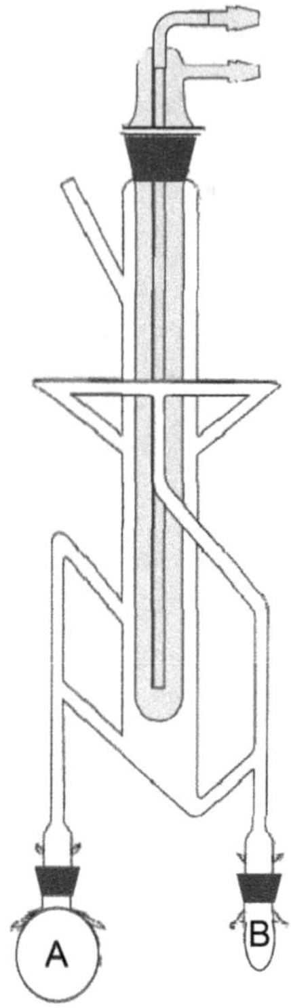

Figure 4.4. Lickens–Nikerson apparatus.

4.2.1.3 *Electric charge*

When charged, an aroma compound becomes too polar to be extracted in an organic phase. On the basis of this principle, modifying the pH of the aqueous phase helps to optimise the extraction of some ionisable compounds, such as phenols, carboxylic acids, pyrazines, and thiazoles, from complex matrices, especially ones rich in lipids (chocolate, cheese, etc.). A first dichloromethane extraction after acidification of the aqueous sample (to protonate pyrazines and thiazoles, for example) or its basification (to deprotonate acids and phenols, for example) helps to get rid of lipids (which move into the organic phase while the compounds of interest remain in the aqueous phase). The pH is then readjusted in order to neutralise the aroma compounds so that they can be extracted with the organic solvent [3, 12, 18, 19].

Hyphenated to SPE, this procedure further improves the extraction specificity (for example, deprotonation of sotolon and abhexon to better distinguish them from wine or beer lipophilic aromas on an XAD-2 column [20]).

4.2.1.4 *Specific interactions*

Aroma compounds containing a thiol function (SH-) can interact strongly with metal ions of mercury (Hg) or silver (Ag). This singular property can be used advantageously to extract polyfunctional thiols, specifically from natural media (where they are usually present at only nanogram-per-litre levels, thus requiring high selectivity, cleaning and concentration). The methods used include the *para*-hydroxymercuribenzoic acid (pHMB) procedure and Ag-ion solid-phase extraction (Ag SPE) [21–24].

Figure 4.5. illustrates the Ag SPE (loading of dichloromethane extract onto the active Ag sites in the cartridge) and pHMB procedures (additional liquid/liquid extraction of the dichloromethane extract before loading the aqueous phase on an anion exchange resin which binds the free pHMB carboxylic ions). In both methods, after rinsing of the solid phase, polyfunctional thiols are eluted with an excess of cysteine and then extracted with dichloromethane before concentration (detailed in a following section).

Besides being less time-consuming and more eco-friendly than the pHMB method, Ag SPE can be applied to smaller samples. One should note, however, that the distribution of extracted polyfunctional thiols can vary slightly according to the method (better recovery of most sulphanylalkyl esters with Ag SPE but lower recovery of 2-sulphanylethyl acetate, 3-sulphanylpropanol and 3-sulphanylpropyl acetate, as discussed in a following section).

4.2.2 *Concentration techniques*

As already mentioned, since many aromas are present only at trace levels in most studied matrices and since most rough extraction techniques yield a dilute aroma-containing solution, a concentration step is often required to reach detectable levels in GC. Among the low-boiling-point solvents, dichloromethane (40°C) and diethyl ether (35°C) are very often selected for the last step of the extraction procedure. When a high concentration factor is targeted, a bidistilled solvent is required to minimise artefacts in GC. Otherwise, impurities found at trace level in the original solvent would be highly concentrated, leading to unexpected peaks on GC chromatograms.

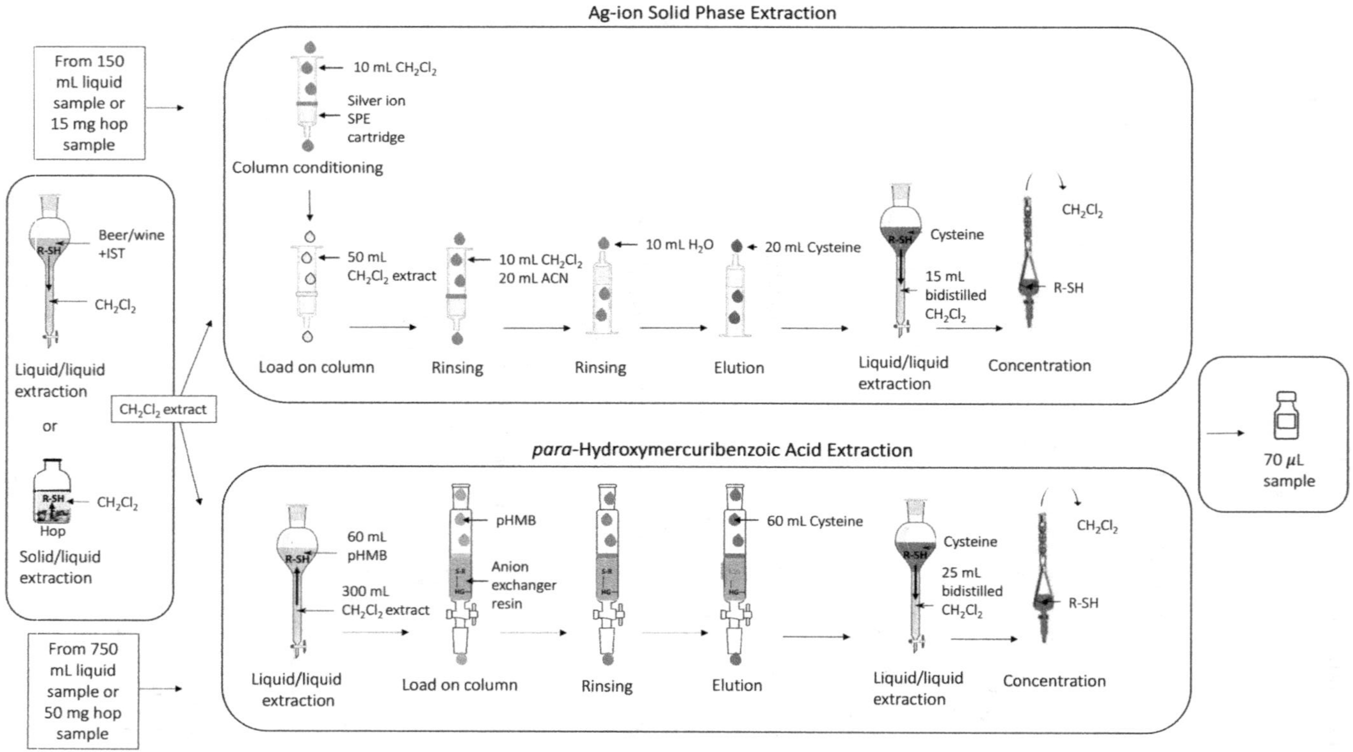

Figure 4.5. Specific extraction of free thiols (from solid matrices such as hop or liquid ones such as beer or wine) using Ag-ion SPE or *para*-hydroxymercuribenzoic acid.

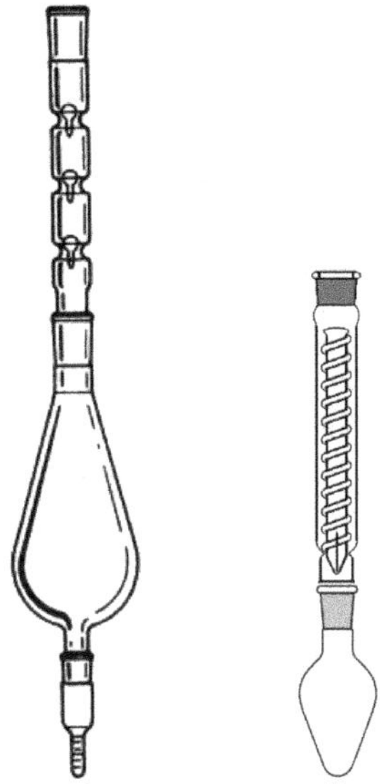
(a) (b)

Figure 4.6. (a) Danish–Kuderna and (b) Dufton reflux systems for volatile concentration.

The classic techniques used in organic chemistry (vacuum rotary evaporation for organic solvents and freeze-drying for aqueous samples) are often not recommended for volatiles, as they could be lost. Slower reflux-concentration techniques are often preferred, including the so-called "Danish–Kuderna" technique (Figure 4.6(a)), which can yield a final volume down to 250 μL without significant losses. To push solvent evaporation further, a similar but tinier apparatus called the "Dufton column" (Figure 4.6(b)) can be used (down to 70 μL extract). Assuming a 700 mL starting volume of complex medium, this means a concentration factor of 10,000 (for example, 1 ng/L thiol in a fermented medium becomes 10 μg/L injected into a GC-PFPD system; 1 μg/L phenols becomes 10 mg/L injected into a GC-MS system, etc.).

4.2.3 *Gas chromatography*

To separate each volatile from the others, GC is the most usual technique. Nowadays, efficient commercially available capillary columns (polar or apolar, sometimes with a chiral phase, length: 5–75 m, and internal diameter: 0.10–0.32 mm) allow reaching sufficient resolution whatever the compound. Of course, the operating conditions must be carefully optimised (including carrier gas flow favouring the highest number of theoretical plates, N_2 preferred to He (itself better than H_2), a sufficiently low oven temperature at the beginning of the separation, and the use of the solvent effect with a splitless or on-column injector).

GC, like any chromatography technique, is based on the respective affinities of the compounds to be solubilised in either the mobile or the stationary phase. The sample is injected into the column as a gas (or as a liquid which is first volatilised) and the compounds distribute differentially between the thin stationary liquid phase and the gas eluent. GC can be hyphenated to different detectors.

4.2.4 *Detection and quantitation techniques*

Thousands of aroma compounds can be identified in fermented media, but only a few are really essential to the aroma profile of a product. The techniques described in the following not only offer accurate quantitation but can also help determine which volatiles are actual key contributors.

4.2.4.1 *The universal flame ionisation detector (FID)*

The effluents from the GC column are directly introduced into an air-hydrogen flame where ionisation occurs (where cations are quantitated at the FID cathode). The response coefficient (the slope of area-versus-concentration in the extract curve) depends on the nature of the aroma (for instance, alcohols are often less sensitive than esters) but rarely allows measuring concentrations below the milligram-per-litre level at injection (1 μL is usually injected). Moreover, except for the retention index (or Kovats index) obtained by comparison with the retention times of a series of alkanes, the FID method does not allow identifying new investigated flavours.

4.2.4.2 *The pulsed flame photometric detector (PFPD) and other specific detectors*

A wide variety of specific detectors can be used to increase both selectivity and sensitivity: the NPD for nitrogen aromas (e.g. pyrazines), the FPD, PFPD and Sievers detector for sulphur-containing odorants (e.g. polysulphides, thioesters and thiols) and the ECD for halogens (e.g. trichloroanisole) and dicarbonyls (e.g. diacetyl).

PFPD and FPD work similarly: combustion of sulphur compounds in a hydrogen-rich flame creates excited-state S_2^*, which returns to its ground state by emitting at 394 nm. The relationship between response and concentration, which is exponential in a classic FPD detector, becomes linear (and within the same chemical family, it is even equimolar) in the pulsed-flame version (ignition occurs a few milliseconds after each C_2^*, CH* and OH* emission). This version is also much more sensitive, making it possible to quantitate compounds present at microgram-per-litre levels at injection. Another equimolar sulphur detector is the Sievers detector, which is highly sensitive for all S-compounds except thiols.

4.2.4.3 *Electronic impact mass spectrometry (EI-MS)*

This identification and quantitation technique uses 70 eV electrons to ionise GC-eluted compounds under high vacuum (10^{-2}–10^{-5} Pa). The molecular ion is then fragmented into smaller cations that can be distinguished according to their mass/charge ratios (electric and/or magnetic field). The resulting plot of the ion signal as a function of the mass/charge ratio is called the mass spectrum.

This spectrometer can be used in full-scan mode, collecting successively all the ions, or in the single-ion monitoring mode (SIM), making possible a 100-fold sensitivity increase (1–10 μg/L at injection is often quantifiable).

4.2.4.4 *Olfactometry*

In olfactometry, the human nose is used to detect odorants. To be sure that compounds, once separated, are not remixed in the detector, humidified air is sent, at 10–20 mL/min, into the funnel located at the outlet of the chromatographic column.

The odour intensity of each compound can be assessed through aroma extract dilution analysis (AEDA), which consists of diluting the extract until the aromas are no longer detected by the panellist. For each compound, the flavour dilution (FD) factor is defined as the highest dilution at which it is still detected:

$$FD = 2^n, \tag{4.2}$$

where $n + 1$ is the number of dilutions applied until no odour is perceived.

Other alternative procedures have been proposed in the literature (such as CHARM, OSME, Finger Span, and Rheostat) [25].

4.2.4.5 *Quantitation*

For each compound, the linear relation (or exponential relation for FPD) between peak areas and concentrations in the GC extract gives the response coefficient:

$$\text{Area }(x) = \text{CR}_x \times \text{Conc }(x), \tag{4.3}$$

where Area (x) is the area for X, Conc (x) is the concentration of X in the injected sample and CR_x is the response coefficient of X at the detector.

The concentrations in the extract are determined using a calibration curve. To obtain the concentration in the complex matrix instead of in the GC-extract, the recovery factor of each compound (RF_x) and the concentration factor between sample and extract (CF) must also be taken into account (although both are unknown in most cases):

$$\text{Area }(x) = \text{CR}_x \times \text{RF}_x \times \text{CF} \times \text{Conc }(x). \tag{4.4}$$

Therefore, the only way to get accurate quantifications is to apply either the time-consuming standard addition method (obtained slope = $\text{CR}_x \times \text{RF}_x \times \text{CF}_x$, which is then used for all similar samples) or the isotopic dilution procedure (in which RF_x and CF are the same for a compound and its deuterated analogue is used as an internal standard, IST).

Even with the traditional standard addition method, it can be advisable to use an IST (or an external standard, EST) to minimise the variation coefficients (similar extraction recovery and same concentration factor). In this case, the quantitation equation becomes:

$$\frac{\text{Surf}(x)}{\text{Surf (IST)}} = \left[\frac{\text{CR}_x}{\text{CR}_{\text{IST}}} \times \frac{\text{RF}_x}{\text{RF}_{\text{IST}}}\right] \times \frac{\text{Conc }(x)}{\text{Conc (IST)}}, \tag{4.5}$$

where $[\text{CR}_x \cdot \text{RF}_x/\text{CR}_{\text{IST}} \cdot \text{RF}_{\text{IST}}]$ is the slope determined by standard addition.

Sometimes, concentrations are given in IST equivalents, indicating that the RF is considered to be the same for all aromas and the IST (no standard addition needed). When commercially unavailable compounds are investigated, a similar simplification can be done: $CR_x = CR_{IST}$ (no calibration curve needed).

When using an equimolar detector (PFPD or Sievers), one can use the molecular weight ratio (MW_{IST}/MW_x) instead of determining the response coefficient ratio with a calibration curve.

Once the actual concentration is obtained, it can be compared to the threshold value to distinguish those molecules which have a real odour impact from the others (concentration above or below the threshold, respectively, flavour units, or AOV, > or < 1). Yet, as synergistic effects may sometimes occur, analytical results should always be compared to sensory analyses performed on the complex matrix.

4.3 Case Study of Odorants in a Yeast-Fermented Medium

Fermented media (beer, wine, etc.) are characterised by a broad spectrum of O, S and N volatiles, most of them being present at trace levels (ranging from mg/L to ng/L). As some sensory thresholds are also close to the nanogram-per-litre level, specific methods of extraction, concentration and GC detection are needed, depending on the nature of the targets.

Most of the volatiles found in such media do not directly arise through dilution of raw materials but are produced, biotransformed or enzymatically released by yeast from non-volatile molecules during the fermentation process. The results presented here illustrate how these reactions can be strongly dependent on the yeast strain.

In order to screen 11 yeasts (*Saccharomyces cerevisiae, S. pastorianus* and *S. diastaticus)* for the ability to produce (pleasant) fermentation flavours (esters, fusel alcohols, phenols, sulphur compounds, etc.) or to release them from bound precursors (e.g. polyfunctional thiols from S-conjugates and terpenols from glycosides), three complementary extraction procedures and two detection techniques were applied.

4.3.1 *Analysis of esters, fusel alcohols, phenols and terpenols*

4.3.1.1 *Esters, fusel alcohols and phenols by SAFE extraction — GC-MS (full scan) quantitation*

Esters are the biggest contributors of fruity aromas in most foodstuffs and natural products (including fermented media). Along with fusel alcohols, esters are produced by yeast according to parameters such as yeast growth, oxygen, temperature, wort density and lipid content.

Other wine or beer flavours are more raw-material-dependent, but the yeast strain still plays a key role, as the corresponding compounds undergo transformations catalysed by yeast during fermentation. For example, odorant volatile phenols are issued from hydroxycinnamic acids present in malt, but yeast is required for decarboxylation (to vinyl derivatives) and possibly reduction (to ethyl derivatives).

Screening of the 11 yeasts first focused on the production of the most common compounds in a fermented medium. Two hundred and fifty millilitres of a 12°P unhopped wort pitched with each yeast (0.5 g/L) and kept for 10 days at 22°C (primary fermentation) and for one day at 4°C (maturation) were further extracted by the SAFE procedure (route B, Figure 4.7). The following steps were carried out: 100 mL fermented medium + 0.025 mg/L 2-acetylthiophene as IST introduced into a SAFE apparatus; 3 × 30 mL dichloromethane extractions of the distillate; washing of the dichloromethane extract with 3 × 30 mL water; and concentration to 70 μL in a Kuderna–Danish and a Dufton apparatus. The resulting extract was then injected into a GC-MS system operating in full-scan mode. Compounds were identified by comparison with the retention times of commercially available standards and by checking the mass spectra against the NIST library.

Since the idea of this first part of the screen was only to compare the yeast metabolites produced rather than to quantitate selected compounds accurately, a simple quantitation in IST equivalents was carried out (relative to the IST RF and CR ratios set at 1 for all compounds) (Table 4.1).

Figure 4.7 shows the chromatograms obtained with yeasts of three different species. As expected, some compounds clearly appeared to be strain- or species-dependent, while some appeared quite common to all the fermented media.

Among the main esters, isoamyl acetate (odour descriptor: banana, candy; perception threshold: 0.5 ppm) is a key ester present in all fermented media. The so-called "apple esters" ethyl caproate (or hexanoate; odour descriptor: apple, rhum; perception threshold: 0.2 ppm) and ethyl caprylate (or octanoate; odour descriptor: pineapple; perception threshold: 0.9 ppm), are also commonly found in such media, as well as 2-phenylethyl acetate (odour descriptor: roses, honey; perception threshold: 0-25 ppm) [26]. *S. pastorianus* W3470, *S. cerevisiae* S33 and *S. diastaticus* WB06 appeared as the best ester producers (up to 1.5 ppm for all of them

Table 4.1. Concentrations (determined in IST equivalents) of four odorant esters, two fusel alcohols and a phenol in a model fermented medium (12°P wort, 10 days at 22°C and one day at 4°C).

	Saccharomyces										
	pastorianus			***cerevisiae***						***diastaticus***	
mg/L IST equivalent	**S189**	**W3470**	**S23**	**S04**	**S33**	**US05**	**K97**	**BE256**	**T58**	**BE134**	**WB06**
Isoamyl acetate	0.05	0.07	0.01	0.29	0.33	0.08	0.35	0.06	0.18	0.07	**0.65**
Ethyl hexanoate	0.08	0.19	0.02	0.13	**0.46**	0.15	0.14	0.01	0.05	*0.17*	0.32
Ethyl octanoate	0.17	**0.36**	0.01	0.13	0.18	0.32	0.80	0.03	0.01	*0.35*	0.27
2-Phenylethyl acetate	0.21	**0.82**	0.05	0.42	0.27	0.09	0.13	0.03	0.10	*0.69*	0.30
3-Methylbutanol	57.00	57.00	21.00	91.00	83.00	70.00	65.00	9.00	31.00	*91.00*	**129.00**
3-Ethoxy-1-propanol	0.01	0.01	0.01	0.29	0.11	**0.71**	0.01	0.01	0.01	*0.13*	0.60
4-Vinylguaiacol	0.03	0.06	0.01	0.14	0.14	0.19	0.11	0.01	**0.78**	*2.00*	**2.40**

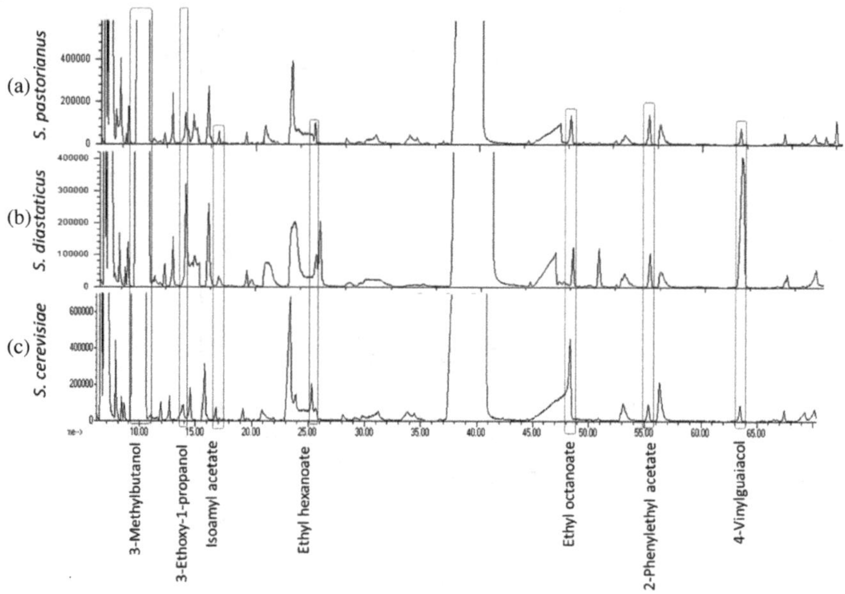

Figure 4.7. GC-MS full-scan chromatograms of SAFE extracts (route B) obtained from media fermented by (a) *S. pastorianus* (S189), (b) *S. diastaticus* (WB06) and (c) *S. cerevisiae* (K97).

together), while *S. pastorianus* S23 and *S. cerevisiae* BE256 hardly reached 0.1 ppm. Since fermentation alcohols can result from Ehrlich degradation of amino acids, it was no surprise to observe a high level of 3-methylbutanol (odour descriptor: alcohol; threshold: 70 ppm) in all trials as its precursor, leucine, is one of the most abundant amino acids in wort.

3-Ethoxy-1-propanol (odour descriptor: blackcurrant; threshold: 0.1 ppm) is another interesting alcohol also believed to be produced through the Ehrlich pathway from ethyl homoserine. This compound was already known to be highly dependent on the yeast strain and species, having been found at levels twice as high in beers fermented by ale yeasts than in beers fermented by lager yeasts [27]. The same trend was observed here, with the top three 3-ethoxy-1-propanol producers being two *S. cerevisiae* strains and one *S. diastaticus* strain.

As mentioned above, 4-vinylguaiacol (4-VG; odour descriptor: clove, curry; threshold: 0.3 ppm) is obtained through the decarboxylation of ferulic acid (one of the most abundant hydroxycinnamic acids). This conversion can either be thermal or enzymatic. In the latter case, only specific yeast strains with the phenotype known as "phenolic off flavour" (POF+) are able to catalyse this reaction. Accordingly, the highest 4-VG levels were found in media fermented with three POF+ yeasts (T58, BE134 and WB06).

4.3.1.2 *Terpenol release activity determined by SAFE extraction — GC-MS SIM quantitation mode*

The terpenoids found in fermented media such as beers or wines are highly dependent on the hop or grape variety, although most of them result from chemical and enzymatic transformations (oxidation, hydrolysis, isomerisation, esterification and enzymatic cleavage) of endogenous precursors. Geraniol (odour descriptor: floral, rose; threshold: 20–400 ppb) and linalool (odour descriptor: floral, lavender, cilantro; threshold: 2–80 ppb) are usually the main contributors of fruity, citrus, floral notes. Both compounds can be released from glycosidically conjugated flavour-inactive precursors (glycosides). Glucoside cleavage can either be chemical or enzymatic, the latter depending on yeast activity [7, 14, 26].

In order to assess the release of terpenoids from glucosides by our 11 yeasts, unhopped worts were spiked with a terpenoid glucoside (5 mg/L geranyl-β-D-glucose) before fermentation (same procedure as described in the previous section). The SAFE extracts were here analysed by GC-MS using the SIM method (focusing on m/z 69 and 93 for geraniol and on m/z 111 and 126 for IST detection). Release activities (expressed as percentages of the added amount of precursor) are depicted in Table 4.2. and Figure 4.8.

Here again, the activity appeared strongly dependent on the yeast strain, with a huge variability within the same yeast species (5.5–41.3% released geraniol). *S. pastorianus* S23 produced flavour from almost half of the hidden potential and *S. cerevisiae* BE256 to a lesser extent.

4.3.2 *Polyfunctional-thiol-specific analysis*

Polyfunctional thiols (PFTs), made of a carbon chain carrying both a thiol function and an alcohol, aldehyde, ester or ketone moiety (usually, three carbons apart from each other), are of prime interest because of their extremely low thresholds (1–100 ng/L) and characteristic odours. While only ppb levels of free PFTs have been found in grapes and hops (e.g. 3-sulphanylhexanol, 3SHol, odour descriptor: grapefruit, rhubarb,

Table 4.2. Percentage of geraniol released from its glucoside in a model fermented medium (12°P wort, 10 days at 22°C, one day at 4°C). No geraniol found in the unspiked-control.

	Saccharomyces										
	pastorianus			***cerevisiae***					***diastaticus***		
	S189	**W3470**	**S23**	**S04**	**S33**	**US05**	**K97**	**BE256**	**T58**	**BE134**	**WB06**
Released geraniol (%)	6.0	5.5	**41.3**	22.4	6.1	22.0	8.1	35.6	16.2	13.8	5.9

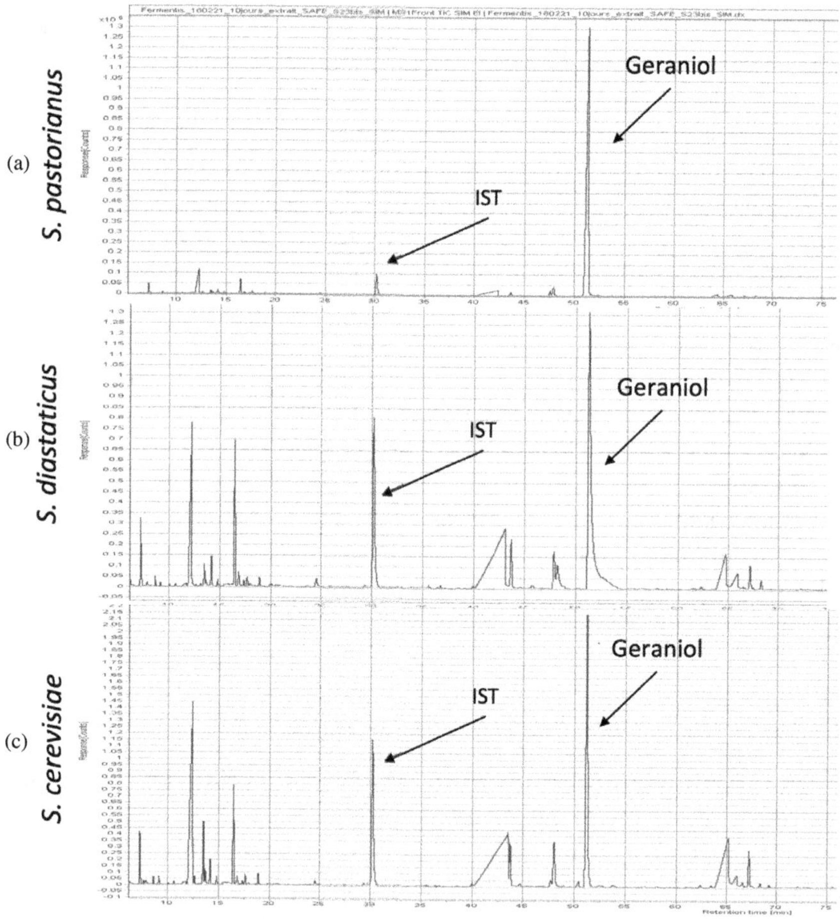

Figure 4.8. GC-MS SIM (m/z 111 and 126 for IST; m/z 69 and 93 for geraniol) chromatograms of SAFE extracts (route B) obtained from media fermented by (a) *S. pastorianus S23*, (b) *S. diastaticus WB06* and (c) *S. cerevisiae K97*.

passion fruit, threshold: 55 ng/L_{beer}; 3-sulphanyl-4-methylpentanol, 3S4MPol, odour descriptor: grapefruit, threshold: 70 ng/L_{beer}; 3-sulphanyl-pentanol, 3SPol, odour descriptor: citrus, threshold: 620 ng/L_{water}), to some extent, their flavour-inactive precursors can be degraded by yeast through fermentation, namely their cysteinylated conjugates (Cys-S-) and glutathionylated (G-*S*-) conjugates (e.g. Cys- and G- 3SHol, -3S4MPol, -3SPol) [14, 26–32]. Our 11 yeast strains were therefore also compared in terms of their ability to release free 3SHol, free 3S4MPol and free 3SPol from added Cys- and G-adducts. The main conclusions have been published in Chenot *et al.* 2021 [21] and 2022 [33], but in this chapter, the focus is on the complementarity of three extraction and two detection techniques which can be used for this purpose.

4.3.2.1 *SAFE vs. pHMB extraction: GC-PFPD quantitation*

Our first strategy was to inject the nonspecific SAFE extracts into a GC column hyphenated to the specific PFPD. SAFE extraction (route B in Figure 4.2) was applied to 100 mL fermentation medium (unhopped wort spiked with 10 ppm each of Cys-3S4MPol and G-3SHol before yeast inoculation; insignificant amounts of both free thiols were found in the control without spiking).

As depicted in Figure 4.9, a large number of sulphur compounds occurred in the fermented worts, only a few of which were thiols (among many sulphides, polysulphides, methional and methionol). Yet, this unspecific quick method allowed distinguishing yeast K97 for its ability to better release 3S4MPol from its Cys- adduct (0.26% vs. 0.02–0.06%) (Table 4.3). Five of the 11 yeasts generated a detectable peak of free 3SHol, confirming the ability of yeast to release thiols from G-adducts as well. However, SAFE extraction did not provide sufficiently clean and concentrated extracts to accurately quantitate the free thiol released from G-3SHol (free 3SHol below the PFPD quantitation limit).

The much more time-consuming thiol-specific pHMB extraction method was therefore applied to the same fermentation media, ensuring a higher recovery factor and a higher concentration factor (values relative to the IST recovery factor set at 1).

Extraction was performed on 750 mL fermented medium (+ 1.3 μg/L of 4-methoxy-2-methylbutane-2-thiol as IST). After extraction with 300 mL dichloromethane and 2 × 30 mL pHMB solution, the aqueous layers were loaded on Dowex resin. Elution was done with 60 mL cysteine solution, and the eluate was further extracted with 1 × 10 and 1 × 15 mL bidistilled

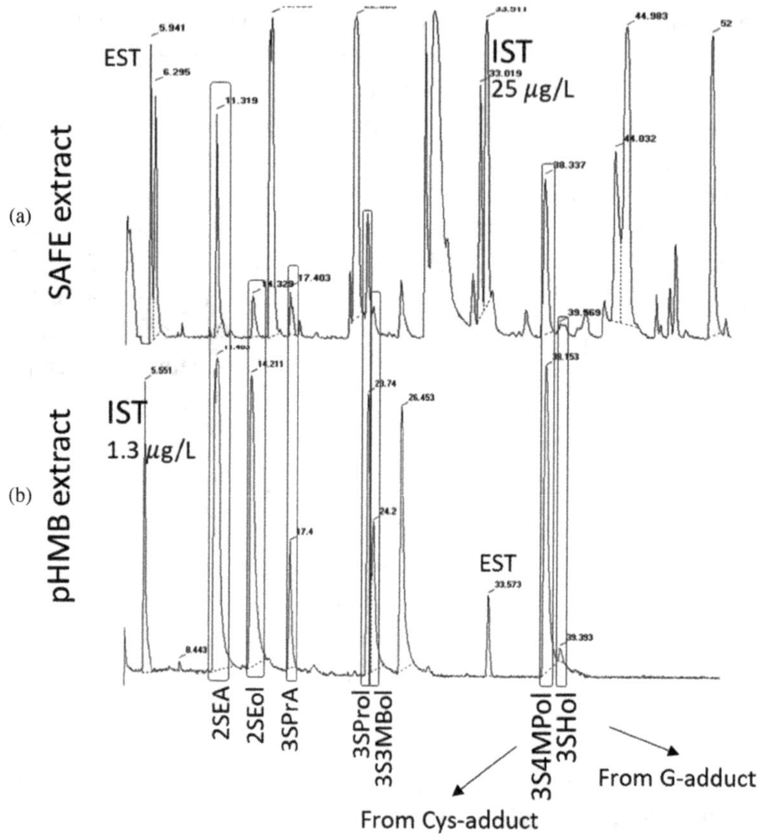

Figure 4.9. Comparison of GC-PFPD (FFAP column) chromatograms from (a) SAFE extract and (b) pHMB extract resulting from the same fermentation trial (12°P wort spiked with Cys-3S4MPol and G-3SHol at 10 ppm each before fermentation by K97 yeast).

Table 4.3. Efficiencies of thiol release from Cys-3S4MPol and G-3SHol added at 10 ppm to a 12°P wort fermented for 10 days at 22°C and for one day at 4°C.

	Saccharomyces										
	pastorianus				*cerevisiae*					*diastaticus*	
% released	S189	W3470	S23	S04	S33	US05	K97	BE256	T58	BE134	WB06
3S4MPol	0.03	d	0.06	0.05	0.02	0.02	**0.26**	0.06	d	0.05	0.02
3SHol	nd	nd	d	d	nd	nd	d	d	nd	d	nd

Note: nd, undetected; d, detected at a trace level.

dichloromethane before concentration to 70 μL (concentration factor: around 11,000 instead of 1,450 for the SAFE procedure) (Figure 4.5). As illustrated in Figure 4.9, the chromatogram of the pHMB extract was also cleaner than the previous one. It allowed accurate quantitation of 3SHol released from G-3SHol by K97 (0.01% in the selected conditions).

4.3.2.2 *pHMB vs. Ag-ion extraction: GC-PFPD quantitation*

In this last section, another thiol-specific extraction, using silver instead of mercury, was applied to our fermented media.

Instead of studying release from only one Cys-adduct and one G-adduct, we spiked worts either with three Cys-adducts (Cys-3S4MPol, Cys-3SHol and Cys-3SPol) or with their glutathionylated counterparts (G-3S4MPol, G-3SHol and G-3SPol). The fermentation parameters were also modified in order to better evidence sulphanylalkyl acetates produced by yeast from released sulphanylalkyl alcohols (15°P wort, 7 days of primary fermentation at 24°C, 3 days at 4°C).

As compared to the pHMB extraction, the specific silver-based procedure was applied to smaller volumes of sample (150 mL instead of 750 mL, so the concentration factor was also divided by five). Ag SPE avoids handling a mercury-containing reagent and uses much less solvent [23]. The extraction was performed as described in Figure 4.5 (150 mL fermented medium + 2 μg/L 4-methoxy-2-methylbutane-2-thiol as IST; extraction with 50 mL dichloromethane; loading on Ag SPE; 30 mL cysteine solution for elution; extraction with 1 × 5 and 1 × 10 mL bidistilled dichloromethane; and concentration to 70 μL).

The recovery factors relative to the IST of the two procedures were compared by standard addition. Compared to pHMB, Ag SPE appeared less thiol-specific, as it also recovered a huge amount of methionol (potential coelution with thiols; e.g. coelution with 3-sulphanyl-3-methylbutanol on CPSil5). Similar recoveries relative to IST were found for most polyfunctional thiols (from 0.8 to 1.2). The Ag SPE procedure emerged as much more efficient for sulphanylalkyl acetates, however, despite its lower concentration factor. As shown in Figure 4.10, three additional peaks corresponding to the pleasant 3-sulphanylhexyl acetate (3SHA; odour descriptor: passion fruit, box tree; threshold: 5 ng/L_{beer}), 3-sulphanyl-4-methylpentyl acetate (3S4MPA; odour descriptor: grapefruit; threshold: 160 ng/L_{beer}) and 3-sulphanylpentyl acetate (3SPA; odour descriptor: peach, passion fruit) were now quantifiable (all resulting from the yeast alcohol acetyl

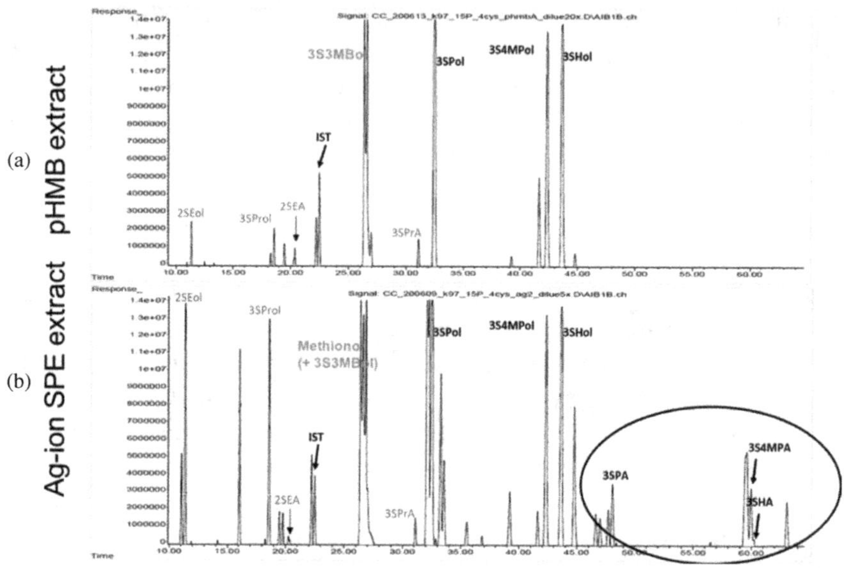

Figure 4.10. GC-PFPD (CPSil 5 column) chromatograms of a wort (15°P) spiked with 5 ppm Cys-3SPol, Cys-3SHol and Cys-3S4MPol (each), fermented by K97 and extracted by (a) the pHMB method or (b) the Ag-ion SPE method.

transferase activity). In a medium spiked with Cys-adducts, these esters could reach up to 0.1% release.

On the other hand, the pHMB method should be preferred when investigating Ehrlich-derived thiols, especially 2-sulphanylethyl acetate (2SEA) and 3-sulphanylpropyl acetate (3SPrA), which showed poor recovery and high variability with the Ag SPE procedure.

References

[1] C. Counet, S. Collin. (2003). Effect of the number of flavanol units on the antioxidant activity of procyanidin fractions isolated from chocolate, *J. Agric. Food Chem.*, 51, 6816–6822.

[2] S. Bailly, V. Jerkovic, A. Meurée, A. Timmermans, S. Collin. (2009). Fate of key odorants in sauternes wines through aging, *J. Agric. Food Chem.*, 57, 8557–8563.

[3] D. Callemien, S. Dasnoy, S. Collin. (2006). Identification of a stale-beer-like odorant in extracts of naturally aged beer, *J. Agric. Food Chem.*, 54, 1409–1413.

[4] C. Scholtes, S. Nizet, S. Collin. (2015). How sotolon can impart a madeira off-flavor to aged beers, *J. Agric. Food Chem.*, 63, 2886–2892.

[5] C. Scholtes, S. Nizet, H. Massart, P. Gerbaux, S. Collin. (2015). Occurrence of theaspirane and its odorant degradation products in hop and beer, *J. Agric. Food Chem.*, 63, 8247–8253.

[6] C. Barba, T. Thomas-Danguin, E. Guichard. (2017). Comparison of stir bar sorptive extraction in the liquid and vapour phases, solvent-assisted flavour evaporation and headspace solid-phase microextraction for the (non)-targeted analysis of volatiles in fruit juice, *LWT — Food Sci. Tech.*, 85, 334–344.

[7] M.-L.K. Cibaka, C.S. Ferreira, L. Decourrière, C.-J. Lorenzo-Alonso, E. Bodart, S. Collin. (2017). Dry hopping with the dual-purpose varieties Amarillo, Citra, Hallertau Blanc, Mosaic, and Sorachi Ace: Minor contribution of hop terpenol glucosides to beer flavors, *J. Am. Soc. Brew. Chem.*, 75, 122–129.

[8] G.A. da Silva, F. Augusto, R.J. Poppi. (2008). Exploratory analysis of the volatile profile of beers by HS-SPME-GC, *Food Chem.*, 111, 1057–1063.

[9] L. Gijs, F. Chevance, V. Jerkovic, S. Collin. (2002). How low pH can intensify beta-damascenone and dimethyl trisulfide production through beer aging, *J. Agric. Food Chem.*, 50, 5612–5616.

[10] C.A. Silva Guimarães Ferreira, E. Bodart, S. Collin. (2019). Why craft brewers should be advised to use bottle refermentation to improve late-hopped beer stability, *Beverages*, 5, 39.

[11] G. Lermusieau, S. Noël, C. Liégeois, S. Collin. (1999). Nonoxidative mechanism for development of trans-2-nonenal in beer, *J. Am. Soc. Brew. Chem.*, 57, 29–33.

[12] C. Counet, D. Callemien, C. Ouwerx, S. Collin. (2002). Use of gas chromatography−olfactometry to identify key odorant compounds in dark chocolate. comparison of samples before and after conching, *J. Agric. Food Chem.*, 50, 2385–2391.

[13] W. Engel, W. Bahr, P. Schieberle. (1999). Solvent assisted flavour evaporation — a new and versatile technique for the careful and direct isolation of aroma compounds from complex food matrices, *Eur. Food Res. Technol.*, 209, 237–241.

[14] C.S. Ferreira, S. Collin. (2021). Fate of hop and fermentation odorants in commercial belgian dry-hopped beers over 2 years of bottle storage: Key-role of oxidation and hop esterases, *J. Am. Soc. Brew. Chem.*, 79, 259–271.

[15] A. Dusart, B. Mertens, E. Van Hoeck, M. Simon, S. Goscinny, S. Collin. (2022). Occurrence of (suspected) genotoxic flavoring substances in belgian alcohol-free beers, *Food Chem.*, 369, 130917.

[16] A. Bouseta, S. Collin. (1995). Optimized likens-nickerson methodology for quantifying honey flavors, *J. Agric. Food Chem.*, 43, 1890–1897.
[17] G. Lermusieau, S. Collin. (2001). Varietal discrimination of hop pellets. II. Comparison between fresh and aged samples, *J. Am. Soc. Brew. Chem.*, 59, 39–43.
[18] M.-F. Herent, S. Collin. (1998). Pyrazine and thiazole structaral properties and their influence on the recovery of such derivatives in aroma extraction procedures, *J. Agric. Food Chem.*, 46, 1975.
[19] C. Scholtes, S. Nizet, S. Collin. (2014). Guaiacol and 4-methylphenol as specific markers of torrefied malts. Fate of volatile phenols in special beers through aging, *J. Agric. Food Chem.*, 62, 9522.
[20] C. Scholtes, S. Nizet, S. Collin. (2012). Occurrence of sotolon, abhexon and theaspirane-derived molecules in gueuze beers. Chemical similarities with 'yellow wines,' *J. Inst. Brew.*, 118, 223–229.
[21] C. Chenot, E. Thibault de Chanvalon, P. Janssens, S. Collin. (2021). Modulation of the sulfanylalkyl acetate/alcohol ratio and free thiol release from cysteinylated and/or glutathionylated sulfanylalkyl alcohols in beer under different fermentation conditions, *J. Agric. Food Chem.*, 69, 6005–6012.
[22] J. Gros, S. Nizet, S. Collin. (2011). Occurrence of odorant polyfunctional thiols in the super alpha Tomahawk hop cultivar. Comparison with the thiol-rich Nelson Sauvin bitter variety, *J. Agric. Food Chem.*, 59, 8853–8865.
[23] K. Takazumi, K. Takoi, K. Koie, Y. Tuchiya. (2017). Quantitation method for polyfunctional thiols in hops (*Humulus lupulus* L.) and beer using specific extraction of thiols and gas chromatography-tandem mass spectrometry, *Anal. Chem.*, 89, 11598–11604.
[24] T. Tominaga, M.-L. Murat, D. Dubourdieu. (1998). Development of a method for analyzing the volatile thiols involved in the characteristic aroma of wines made from *Vitis vinifera* L. cv. Sauvignon blanc, *J. Agric. Food Chem.*, 46, 1044–1048.
[25] M. Brattoli, E. Cisternino, P.R. Dambruoso, G. de Gennaro, P. Giungato, A. Mazzone, J. Palmisani, M. Tutino. (2013). Gas chromatography analysis with olfactometric detection (GC-O) as a useful methodology for chemical characterization of odorous compounds, *Sensors (Basel)*, 13, 16759–16800.
[26] S. Holt, M. Miks, B. Carvalho, M. Foulquie-Moreno, J. Thevelein. (2018). The molecular biology of fruity and floral aromas in beer and other alcoholic beverages, *FEMS Microbiol. Rev.*, 43.
[27] A.J. Irwin. (1992). 3-Ethoxy-1-propanol: A strain- and species-dependent yeast metabolite, *J. Inst. Brew.*, 98, 427–431.

[28] C. Chenot, R. Robiette, S. Collin. (2019). First evidence of the cysteine and glutathione conjugates of 3-sulfanylpentan-1-ol in hop (*Humulus lupulus* L.), *J. Agric. Food Chem.*, 67, 4002–4010.

[29] C. Chenot, S. Collin. (2021). Ability of the mandarina bavaria hop variety to release free odorant polyfunctional thiols in late-hopped beers, *J. Inst. Brew.*, 127, 140–148.

[30] J. Gros, F. Peeters, S. Collin. (2012). Occurrence of odorant polyfunctional thiols in beers hopped with different cultivars. First evidence of an S-cysteine conjugate in hop (*Humulus lupulus* L.), *J. Agric. Food Chem.*, 60, 7805–7816.

[31] M.-L. Kankolongo, L. Decourrière, C.-J. Lorenzo-Alonso, E. Bodart, R. Robiette, S. Collin. (2016). 3-sulfanyl-4-methylpentan-1-ol in dry-hopped beers: First evidence of glutathione S-conjugates in hop (*Humulus lupulus* L.), *J. Agric. Food Chem.*, 64, 8572–8582.

[32] R.E. Miracle, S.E. Ebeler, C.W. Bamforth. (2005). The measurement of sulfur-containing aroma compounds in samples from production-scale brewery operations, *J. Am. Soc. Brew. Chem.*, 63, 129–134.

[33] C. Chenot, W. Donck, P. Janssens, S. Collin. (2022). Malt and hop as sources of thiol S-conjugates: Thiol-releasing property of lager yeast during fermentation, *J. Agric. Food Chem.*, 70, 3270–3279.

Bibliography

Belgiorno, V., Naddeo, V., Zarra, T. (2013). *Odour Impact Assessment Handbook*, (John Wiley & Sons, West Sussex).

Collin, S. (2022). *Traité de Brasserie, Tome 1: La Bière et ses Matières Premières & Tome 2 : Etapes du Procédé Brassicole*, (Dunod, Paris).

Parker, J. K., Elmore, S., Methven, L. (2014). *Flavour Development, Analysis and Perception in Food and Beverages*, (Elsevier, Cambridge).

Reineccius, G. (1994). *Source Book of Flavors*, (Aspen Publishers, Inc., Maryland).

https://doi.org/10.1142/9781800617780_0005

Chapter 5

Manufacturing Natural Ingredients*

Bernard Blerot[†,‡] **and Marine Goubet**[†,§]

[†] *18/20 Av. Joseph Honoré Isnard, Grasse, France*

[‡] *bernard.blerot@iff.com*
[§] *marine.goubet@iff.com*

Abstract

Fragrant compounds have been extracted from plants for centuries. The flavour and fragrance industry now has access to modern extraction and purification technology; however, ancestral methods, such as distillation, are still in widespread use. After working with the principles of agronomy, to fully understand the importance of several specific botanical parameters and their influence on the organoleptic quality of the end ingredient — such as the genetic material of plants, farming conditions and location and post-harvest treatments — we move on to considerations related to processes. This chapter contains a review of all the classic extraction methods currently used in the flavour and fragrance industry. We present their principles, their specificities and the advantages and disadvantages of each method. We illustrate the information provided with examples based on the bitter orange tree, a very generous plant in the eyes of the flavour and fragrance industry, which yields its fruit, flowers, twigs and leaves to extraction.

*Translated from French by Amanda Ménage.

Initially, we discuss primary extraction methods, which are applied directly to the plant matter and which very often occur near the site of harvest: cold pressing for citrus fruits, steam distillation and hydrodistillation, organic solvent extraction and supercritical CO_2 extraction. We then discuss secondary extraction or purification processes, such as molecular distillation, fractional distillation and the refinement of absolutes. These practices involve treating the primary extract a second time to modify its composition to comply with restrictions imposed by its downstream use as an ingredient, such as to render it soluble in alcohol, to remove specific compounds or to increase the concentration of other compounds.

5.1 Introduction

Natural fragrance ingredients are manufactured using various methods and techniques that can be divided into two main categories: primary extraction techniques and processes, and secondary extractions or purifications.

Primary extraction methods involve moving compounds from one phase (plant matter) to another phase (the solvent) according to their differences in chemical potential. The main techniques that will be discussed in this chapter are cold pressing, steam distillation and hydrodistillation, organic solvent extraction and supercritical CO_2 extraction (Figure 5.1). The resulting primary extract can then be used as is or subjected to further purification using one or more secondary extraction methods, with the aim of modifying its physicochemical properties (solubility in alcohol, colour, etc.) or its olfactory properties (deterpenation, removal of fractions that are undesirable due to their olfactory profile or for toxicological reasons, etc.).

Primary extraction, which requires a significant quantity of biomass, is quite often carried out in close proximity to the sites where the plants are grown, both to preserve all the olfactory qualities of the plant matter and to avoid complicated logistics.

Indeed, for perfumery, our intention is to extract volatile metabolites present in plants at very low concentrations (0.01–2% on average). These metabolites often fulfil an important biological function for the plant, such as attracting pollinators, repelling herbivorous insects, communicating or increasing resistance to biotic and abiotic stress factors [1]. The concentration factor therefore ranges from 50 to 10,000×, which explains the importance of being located as close as possible to the growing site.

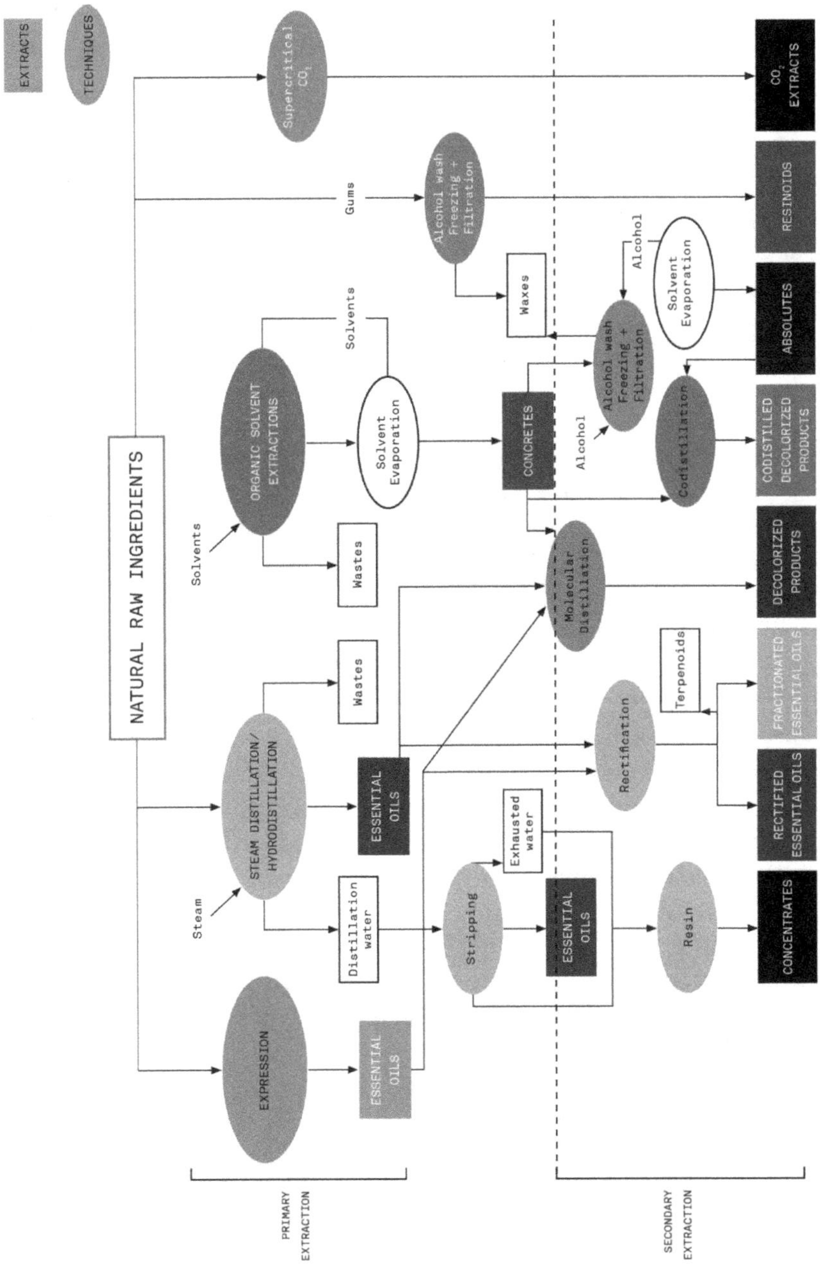

Figure 5.1. Main extraction and purification processes.

The strategy employed by industrial manufacturers is to move closer and closer to primary extract producers in the geographic areas where aromatic plants are grown by establishing long-term partnerships. The goal of such partnerships is twofold.

The first objective is to constantly improve control over quality and to ensure that production is sustainable, in keeping with international standards. This is a growing concern both for the clients of these manufacturers and for end consumers. Transparency is therefore increasing in the supply chains, which may now qualify for CSR certification from independent auditors.

The second objective of these partnerships is innovation. While innovation may be difficult for a secondary extract — since manufacturers have perfected more or less the same purification technologies — improvements and innovations at the source remain numerous. To distinguish themselves from the competition, manufacturers are investing in research programmes at both the agricultural and primary extraction levels.

Biology and agronomy experts aim to improve the genetic material of plants through selective breeding, to increase biomass production and yield per surface and to boost quality by implementing best farming practices and mechanising harvests, for example. This approach and an understanding of biology are helping researchers to make significant improvements in their supply chains and extraction methods.

The second focal point of innovation within supply chains is to improve primary extraction methods with the goal of reducing their inherent energy consumption (for distillation in particular) or to develop an alternative to petrochemical extraction solvents, which are highly toxic to humans and the environment.

Manufacturing a primary extract at the source is extremely important and greatly influences the final quality of the ingredient. For this reason, factors that contribute to ingredient quality (Section 5.2) will be addressed before the extraction methods are exposed in more detail (Section 5.3).

Secondary extracts, which are derived from primary extracts, require the use of more advanced technology (high vacuum, liquid nitrogen, etc.) and greater technical skill. The volumes processed are much smaller, and the equipment is often centralised in a factory that processes only natural ingredients in order to prevent cross-contamination and to pool resources (costly machinery).

Generally, the various advancements in extraction methods are often the result of small improvements at different stages in the process and are therefore difficult to protect by patents. Manufacturers work in closely guarded secrecy, each operating with their own techniques, which are closely linked to their equipment and are responsible for the differences in quality when compared to the competition.

The fragrance industry grows and processes approximately 120–150 botanical species to extract between 600 and 1,000 different qualities and grades of ingredients. Indeed, one botanical species can give rise to an array of different extracts according to the primary extraction method (distillation for essential oil or organic solvent extraction for concrete) and multiple refinement options.

This chapter discusses the production of natural ingredients using traditional methods, which still account for the vast majority of processing currently performed by industrial manufacturers. There are many R&D initiatives that aim to intensify extraction processes (e.g. pressurised fluid extraction, ultrasound- or microwave-assisted extraction) or to use green solvents (liquefied gases, supercritical fluids, agro-based solvents, ionic liquids, deep eutectic solvents, etc.), but these initiatives are not always appropriate in perfumery or economically realistic. They are not widespread at this point and will be discussed in Chapter 2.

To make it easier to understand the various methods discussed, we illustrate the techniques described with examples of extracts obtained from bitter orange (*Citrus aurantium* L.), a very generous tree in perfumery, yielding extracts from its flowers, fruits, twigs and leaves (see Figure 5.2, Tables 5.1–5.2 for chemical analysis of the extracts). We will go deeper in the analysis of the compositions provided in Table 5.1 later in this chapter.

5.2 Factors Influencing Ingredient Quality

Consistent quality is required when making natural ingredients. Due to the complexity and olfactory intensity of natural ingredients, variations in their quality can have a major impact on the final composition in which they are used. Stable quality requires masterful extraction, but skill is far from sufficient. Natural ingredients are unique precisely because they are derived exclusively from biological raw materials — materials that, by definition, change and are influenced by many factors prior to extraction.

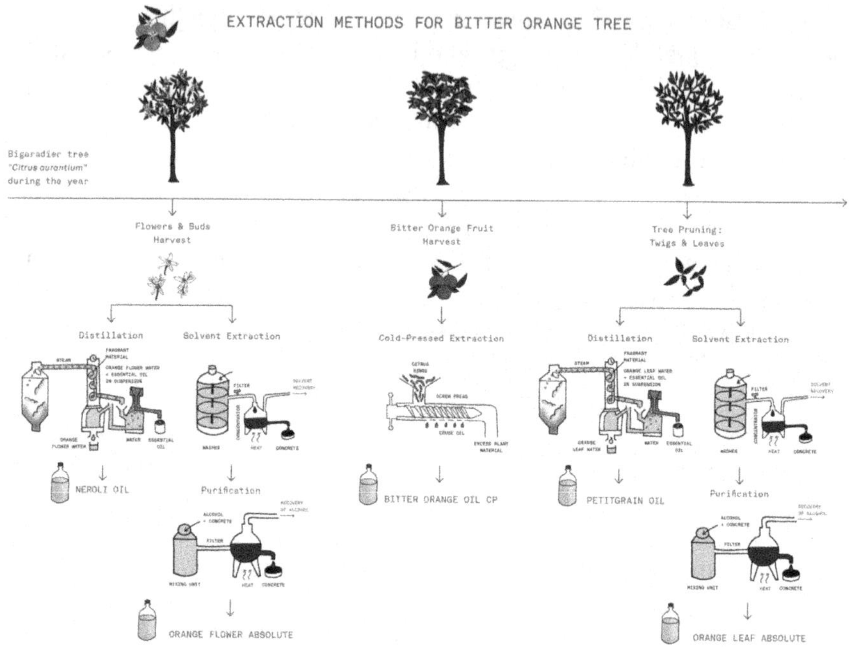

Figure 5.2. Overview of the various extracts that can be obtained from bitter orange tree.

The influencing factors have a major impact on the final quality of an ingredient and merit a brief discussion.

5.2.1 *Plant matter*

The quality of a natural ingredient is directly related to the quality of the raw material from which it is extracted. Perfumery calls upon a vast array of open-pollinated plant species and hybrid cultivars, which can be very closely related both genetically and morphologically while featuring completely different chemical profiles. The different breeds are also known as chemotypes or chemical phenotypes. For example, two hybrid cultivars of *Pelargonium* "Rosat" can produce very different essential oils (EOs) [2]. Some varieties of basil (*Ocimum basilicum* L.) yield highly divergent EOs (linalool chemotype, citral chemotype, methyl chavicol chemotype, etc.) and may cross-hybridise. In such cases, it is important to ensure that the fields are spaced far enough apart to prevent cross-pollination between species. This would cause shifts in quality due to seeds resulting from uncontrolled cross-breeding. It is crucial to have a thorough genetic characterisation of the plant from which the essence is extracted. This prevents deviations in quality, which could occur as a shift to a morphologically similar chemotype producing a different EO [3].

5.2.2 *Growing conditions*

The environmental factor and human intervention (growing practices) both have a significant impact on the final quality of the ingredient. If two genetically identical plants are cultivated in two different climates by two different people, the extracts will also be different.

The expression of genes coding the production of enzymes involved in the various biosynthesis pathways of fragrant molecules is very delicately regulated by mechanisms that remain poorly understood [4, 5]. The enzymes themselves have optimal functioning temperatures. They are more or less effective at converting one molecule to another depending on these conditions [6]. Perceptible differences in scent profile from one crop to the next are caused by the combination of all these conditions acting at the molecular level. Thus, in warm weather, we know that geraniol is converted to citronellol more efficiently in *Pelargonium* or *rose*. As a result, in warm years, we observe considerable variations in the proportion of these compounds in the EO, with a rise in citronellol in relation to geraniol (unpublished results).

The concept of *terroir* represents a certain consistency in conditions, such as temperature or day/night temperature differential, luminosity, rainfall and soil. Terroir is often limited to a specific geographic area and exerts a major influence on the quality of the final extract.

Human intervention is also a determining factor due to variations in farming practices. The way a shrub is pruned determines how many young leaves will sprout, and young leaves have a different chemical composition than older ones. The molecules in young leaves are produced further upstream of the biosynthesis pathway (geraniol), while those in older leaves are produced further downstream of the biosynthesis pathway (citronellol, etc.). Excessive or insufficient irrigation supplied by the farmer also affects EO quality [7].

5.2.3 *Harvest and post-harvest treatments*

The steps involved in harvesting and post-harvest treatment are also critical to the final quality of a natural ingredient. The way the raw material is harvested, how quickly it is transported to the factory and any pre-treatments (such as shredding, maceration or heat treatment) will influence not only yield but also the organoleptic qualities of the extract.

To take orange blossom as an example, harvesting practices are different between the two countries responsible for the bulk of production: Tunisia and Egypt. In Tunisia, the flowers are harvested by hand when they are big buds (just before the blossoms unfurl). In Egypt, on the other hand, the harvest involves knocking the trees with poles to cause the flowers to drop, resulting in a mix of blossoms at different stages of development (small bud, big bud or open flower). The quality of the flower essence is not the same between a small bud and an open flower (Table 5.3). As a result, the difference in quality between Tunisian and Egyptian orange blossom extracts is partly due to the method used to harvest the flowers. It is important to note that even within a single origin, quality also varies between the beginning of a harvest, when small buds account for the majority of flowers, and the end, when unfurled blossoms are proportionately more numerous. The resulting EOs will therefore smell fresher (linalool and linalyl acetate) if they are extracted early on in the harvest, or they will smell more intense (indole and methyl anthranilate) if taken closer to the end of the harvest.

The raw material may still be subject to several additional qualitative changes even after the harvest. While some raw materials are able to withstand long transportation times (wood, bark, dry leaves, etc.), which results in comparatively stable extract quality, fresh raw materials can yield extracts with higher variations in quality if storage conditions are poor or if the plant matter is processed quickly. Generally, raw materials with a high water content may be affected by a biological reaction with microorganisms, leading to changes in the aromatic compounds (methylation, oxidation, etc.). If they are controlled properly, these reactions can be valuable for the fragrance industry.

Indeed, agarwood infected with fungi results in specific animalic notes caused by carboxylic acids (hexanoic, heptanoic, caprylic, nonanoic,

Table 5.1. Chemical composition of orange flower extracts obtained using different extraction processes.

	Orange flower - study of the impact of the primary extraction process on extract quality (steam distillation vs. hydrodistillation vs. hexane extractions in two regions)				
		Egypt	**Tunisia**	**Egypt**	**Tunisia**
	Material	**Flowers**			
	Process	**Steam distillation**	**Hydro-distillation**	**Hexane extraction followed by purification with alcohol**	**Hexane extraction followed by purification with alcohol**
HEADS	α-Pinene	0.3	0.8	/	/
HEADS	β-Pinene	4.7	11.5	0.4	0.8
HEADS	Sabinene	0.5	1.1	0.1	0.2
HEADS	Myroene	1.8	2.0	0.1	0.2
HEADS	Limonene	8.0	13.0	1.7	1.7
HEADS	Cis-β-Oeimene	0.8	0.8	0.2	0.2
HEADS	Trans-β-Ocimene	4.8	6.9	1.0	1.4
	Linalool	39.5	38.4	40.7	38.2
	Linalyl Acetate	16.0	5.1	12.8	21.3
	β-Caryophyllena	0.7	0.7	0.2	0.1
	α-Terpineol	5.5	4.2	0.6	1.8
	Neryl Acetate	2.4	1.5	0.2	0.6
	Geranyl Acetate	4.6	2.8	0.3	1.1
	Nerol	1.2	0.9	0.2	0.4
	Geraniol	3.4	2.5	0.4	0.9
	Trans-Nerolidol	1.6	1.8	2.2	4.1
TAILS	Methyl Anthranilate	0.1	0.1	0.3	2.1
TAILS	Trans-Trans-Farnesol	1.2	1.4	2.3	5.9
TAILS	Indole	0.1	0.1	0.2	1.5
	Total	97.0	95.6	63.8	82.4

Table 5.2. Chemical composition of various extracts obtained by processing different parts of the bitter orange tree.

Orange flower - how the primary extraction process and the parts of the tree chosen for processing affect quality				
	Tunisia	**Tunisia**	**Egypt**	**Tunisia**
Material	**Flowers**	**Flowers**	**Fruits**	**Twigs & leaves**
Process	**Hydrodistillation**	**Hexane extraction followed by purification with alcohol**	**Cold-pressed extraction**	**Hydrodistillation**
α-Pinene	0.8	/	0.5	0.2
β-Pinene	11.5	0.8	0.4	2.2
Sabinene	1.1	0.2	0.2	0.4
Myrcene	2.0	0.2	2.0	2.0
Limonene	13.0	1.7	93.5	4.7
cis-β-Ocinwne	0.8	0.2	0.0	0.8
Trans-β-Ocimene	6.9	1.4	0.3	2.5
Linalool	38.4	38.2	0.3	15.8
Linalyl Acetate	5.1	21.3	0.5	52.8
β-Caryophyllene	0.7	0.1	0.1	1.1
α-Terplneol	4.2	1.8	0.1	3.4
Neryl Acetate	1.5	0.6	0.0	1.7
Geranyl Acetate	2.8	1.1	0.1	3.0
Nerol	0.9	0.4	0.0	0.7
Geraniol	2.5	0.9	0.0	2.0
Trans-Nerolidol	1.8	4.1	0.0	0.1
Methyl anthranilate	0.1	2.1	/	/
Trans-Trans-Farnesol	1.4	5.9	/	/
Indole	0.1	1.5	/	/

capric and 3-methyl-valeric acids). In China, osmanthus flowers are steeped in brine for several weeks before extraction takes place. This pre-treatment induces the development of the microorganisms responsible for

Table 5.3. Orange flower: Study of volatile compounds in the flowers at different stages of development.

Orange flower: Study of volatile compounds in the flowers at different stages of development 			

	Egypt		
(% composition)	**Small buds**	**Flower stage (big buds)**	**Flowers**
β-Pinene + Sabinene	0.55	0.34	0.38
Myrcene	0.17	0.16	0.16
Limonene	0.87	0.77	0.71
Trans-β-Ocimene	0.41	0.42	0.33
Linalool	61.62	54.17	42.43
Nerol	0.11	0.12	0.18
Phenylethyl Acetate	0.09	0,17	1.04
Linalyl Acetate	7.29	6.49	5.30
Indole	0.97	3.37	5.03
Methyl Anthranilate	0.17	0.43	3.36
β-Caryophyllene	0.05	0.11	0.05
α-Humulene	0.03	0.06	0.13
Trans-Trans-α- Farnesene	0.01	0.04	0.07
Trans-Nerolidol	0.18	0.97	3.33
Trans-Trans-Farnesol	0.56	4.05	9.89
Total	73.07	71.67	72.37

the "animalic" scent of the final extract, which is high in skatole, indole and an array of acids (valeric and butyric acids, as well as derivatives).

Other forms of processing, such as heat treatment, are also widely used to induce changes in raw materials to give them a specific character. For example, oakwood only develops its vanillic notes after being treated with heat [8], and orris also needs to be oxidised — naturally (atmospheric ozone), enzymatically or chemically — to generate the precious irones derived from iridales [9, 10].

5.2.4 *Primary and secondary extraction processes*

The fragrance industry uses many different raw materials, including fresh flowers, fresh or dry leaves, fresh or dry fruits, woods, barks, beans, roots and seeds. While, as we will see later on, certain methods are preferred for processing fresh raw materials (those that contain water), it is also possible to apply different processes to the same raw material to obtain extracts with different olfactory qualities.

These different primary extraction techniques rely on different physical and/or chemical properties and are more or less suitable depending on the plants to be processed.

Several raw materials can be processed using different methods: ginger can be distilled or subjected to supercritical CO_2 extraction; rose can be hydrodistilled or solvent-extracted; and orange blossom can be distilled (neroli) or solvent-extracted (orange flower absolute). When carried out under different conditions (temperature, pressure, contact with water, etc.), the processes yield different results and are used by perfumers in different ways, depending on their goals and creative needs. In the fourth section of this chapter, we will see that secondary extraction processes, or purification, also have a major impact on the quality of the final ingredient.

5.3 Primary Extraction Processes

"Primary extraction" refers to the first processing step applied to plant matter to extract the volatile odour compounds. Primary extraction is quite often performed near the site where the plants are grown, such that they can be processed without having to transport them over long distances. Lengthy transportation would result in impaired quality due to deterioration.

This is how vetiver EO is produced in Haiti, for example, and how jasmine and orange flower concretes are respectively produced in India and in Northern Africa.

5.3.1 *Cold pressing*

Expression, also known as cold pressing, is a technique used specifically to extract EOs from citrus fruit rinds. The EOs are contained in small sacs inside oleiferous glands at varying depths within the flavedo. The flavedo is the outer skin of a citrus rind, colourful and waxy, as opposed to the

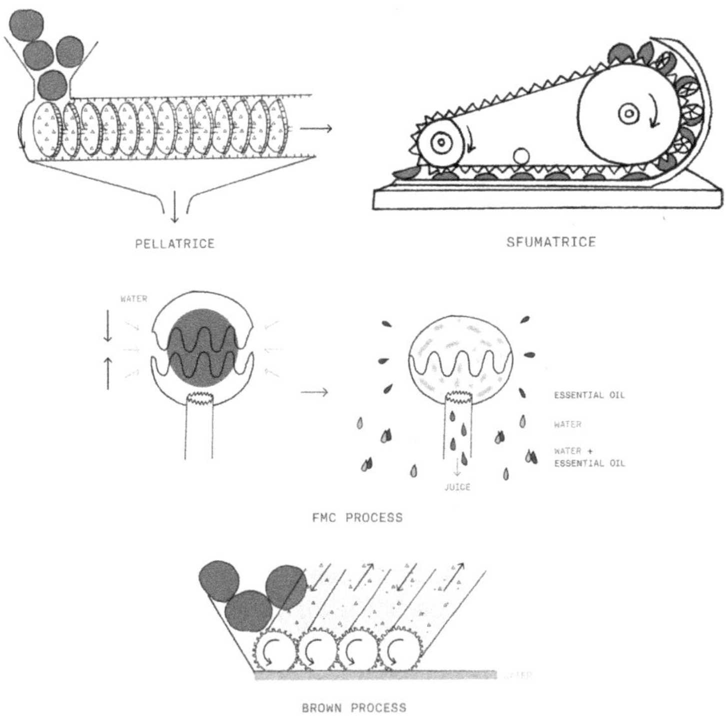

Figure 5.3. Diagrams of four types of cold-pressing equipment (Pellatrice, Sfumatrice, FMC Process and Brown Process).

albedo, which is the white, spongy, inner pith of the rind. Simply lacerating or pressing the flavedo releases the EOs by piercing the small pockets that contain it. Only citrus skins contain a concentration of EO high enough to be harvested mechanically. Several industrial techniques have been developed to extract citrus oils. The following provides a brief overview (see Figure 5.3 for diagrams).

Pellatrice: Whole fruits are sent through an abrasive horizontal tube fitted with an endless screw, also abrasive, which pushes the fruits through the tube. The fruits are sprayed with water inside the tube while they are being peeled. The water carries away the EO released, as well as the zest. Solid matter is filtered out, and then the EO is separated from the water in a centrifuge. The water is recovered and sent back into the tube. When they reach the end of the tube, the fruits are peeled and ready to be processed for juice and pulp extraction. The EO collected is never in contact with the acidic juice of the fruits.

***Sfumatrice*:** The fruits are sliced in half, and the juice is extracted from both halves by rotating pressure, like in a traditional citrus juicer. The half-rinds are then carried down a conveyor belt where they are compressed by a metal chain located directly above them. The pressure exerted on the rinds causes the small sacs containing the EO to burst open. As with the peeler, sprayed water carries the EO to centrifuges where the oil and water are separated. With this device, the EO may temporarily come into contact with a small amount of the acidic fruit juice that remains on the rinds after the first pressing. The profile of the resulting oil may be slightly different in comparison to the EO extracted with a peeler since hydrolysis may occur due to contact with the acidic juice.

***FMC Process*:** This process extracts the juice and the EO practically at the same time, without allowing them to come into contact with each other. The fruits fall into cup-like cavities that are then covered with another cup to envelop the fruit. The cup-shaped cavities do not have smooth rims; they are scalloped, with alternating metal tabs and empty spaces, which interlock when the two cups are facing each other. In the bottom of the lower cup, a knife cuts a circle in the rind of the fruit. At the same time, the upper cup moves downwards, gradually compressing the fruit. Under the effect of the compression, the juice and pulp are ejected through the circular hole at the bottom of the fruit. At this point, only the emptied fruit rinds remain. They continue to experience the growing pressure from the cups closing in on each other, which causes the sacs containing the EO to burst. The EO released is projected outwards by powerful sprays of water, and the water/oil emulsion is then separated in a centrifuge.

***Brown Process*:** Whole fruits are placed on a kind of conveyor belt featuring a succession of rollers moving in a circular and lateral pattern. The rollers are covered in spikes that pierce the epicarp of the fruits as they move along. The rollers are submerged in a thin layer of water, in which the EO exuded from the fruit trickles along. This EO-rich water flows in the opposite direction of the conveyor belt and is sent continuously into a centrifuge so that the oil can be isolated after the suspended solids have been removed.

Cold pressing is the preferred extraction method for citruses for the following reasons:

- It works well with the physiognomy of the fruits, whose EO pockets are located on the surface of the rind. This is the age-old method of citrus EO extraction; the pressing and collection techniques of the past have simply been mechanised and optimised on a larger scale.
- Today, the EO of citrus fruits is often a by-product of the fruit juice industry. Its extraction must therefore be compatible with juice extraction, a condition that is met quite well with cold pressing in modern industrial facilities.
- Citrus EOs contain a wealth of top notes (zesty, fresh scent) delivered by highly volatile compounds, which would be lost or broken down by heat-based extraction methods.

That said, distillation is sometimes used with spent citrus rinds that have already undergone cold pressing in order to extract a small amount of residual EOs and increase yield. However, the olfactory profile of the EOs obtained via distillation is very different and often less desirable than with cold pressing.

The EO accessed through cold pressing is practically identical to that secreted by the fruit, the only difference being that it passes through water during the industrial process. This is not the case with extraction techniques involving distillation or volatile solvents, which we will discuss later on.

Table 5.4 lists the pros and cons of cold pressing.

Table 5.4. Advantages and disadvantages of cold pressing.

Advantages of cold-pressing	Disadvantages of cold-pressing
• Essential oil intact and very close to the quality secreted by the citrus fruit • Low-cost extraction technique (high yield and often coupled with juice extraction) • Limited energy consumption since no heating is required	• Coloured essential oil • Furocoumarins extracted (EO in need of refinement)

For applications Table 5.2 in perfumery, cold-pressed citrus EOs are often modified. Indeed, the crude oils contain heavy, non-fragrant molecules, such as pigments and furocoumarins, which must be removed. The pigments colour the EO and render it unstable, and furocoumarins are photo-sensitising compounds that may only be present in low concentrations in cosmetic products designed to be applied to the skin. The heaviest compounds must therefore be isolated and removed, which is facilitated by their low volatility.

Furthermore, as shown by the composition of sweet orange EO presented in Table 5.2, citrus EOs can be very high in limonene (>93% in this case). The remaining few percentage points give the oil its character. Since limonene is among the most highly volatile compounds, it is common to remove a portion of it through rectification (partial distillation of the EO, causing the heads to evaporate). This concentrates the heavier volatile molecules — those that are responsible for giving a citrus its distinctive identity as compared with other citruses. In this second type of modification, the goal is to remove the most highly volatile compounds (heads). Both types of modification can be applied to the same EO.

5.3.2 *Steam distillation and hydrodistillation*

Distilling plant matter to extract the EO is carried out in four main steps:

- The plant cells that contain the EO are broken down upon contact with the hot water, whether it is in liquid form or as steam, releasing the EO into the still. Highly ligneous plant matter, such as wood or roots, may be distilled under two to three bars of pressure. This results in a steam temperature of between 120°C and 130°C (248 –266°F), which breaks down plant cells more quickly. Pressurised distillation can, however, result in notes of warmth or smokiness that may be undesirable from an olfactory perspective. The EOs obtained through pressurised distillation are generally flatter and longer-lasting than those obtained via traditional distillation. Hotter steam can carry heavier molecules, but it also destroys the most highly volatile compounds.
- Azeotropic distillation of the mixture (water + EO) contained in the still: since water and EO are not miscible, their respective steam pressures (P_{water} and P_{oil}) remain unchanged. The mixture vaporises when the total steam pressure $P_{tot} = P_{water} + P_{oil}$ is higher than the pressure in

the still (atmospheric pressure, in general). Since P_{tot} is always higher than both P_{water} and P_{oil}, this means that the boiling point of the mixture (water + EO) will always be lower than that of water alone or the EO alone. Distillation therefore lifts the EO out of the plant matter at temperatures below its boiling point.
- Condensation of the mixture (water + EO) along a cooling element.
- Separation of the EO from the hydrosol via decantation in a Florentine flask. In the vast majority of cases, the EO floats on top of the hydrosol.

Depending on the plant and the distillation equipment used, the hydrosol may be reintroduced into the still during distillation (known as *cohobation*), recovered and used as floral water, reprocessed to extract the water-soluble volatile compounds (volatile solvent extraction or application to a resin, which is then desorbed) or discarded.

It is important to note that EO obtained through distillation does not generally have a composition strictly identical to that produced by the plant from which it is extracted for the following reasons:

- Highly volatile compounds are not condensed and are therefore lost (e.g. *cis*-3-hexenol); heavier compounds are not fully extracted (e.g. indole or methyl anthranilate).
- Water-soluble volatile compounds are absent from, or only partially present in, the EO because they are preferentially trapped in the hydrosol (e.g. alpha-terpineol, geraniol).
- Compounds may be altered by heat or when they come into contact with water (hydrolysis, oxidation, etc.) (e.g. linalyl acetate).

In steam distillation, bursts of water vapour generated outside the still pass through the plant matter, whereas in hydrodistillation, the plant matter is submerged in boiling water, as shown in Figure 5.4. Steam distillation presents the advantage of modifying the composition of the EO to a lesser degree than with hydrodistillation, the latter causing more volatile compounds to break down due to hydrolysis. But steam distillation is not recommended for "soft" plant matter, such as flowers, since they would fall to the bottom of the still and form an uneven mat through which the steam would pierce in preferential channels, significantly reducing the extraction yield. The preferred method for flowers is hydrodistillation, which delivers a more homogeneous extract.

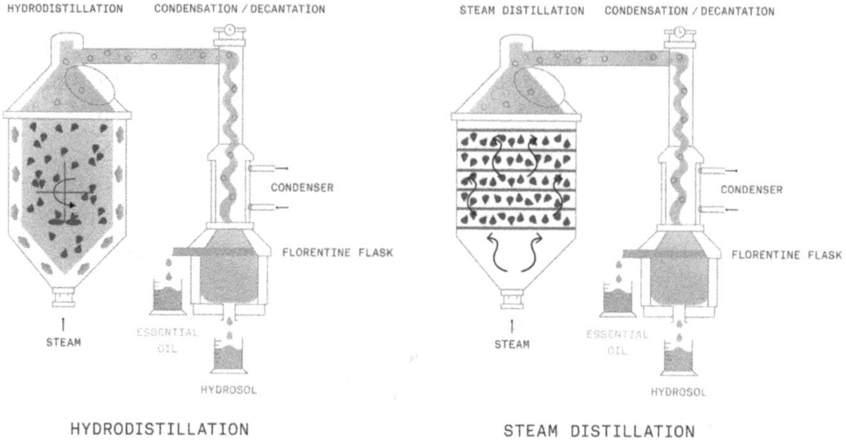

Figure 5.4. Diagrams of the hydrodistillation and steam distillation processes.

Orange blossoms, however, may be processed with both methods since the flowers are usually harvested at the bud stage. The plant matter is firm enough to hold up well in the still, and it loses some volume during distillation. The flowers are spread out on several stacked trays. This prevents the formation of a thick layer of flowers, which could interfere with distillation due to crushing, which in turn would create preferential channels for the steam and result in poor homogeneity. The composition of the EOs obtained varies slightly according to the distillation technique used. Table 5.5 shows the pros and cons of distillation.

The impact of the process on oil quality is apparent in the comparison between the volatile compound compositions of orange blossom EOs (neroli) obtained via hydrodistillation versus steam distillation, as illustrated previously in Table 5.1.

The EO obtained through hydrodistillation contains significantly less linalyl acetate, which breaks down more quickly into linalool via the following hydrolysis reaction (Figure 5.5) upon contact with water at 100°C (212°F).

Therefore, we would expect to find a higher concentration of linalool in the Tunisian hydrodistilled extract than in the Egyptian steam-distilled extract. However, this is not the case (see Table 5.1). This can be explained by other studies (unpublished results, data internal to IFF-LMR) performed on the chemical composition of Tunisian and Egyptian flowers

Table 5.5. Advantages and disadvantages of distillation.

Advantages of distillation	Disadvantages of distillation
• High-quality essential oil is obtained	• High distillation temperature (100°C) that may break down heat-sensitive compounds
• A simple technique, easy to set up in the field, even in rudimentary conditions	• Energy consumption that may be significant for raw materials requiring long processing times
• Product with little colour, usable without further purification	• Risk of hydrolysis due to the long duration of distillation
• No use of solvents that are toxic to humans and/or the environment	• The most highly volatile compounds are lost (no trap)
	• Water-soluble compounds are lost
	• Low distillation yield

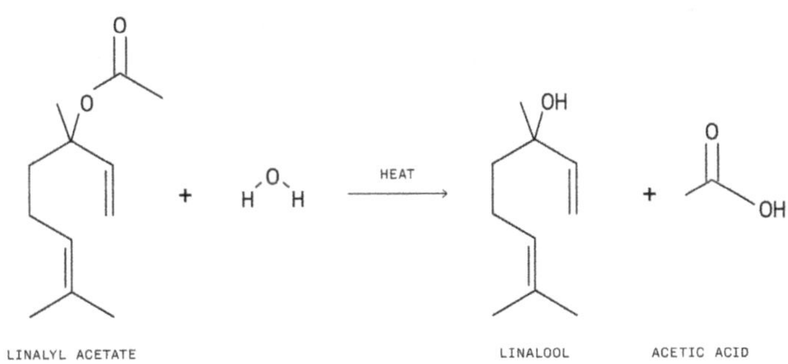

Figure 5.5. Hydrolysis of linalyl acetate into linalool.

just before they are distilled. The studies showed much lower levels of linalool in Tunisian flowers. Linalyl acetate hydrolysis does increase the amount of linalool in the Tunisian hydrodistilled extract but not enough to compensate for the lower concentration in the source material.

5.3.3 *Solvent extractions*

The solvent, very often hexane or ethanol, plays two roles in volatile solvent extraction. It starts by dissolving the plant cell membranes or waxy extracellular storage walls (cuticles) to release the EO. The solvent then extracts the volatile compounds that migrate, through affinity, from the plant matrix to the organic solvent. The solvent also dissolves other elements of the plant: odourless waxes. In practice, the plant matter (shredded or not, depending on its nature) is placed in an extractor in the presence of the volatile solvent. Extraction may take place under heat or at room temperature, with or without agitation, depending on the equipment. After one to two hours of infusion, the saturated solvent is removed from the extractor and replaced with clean solvent. Usually, two or three extraction cycles (or, equivalently, two or three "washings") are carried out until the plant matter is spent. The various extraction solvents are combined and evaporated off in a heated vacuum chamber. The evaporated solvent is then condensed for reuse in a future extraction of the same plant. Indeed, it contains residual plant compounds, which would contaminate future extracts from other plants containing other volatile organic compounds. Once all the extraction solvent has evaporated, only the compounds extracted from the plant remain. These volatile compounds and waxes take the form of a more or less solid paste known as *concrete* when the source is fresh plant matter (usually fresh flowers) processed with one or several solvents (e.g. hexane), or *resinoid* in the case of dry plant matter, gums, or resins processed with one or several solvents (e.g. ethanol), as defined in ISO standard 9235:2021. Figure 5.6 presents a simplified illustration of the solvent extraction process, and Table 5.6 points out its advantages and disadvantages.

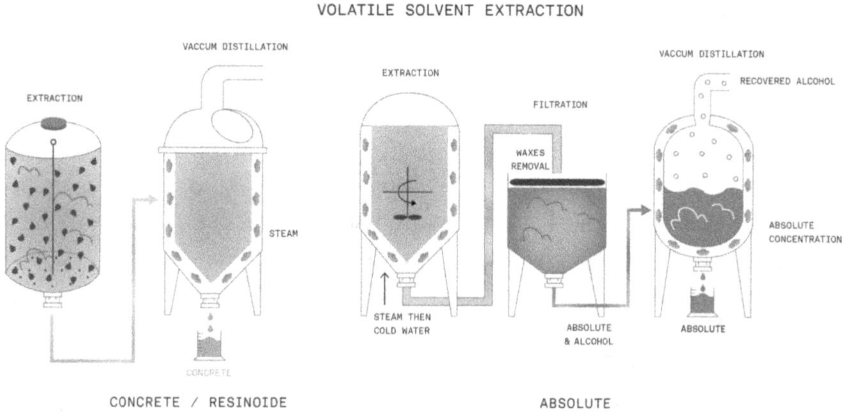

Figure 5.6. Diagrams of the volatile solvent extraction process.

Table 5.6. Advantages and disadvantages of volatile solvent extraction.

Advantages of volatile solvent extraction	Disadvantages of volatile solvent extraction
• Relatively low extraction temperature (<50°C or 120°C)	• Residual traces of solvent in the extract
• A range of solvents to choose from diversity of extract profiles	• The most volatile compounds are lost during solvent evaporation
• Low solvent consumption thanks to recycling	• Potential reduction in quality during solvent evaporation (heat causing breakdown)
	• Lower extraction yields than with other, less specific solvents

5.3.4 *Supercritical CO_2*

Supercritical CO_2 is already in widespread use by industrial manufacturers in a variety of applications, including decaffeination, soil decontamination, physical transformation (particle generation, liposome production, impregnation, functionalisation, etc.), polymer fractionation, deodorisation and the sterilisation of medical equipment. In this section, we see that supercritical CO_2 extraction presents several advantages, giving it an edge in natural raw material processing for perfumery and food flavouring [11]. In particular, it is a credible alternative as a green solvent: more environmentally friendly and less toxic.

Along with methane, ethane, propane, N_2O, ethylene and others, CO_2 is one of the gases that can enter a supercritical state at specific temperatures and pressure levels. Unlike other gases, CO_2 presents the advantage of being nontoxic and chemically inert. It also reaches the supercritical state at low temperatures, which is important for preserving thermolabile molecules. The other decisive advantage of supercritical CO_2 is that it is easier to separate from the extract once the extraction process is complete. Indeed, the compounds extracted are separated from the CO_2 through simple depressurisation, which returns the CO_2 to a gaseous state. Thanks to this phase change, there is no residual solvent left in the extract, and there is no need to distil the extraction solvent. The most highly volatile compounds are therefore left intact. Lastly, it is important to point out that CO_2 extracts do not undergo hydrolysis, oxidation, esterification or thermal changes, resulting in quality that is closer to the original substance.

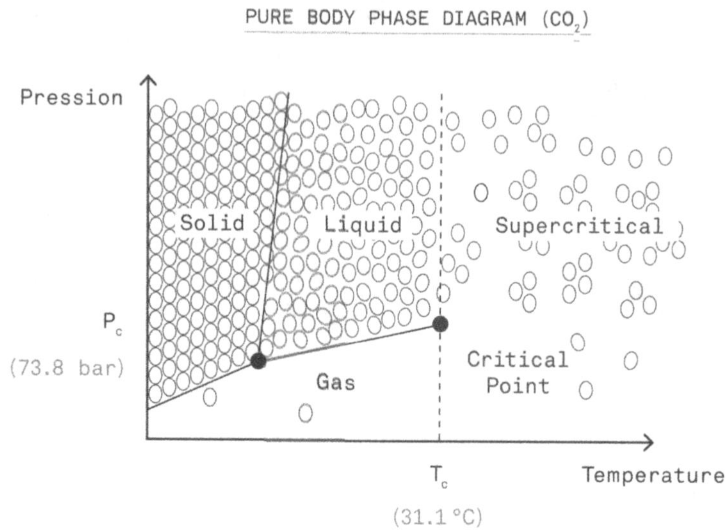

Figure 5.7. Pure CO_2 phases.

To reach the supercritical state, the CO_2 must be subjected to temperatures higher than 31°C (88°F) and pressures greater than 73 bar (Figure 5.7). In the supercritical state, its physical and chemical properties are different from those of a liquid or gas. Supercritical CO_2 has low polarity (close to that of hexane), a molecular weight close to that of liquid CO_2, a viscosity that remains close to that of gaseous CO_2 and intermediate diffusivity [12, 13].

Supercritical CO_2 is an effective solvent thanks to its liquid properties (density) and gaseous properties (viscosity, diffusivity), which endow it with strong capabilities for both extracting and transporting the solutes. It is remarkable that these properties can vary depending on the conditions applied (temperatures and pressure). As a result, CO_2 is a versatile solvent with properties of variable geometry (Table 5.7).

To highlight the advantages of this versatility, the graphs in Figure 5.8 show a schematic and didactic comparison of the various families of compounds extracted during distillation, solvent extraction and CO_2 extraction at different levels of pressure (80–100 and 300 bar). While steam distillation yields an extract composed primarily of volatile compounds, solvent extraction results in a more complete extract, which nevertheless lacks the most highly volatile molecules since they are lost when the solvent evaporates during the concentration phase. Supercritical CO_2 extract, which is

Table 5.7. Density, viscosity and diffusivity characteristics of CO_2 under different temperature and pressure conditions.

	Density	Viscosity	Diffusivity
	ρ (kg/m³)	η (Pa/s)	D (m²/s)
Gas (30°C at 1 atm)	0.6–2	$(1–3) \cdot 10^{-5}$	$(0.1–0.4) \cdot 10^{-4}$
Supercritical CO_2 (Tc & Pc)	200–500	$(1–3) \cdot 10^{-5}$	$0.7 \cdot 10^{-7}$
Supercritical CO_2 (lc, 4Pc)	400–900	$(3–9) \cdot 10^{-5}$	$0.1 \cdot 10^{-7}$
Liquid (30°C at 1 atm)	600–1600	$(0.2–3) \cdot 10^{-3}$	(0.2–2).

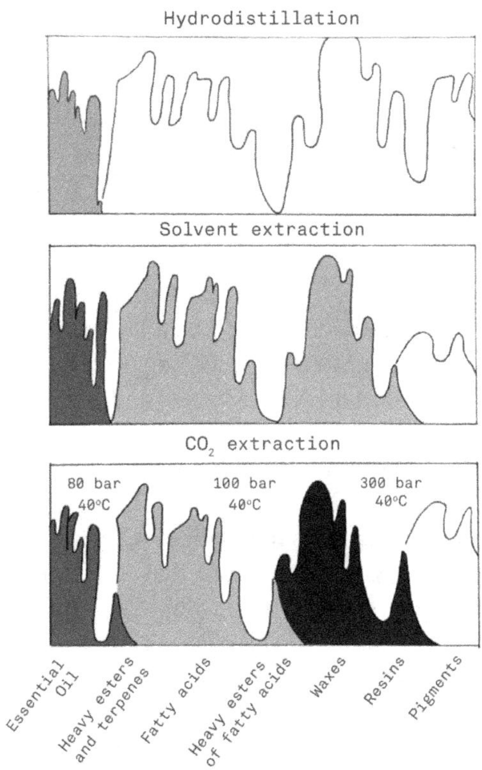

Figure 5.8. Comparison of the families of compounds extracted during distillation, organic solvent extraction and CO_2 extraction at different levels of pressure.

Source: Modified according to Brogle, 1982 [14].

much more versatile, results in a profile similar to an EO at low pressure. As the pressure is increased, other families of compounds can be targeted, including those that are also extracted during solvent extraction (Figure 5.8, modified according to Brogle, 1982 [14]).

5.3.4.1 *Supercritical CO_2 for solvent extraction*

Perfumery uses extraction with supercritical CO_2 to extract solid matrices via discontinuous extraction (the most commonly used technique) or liquid matrices via continuous counter-current extraction.

When withdrawing an extract from solid matter (e.g. ginger rhizomes, oakwood chips, vanilla pods) [15, 16], the raw material can be macerated, ground or freeze-ground into small particles to facilitate extraction. The extractable is then loaded into an extraction cylinder, which is fitted with sintered metal at each end to allow for unrestricted movement of the CO_2 and the dissolved substances while retaining the solid raw material within the extractor. The settings for temperature, extraction pressure, number of recirculation cycles and CO_2 flow rate — which is equal to the total volume of CO_2 used in the extraction process ($m_{CO_2}/m_{biomass}$) — are selected for optimal extraction. The pressurisation, extraction and depressurisation stages are illustrated in Figures 5.9 and 5.10.

Quite often, extraction under high pressure or with a co-solvent (EtOH) increases the polarity of the mixture, which broadens the spectrum of families of compounds that can be extracted and therefore increases yield. At high pressure, according to the raw material being processed, the CO_2 will often extract waxes and other molecules that are insoluble in the EtOH. These "waxes" are removed by chilling and then filtering, after which the final extract can then be called an absolute according to ISO standard 9235:2021.

Processing liquids using supercritical CO_2 to obtain an extract requires a device slightly different from that used for the discontinuous extraction of solids. A traditional liquid-liquid extraction column is used, but it must be strong enough to withstand high pressure. The liquid to be processed is injected into the column, where it encounters a counter-current flow of CO_2 in its supercritical state. Here again, the settings for recirculation, temperature and extraction pressure can be adjusted for optimal extraction. This extraction process presents the advantage of being continuous, and a wide variety of pumpable raw materials can be processed, such as EOs, fruit or vegetable juices and spirits (rum, tequila, etc.).

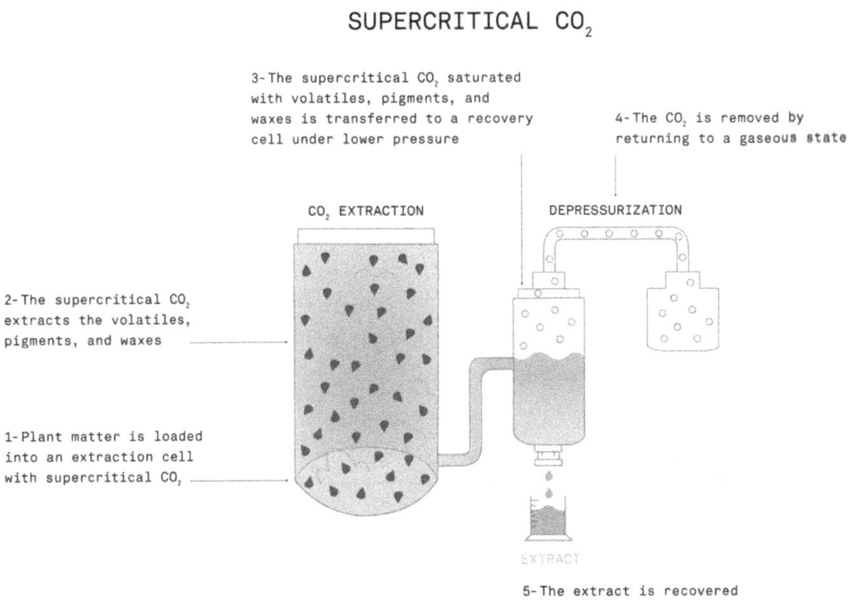

Figure 5.9. Diagram of the supercritical CO_2 extraction process.

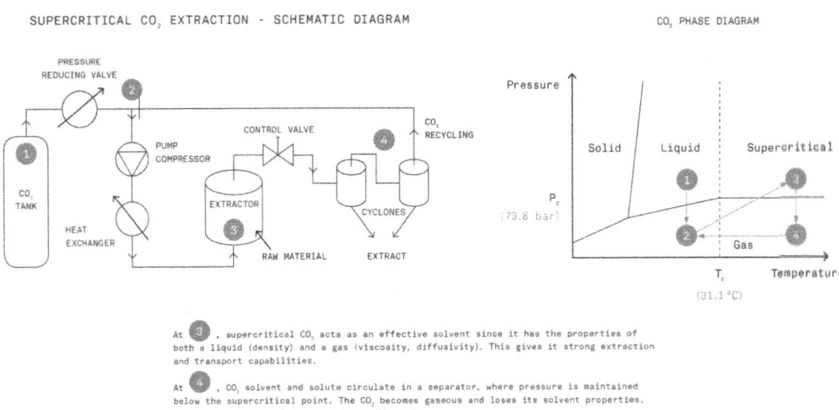

Figure 5.10. Schematic diagram of supercritical CO_2 extraction and CO_2 phase diagram.

Lastly, supercritical CO_2 extraction can itself be combined with other purification techniques, which will be presented in the section on secondary extraction methods [13].

The advantages and disadvantages of supercritical CO_2 extraction are summarised in Table 5.8.

Table 5.8. Advantages and disadvantages of supercritical CO_2 extraction.

Advantages of supercritical CO_2 extraction	Disadvantages of supercritical CO_2 extraction
• Good organoleptic quality (no high T°) • Abundantly available • Tasteless, colourless, odorless solvent • Green solvent (nonflammable, nontoxic) • Inoxidisable, chemically inert • High-purity extracts with no residual solvent • Selective extraction (P & T°) • Rapid extraction • GRAS (Generally Recognized As Safe) by the Food & Drug Administration (USA)	• High cost of equipment • Cleaning difficult and time-consuming • Limited options for co-solvents • Difficult to extract fresh material • Acidity of the solvent

5.3.4.2 *Supercritical CO_2 for purification*

In this section, we briefly jump forward to the discussion of secondary extraction processes (purification) to come in Section 5.4, simply because it makes sense to discuss all techniques and processes related to supercritical CO_2 together.

Purification via fractionation through supercritical CO_2 is in its early stages for the F&F industries, but it is already in widespread use for other applications, such as the fractionation of polymers and pharmaceutical active compounds. This technique is also promising for applications in fragrance and cosmetics specifically [17, 18].

Researchers have recently been experimenting with the purification of sclareol taken from a starting mixture extracted from clary sage [19], as well as with ginger oleoresin from which an extract high in gingerols is obtained [20]. Citrus extracts are also frequently refined using supercritical CO_2 fractionation to remove terpenes [21], to concentrate a valuable terpene fraction [22] or to remove pesticides.

5.4 Secondary Extraction Processes

"Secondary extraction" refers to the processing of a primary extract to obtain a product that can be used in perfumery (concrete processed to extract an absolute, EO decolourised via molecular distillation, etc.) or to purify a product to remove the fractions that are either undesirable from an olfactory standpoint or problematic for regulatory or toxicological reasons.

5.4.1 *Absolute processing*

To obtain an absolute, concretes or resinoids are processed to remove the plant waxes, which are odourless and insoluble in alcohol, making them undesirable for perfumery and problematic in terms of stability. The concrete or resinoid is dissolved in ethanol at a ratio of one to ten, in general, at a temperature of approximately 120–140°C (248–284°F) with agitation. Next, the solution is chilled to between 0°C and −10°C (14–32°F) and filtered through paper at that temperature (known as *cold filtration*). The waxes solidify and are retained by the filter. The filtrate contains the volatile compounds extracted from the plant in solution in the alcohol, along with other alcohol-soluble compounds if present. The solution is then heated and depressurised, which causes the alcohol to evaporate off. The product obtained after the alcohol has evaporated is called an *absolute*. In general, up to 2% of residual alcohol is considered acceptable within the absolute. It is important to point out that the most highly volatile compounds are lost when the solvent evaporates (during primary extraction and when the concrete is processed to obtain the absolute). An EO distilled from a given plant will contain more heads than an absolute extracted from the same material.

The impact of the extraction process on the olfactory profile of an extract can be appreciated when comparing the various types of flower extracts: hydrodistillation versus volatile solvent extraction (followed by separating the absolute from the concrete) versus volatile solvent extraction of the aqueous distillate.

Yields from absolute extractions are usually higher than for EOs extracted from the same plant, as the solvent also extracts non-volatile compounds. EO, which is made up of 95% volatile compounds on average versus approximately 60% in the case of an absolute, is often considered to be the most noble form of extraction and is sold at higher prices. Neroli breaks this rule, as the EO contains a high proportion of heads (14% beta-pinene and 12% limonene) that are lost in the process of obtaining the absolute via solvent evaporation. Extraction yields are approximately 0.11% in both cases.

Despite containing fewer head compounds than the EO, orange flower absolute has an olfactory profile that more closely resembles the scent of the fresh plant. Chemically, absolute extraction breaks down the volatile compounds to a lesser degree (all of the linalyl acetate from the plant matter is preserved, whereas it is reduced by a factor of three in the EO since it is hydrolysed into linalool) and retains more volatile compounds overall. Indeed, distillation does not retain the most highly volatile compounds, which are lost since they do not condense; the least volatile compounds, which cannot be distilled; or the water-soluble compounds, which remain trapped in the aqueous distillate (the hydrosol). This is why an absolute contains a higher percentage of heavy compounds (indole or methyl anthranilate, for example) than an EO. The alcohols largely remain in the hydrosol, along with the oxides, due to their high solubility in water. To illustrate this, Figure 5.11 compares the volatile compound compositions of an EO, an absolute, and a hydrosol, all extracted from bitter orange (*Citrus aurantium* L.) flower material in Tunisia. A similar observation can be

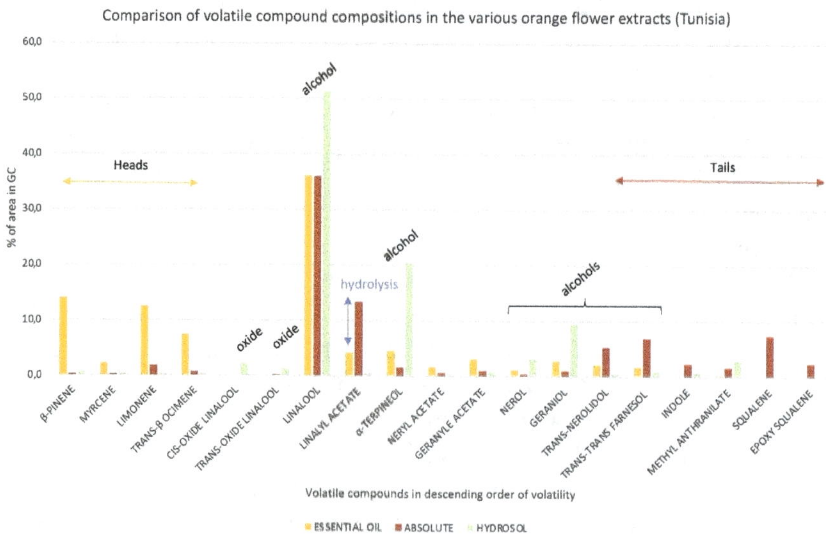

Figure 5.11. Comparison of volatile compound compositions in the various orange flower extracts (Tunisia).

made with the chemical composition of Egyptian orange flower EO and absolute, as shown in Table 5.1: there are fewer heads and more tails in the absolute than in the EO, regardless of the geographical origin of the flower (see the yellow- and green-coloured boxes for Egypt and Tunisia, respectively).

The EO presents zesty and fruity facets, delivered respectively by the head compounds and by ocimene, while the facets of the absolute are more animalic and floral, underscored namely by indole, which is very characteristic even at low concentrations. Even though neroli EO has a higher content of volatile compounds than orange flower absolute, the olfactory intensity of both extracts is similar since the absolute contains higher amounts of the more potent molecules such as indole. The EO is used in the top notes of a fragrance, while the absolute delivers the more persistent middle notes.

5.4.2 *Molecular distillation*

Molecular distillation (or MD) is a method of continuous distillation in a high vacuum. In the field of natural ingredient production for perfumery, MD is primarily used to decolourise natural extracts (EOs, concretes, and absolutes) without negatively impacting their olfactory qualities. By applying a reasonable heating temperature, as permitted by the high vacuum, to the volatile molecules in contact with the evaporator or heat source for a very short period of time, MD effectively separates heat-sensitive volatile compounds from the compounds with lower volatility, such as pigments and heavy molecules that cannot be distilled even in a vacuum of 10^{-1} or 10^{-2} mbar.

The distillation equipment includes a cylindrical compartment in which a vacuum is created, surrounded by a heated outer shell (or an evaporator), an internal condenser at a cooler temperature and a device with rotating blades that spread the extract to be distilled in a thin layer on the heated inner wall of the cylinder. The extract to be decolourised enters the MD device from above, pure or blended with a "heavy MD solvent" to liquefy it if needed, and it is heated to prevent thermal shock upon its entry into the MD compartment. Gravity pulls the extract down the compartment from top to bottom. As the extract is spread in a thin layer on the evaporator, the volatile compounds vaporise almost immediately, which limits the duration of contact with the heat source.

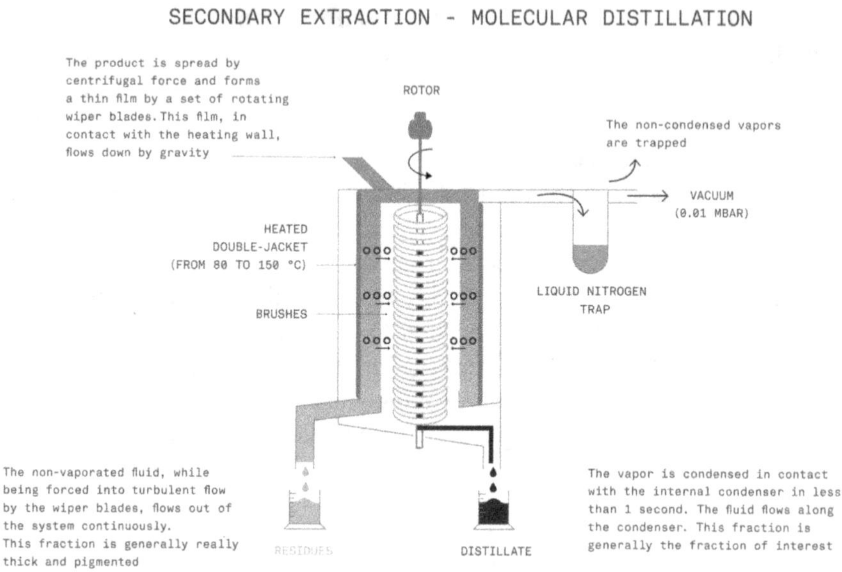

Figure 5.12. Diagram of the molecular distillation process.

Furthermore, the usual space of 1–5 cm between the evaporator and the internal condenser roughly matches the mean free path[1] of the vaporised molecules. The high vacuum within the device drastically reduces the number of residual molecules in the ambient air, such that the molecules vaporised are highly likely to condense directly on the condenser without colliding with other molecules in the environment (from the extract or ambient air). It's almost as if the molecules vaporise/condense one by one in an isolated manner, which is why the process is called molecular distillation.

As shown in Figure 5.12, the MD compartment has three outlets (in addition to the inlet through which the extract to be distilled is introduced):

- a central outlet located just below the condenser through which the distillate (condensed steam) runs out;

[1]Mean free path is the average distance travelled by a moving particle between two successive collisions.

- a peripheral outlet that collects the portion of the extract that isn't vaporised when lowered into the MD compartment and which flows along the heated walls of the MD cylinder to the bottom;
- an outlet at the top of the MD cylinder, designed to collect the most highly volatile compounds that do not condense on the internal condenser — these compounds are then trapped in a liquid nitrogen trap.

Table 5.9 highlights the pros and cons of the molecular distillation process.

Table 5.9. Advantages and disadvantages of molecular distillation.

Advantages of molecular distillation	Disadvantages of molecular distillation
• Low duration of contact between the volatile compounds and the heated surface • High vacuum significantly reducing boiling temperatures (breakdown minimised) • Continuous distillation (the substance to be distilled is introduced in a continuous stream and the distillate separated from the residue is collected likewise) • Can be used to decolourise heat-sensitive substances with no impact on their olfactory profile • The most highly volatile compounds are collected in a nitrogen trap	• Not suitable for precise fractionation • More expensive process than rectification (which can also be used to decolourise an extract but exposes more of the volatile compounds to destructive heat)

5.4.3 *Fractional distillation*

Fractional distillation is a technique for separating volatile compounds from a liquid extract on the basis of their relative volatility, relying on the thermodynamic properties of the liquid/vapour equilibria. Figure 5.13 gives an overview of the whole process. Since natural extracts have a complex composition, fractional distillation is rarely able to isolate specific compounds in pure form. The process can, however, be used to concentrate desirable compounds, to produce "hearts" by removing the "heads" (the most volatile compounds) and the "tails" (the least volatile compounds)

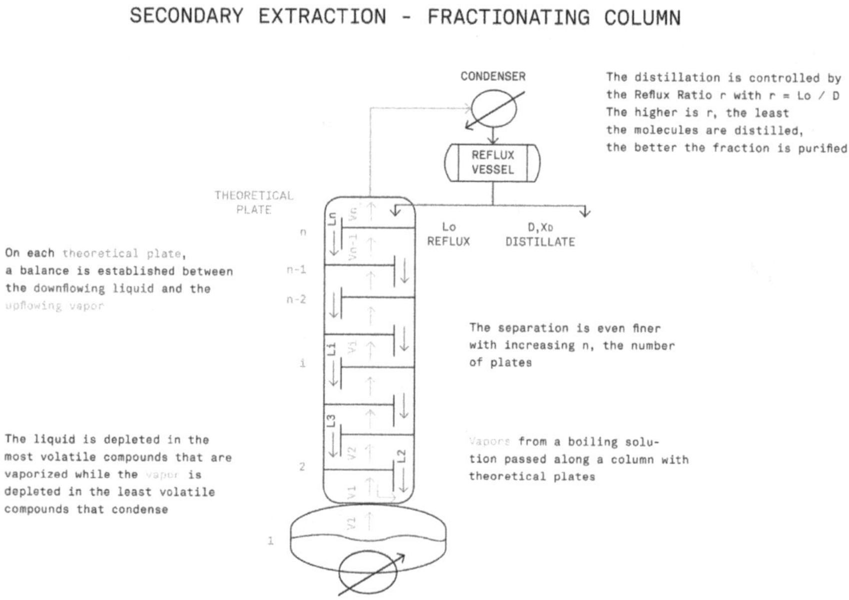

Figure 5.13. Diagram of the fractional distillation process.

from a mixture, or to remove problematic molecules for regulatory or toxicological reasons, such as for methyl eugenol, which must remain under a defined limit in natural extracts due to its carcinogenic, mutagenic and reprotoxic (CMR) properties.

The extract to be refined is placed in the boiler. Under the effect of heat and the intense vacuum created, the most highly volatile compounds vaporise above the surface of the liquid and rise up in the column. Since the heat source is present only at the boiler, the temperature decreases as the column rises. Higher temperatures within the column are caused only by the vapours rising within it. When the temperature in the environment drops, condensation takes place among the least volatile compounds from those that have evaporated. The resulting liquid sinks back down towards the boiler, while the most volatile compounds continue to rise further up within the column. Heat and substance transfers occur along the full length of the column, taking the form of a succession of liquid/vapour equilibria governed by the physical phenomenon of counter-current diffusion between the hot, rising vapours full of volatiles and the cooled, sinking liquid, relieved of the burden of its most volatile compounds.

To prevent the extract from deteriorating due to excess heat, fractional distillation is performed in a vacuum (up to approximately 1 mbar) in order to lower the boiling points of the compounds. This also makes it possible to distil heavier molecules that would not have vaporised at atmospheric pressure.

The vapours that reach the top of the column are condensed by a water refrigerant and are usually retained in part, with one portion kept as the distillate and the other portion sent back into the column. *Reflux* is the term used when the extract is returned to the column. The higher the proportion of vapours returning to the column (called the reflux rate), the better the separation. This also results in a longer duration of fractional distillation. The quality of the separation also depends on how tall the column is (the taller the column, the better the separation), what it is filled with and which type of thermal insulation is used. It is important to note that compounds with more extreme boiling points are easier to separate out.

Since a vacuum is created within the system, the most volatile compounds may not be condensed when they pass through the water refrigerant if it is not cold enough to cause them to liquefy in the conditions of the vacuum. These compounds are intercepted by a liquid nitrogen trap located slightly further along in the structure. They make up the trapped fraction.

Table 5.10 lists the main advantages and disadvantages of fractional distillation.

Table 5.10. Advantages and disadvantages of fractional distillation.

Advantages of fractional distillation	Disadvantages of fractional distillation
• Ability to refine extracts by separating or concentrating fractions with varying degrees of subtlety according to customer preferences or regulatory requirements, by adapting the performance of the column to the desired degree of separation (removal of heads such as terpenes in citrus EOs, allergen removal, concentration in the form of "hearts" or "isolates")	• Costs rise along with the degree of precision when separating fractions • High energy use (heating), applicable only to liquid extracts • Separation based on compound vaporisation temperatures (difficulty in separating compounds with similar boiling points) • Prolonged exposure to heat (in the boiler and then in the column)

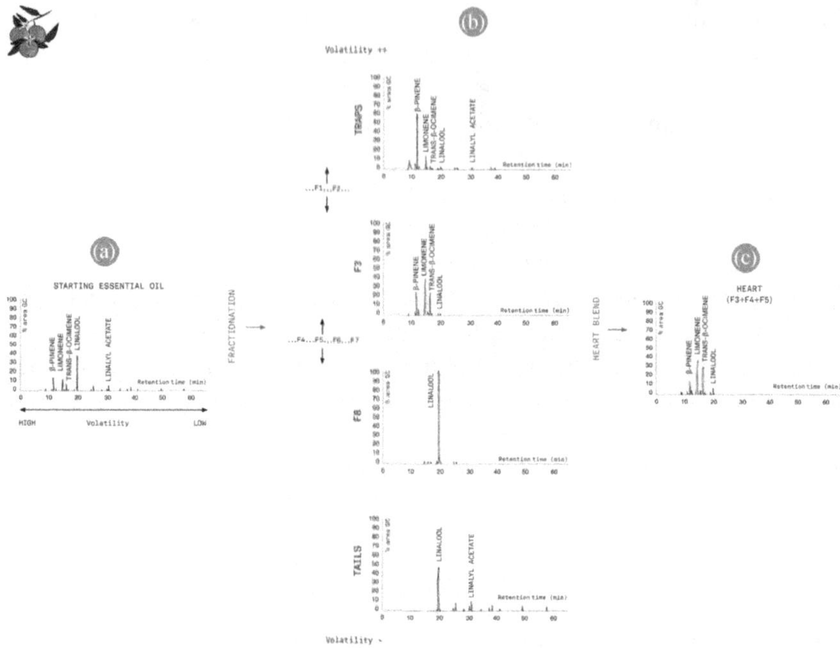

Figure 5.14. Chromatograms of (a) Tunisian neroli EO before fractionation, (b) fractions of Tunisian neroli oil and (c) a blend of three fractions to create a heart.

Such distillation is called *fractional* because the device makes it possible to distil the compounds of a mixture in descending order of volatility and to withdraw distillates with varying compositions during the distillation process. Each portion of the distillate withdrawn is known as a *fraction*. The extract left over in the boiler at the end of the distillation process is known as *tails* and contains the less volatile compounds at a higher concentration than in the initial extract.

$$\text{Traps} > \text{F1} > \text{F2} > \cdots > \text{F}n > \text{Tails}$$

Most volatile compounds ←----------→ Least volatile compounds

The following example is fractional distillation of neroli EO (Figure 5.14 and Table 5.11). Pay attention to the evolution of five compounds in the fractions *Traps*, *F*3, *F*8 and *Tails*.

The most volatile molecules are observed to be present in greater quantities in the first fractions. Conversely, the proportion of less volatile

Table 5.11. Proportional changes in composition of the main volatile compounds found in neroli EO before fractionation, in four fractions, and in the blend of three fractions to form a heart.

VOLATILITY (HIGH → LOW)	% WITHIN THE VOLATILE COMPOUNDS OF EACH FRACTION									
	Neroli EO		Traps	...	F3	...	F8	Tails		Heart F3+F4+F5
β-PINENE	13.65	FRACTIONATION	**60.56**		23.88		-	-	HEART BLEND	13.29
LIMONENE	11.41		13.19		**38.42**		0.01	-		35.85
TRANS-β-OCIMENE	6.95		2.95		**22.54**		0.05	-		29.53
LINALOOL	37.09		1.02		0.02		**99.14**	46.84		6.85
LINALYL ACETATE	4.69		0.01		-		-	**10.42**		-

molecules increases with each additional fraction. The first fractions are enriched with more volatile molecules and contain few of the less volatile molecules in comparison with the initial neroli EO. The opposite is true for the last fractions. Fraction 8 is almost entirely composed of linalool (>99%). The value of fractionation when refining an EO is that specific fractions can be isolated or reassembled to create an olfactory profile different from that of the initial EO. In the case of this development, the perfumers chose to retain three central fractions (F3, F4 and F5) to create what's known as a *heart*. This heart contains fewer head and tail molecules than the initial EO, but higher proportions of heart molecules, as illustrated by the chromatograms in (Figure 5.14(a)) and the accompanying table (Table 5.11).

5.5 Conclusions

Primary extraction techniques and methods were inherited from the past and have evolved only slightly in terms of their design and applications.

The distillation process for extracting EOs has been known and used since the thirteenth and fourteenth centuries. On the other hand, the extraction process for the most delicate flowers has evolved from enfleurage, used by the Egyptians, into the nineteenth century, to organic solvent extraction. Improvements were made to enfleurage in the nineteenth century, when the process of washing the fats with ethyl alcohol to purify them was perfected. When organic solvents became available, enfleurage was definitively replaced. Petroleum ether was followed by benzene and then hexane, which is still in widespread use.

Secondary extraction, on the other hand, which was developed more recently and uses more modern equipment, made it possible for industrial manufacturers to refine their raw extracts to improve their olfactory qualities. These new technologies also allowed them to meet continually stricter and more restrictive regulatory requirements.

The production of natural ingredients is facing a rise in the following three factors that substantially influence prices and, therefore, the future use of these ingredients: the cost of energy, the cost of labour and the cost of land. To counterbalance these structural costs, industrial manufacturers are now becoming involved higher and higher up the supply chain to insert technologies that help them improve biomass production yields (varietal selections, new farming practices, etc.) and also to change primary extraction methods to reduce their impact on humans and the environment. New distillation units, equipped with more efficient boilers and featuring optimised designs, are gradually being installed to reduce the amount of energy required to produce natural ingredients. The use of green, biosourced solvents that are less toxic to humans and the environment is starting to become a viable alternative in the perfumer's organ, although perfumers will need to learn how to use the new extracts since they will always be different from those obtained with hexane.

Major innovations are taking place in the production of both natural and synthetic ingredients, and there is plenty of room for new technologies in the current state of change and reflection. There is no doubt that, driven by consumers, climate change and environmental urgency, the perfumery of the future will be reinvented. The perfumer's palette will change, as ingredients that are too burdensome on the environment disappear, and new ones that can rise to the challenges guiding the world of tomorrow emerge.

References

[1] N. Dudareva, E. Pichersky. (2000). Biochemical and molecular genetic aspects of floral scents, *Plant Physiol.*, 122, 627–633.

[2] B. Blerot, S. Baudino, C. Prunier, F. Demarne, B. Toulemonde, J.-C. Caissard. (2016). Botany, agronomy and biotechnology of Pelargonium used for essential oil production, *Phytochem. Rev.*, 15, 935–960.

[3] M. Kharbach, I. Marmouzi, M. El Jemli, A. Bouklouze, Y. Vander Heyden. (2020). Recent advances in untargeted and targeted approaches applied in herbal-extracts and essential-oils fingerprinting — A review, *J. Pharm. Biomed. Anal.*, 177, 112849.

[4] J. Magnard, A. Roccia, J. Caissard, P. Vergne, P. Sun, R. Hecquet, A. Dubois, L. Hibrand-Saint Oyant, F. Jullien, F. Nicole, O. Raymond, S. Huguet, R. Baltenweck, S. Meyer, P. Claudel, J. Jeauffre, M. Rohmer, F. Foucher, P. Hugueney, M. Bendahmane, S. Baudino. (2015). Biosynthesis of monoterpene scent compounds in roses, *Science*, 349, 81–83.

[5] J.-L. Magnard, A. Bony, F. Bettini, A. Campanaro, B. Blerot, S. Baudino, F. Jullien. (2018). Linalool and linalool nerolidol synthases in roses, several genes for little scent, *Plant Physiol. Biochem.*, 127, 74–87.

[6] H. Trindade, L.G. Pedro, A.C. Figueiredo, J.G. Barroso. (2018). Chemotype and terpene synthase genes in thymus genus: State of the art, *Ind. Crops Prod.*, 124, 530–547.

[7] A. Salata, H. Buczkowska, R. Nurzynska-Wierdak. (2020). Yield, Essential oil content, and quality performance of *Lavandula angustifolia* leaves, as affected by supplementary irrigation and drying methods, *Agriculture*, 10, 590.

[8] M. Courregelongue, S. Shinkaruk, A. Prida, P. Darriet, A. Pons. (2022). Identification and distribution of new impact aldehydes in toasted oak wood (*Quercus petraea*), *J. Agric. Food Chem.*, 70, 11667–11677.

[9] F.-J. Marner, D. Gladtke, L. Jaenicke. (1988). Studies on the biosynthesis of iridals and cycloiridals, *Helv. Chim. Acta*, 71, 1331–1338.

[10] G. Gil, J.L. Petit, J.L. Series. (1992). Process for obtaining irone by enzymatic route, US Patent 5100790A.

[11] A. Capuzzo, M.E. Maffei, A. Occhipinti. (2013). Supercritical fluid extraction of plant flavors and fragrances, *Molecules*, 18, 7194–7238.

[12] V. Rapinel, N. Rombaut, N. Rakotomanomana, A. Vallageas, F. Chemat. (2018). Gaz liquéfiés comme solvants alternatifs appliqués à l'éco-extraction du végétal, *Tech. Ing.*, J2775.

[13] M. Perrut, V. Perrut. (2018). Toward ingredients by combining supercritical fluids with other processes, *J. Supercrit. Fluids*, 134, 214–219.

[14] H. Brogle. (1982). CO_2 in solvent extraction, *Chem. Ind.*, 385–390.

[15] A. Donelian, L.H.C. Carlson, T.J. Lopes, R.A.F. Machado. (2009). Comparison of extraction of patchouli (*Pogostemon cablin*) essential oil with supercritical CO_2 and by steam distillation, *J. Supercrit. Fluids*, 48, 15–20.

[16] M.N. Sovilj, B.G. Nikolovski, M.D. Spasojevic. (2011). Critical review of supercritical fluid extraction of selected spice plant materials, *Maced. J. Chem. Chem. Eng.*, 30, 197–220.

[17] E. Reverchon. (1997). Supercritical fluid extraction and fractionation of essential oils and related products, *J. Supercrit. Fluids*, 10, 1–37.

[18] M. Perrut. (2000). Supercritical fluid applications: Industrial developments and economic issues, Ind. *Eng. Chem. Res.*, 39, 4531–4535.

[19] C. Dufour, C. Crampon, C. Delbecque, P.P. Garry, E. Badens. (2020). Sclareol purification using the supercritical fractionation process: A modeling case study, *J. Supercrit. Fluids*, 159, 104754.

[20] A. Shukla, S.N. Naik, V.V. Goud, C. Das. (2019). Supercritical CO_2 extraction and online fractionation of dry ginger for production of high-quality volatile oil and gingerols enriched oleoresin, *Ind. Crops Prod.*, 130, 352–362.

[21] L. Danielski, G. Brunner, C. Schwänke, C. Zetzl, H. Hence, J.P.M. Donoso. (2008). Deterpenation of mandarin (*Citrus reticulata*) peel oils by means of countercurrent multistage extraction and adsorption/desorption with supercritical CO_2, *J. Supercrit. Fluids*, 44, 315–324.

[22] A. Terada, N. Kitajima, S. Machmudah, M. Tanaka, M. Sasaki, M. Goto. (2010). Cold-pressed yuzu oil fractionation using countercurrent supercritical CO_2 extraction column, *Sep. Purif. Technol.*, 71, 107–113.

Bibliography

Baker Gary, R., F. Lowe Robert, A. Southwell Ian (2000). Comparison of Oil Recovered from Tea Tree Leaf by Ethanol Extraction and Steam Distillation, *Journal of Agricultural and Food Chemistry*, 48(9), 4041–4043.

Bonaccorsi, I., D. Sciarrone, L. Schipilliti, A. Trozzi, H.A. Fakhry, G. Dugo (2011). Composition of Egyptian neroli oil, *Natural Product Communications*, 6(7), 1009–1014.

Bozovic, M., A. Navarra, S. Garzoli, F. Pepi, R. Ragno (2017). Essential oils extraction: a 24-hour steam distillation systematic methodology, *Natural Product Research*, 31(20), 2387–2396.

Burger, P., H. Plainfossé, X. Brochet, F. Chemat, X. Fernandez (2019). Extraction of natural fragrance ingredients: History overview and future trends, *Chemistry & Biodiversity*, 16 (10), e1900424.

Charpentier, B.A., M.R. Sevenants, R.A. Sanders. (1986). Comparison of the effect of extraction methods on the flavor volatile composition of shiitake mushrooms (*Lentinus edodes*) via GC/MS and GC/FTIR, *Developments in Food Science*, 12, 413–433.

Chemat, F. (2010). *Citrus Essential Oils*, Chp 2 "Techniques for oil extraction" pp. 9–36.

Chen, X., T. Rehka, J. Esque, C. Zhang, S. Shukal, C.C. Lim, L. Ong, D. Smith, I. André (2022). Total enzymatic synthesis of cis-a-irone from a simple carbon source. *Nature Communications*, 13, 7421.

Cicile, J.C. (1994). *Techniques de l'ingénieur, traité Génie des procédés*, Distillation, Absorption, Généralités sur les colonnes de fractionnement, J 2 621.

Cu, J., F. Pu, Y. Shi, F. Perineau, M. Delmas, A. Gaset (1990). The chemical composition of lovage headspace and essential oils produced by solvent extraction with various solvents, *Journal of Essential Oil Research*, 2(2), 53–59.

Deterre, S., B. Rega, J. Delarue, M. Decloux, M. Lebrun, P. Giampaoli (2012). Identification of key aroma compounds from bitter orange (*Citrus aurantium L.*) products: Essential oil and macerate-distillate extract, *Flavour and Fragrance Journal*, 27(1), 77–88.

Dugo, G., I. Bonaccorsi, D. Sciarrone, R. Costa, P. Dugo, L. Mondello, L. Santi, H.A. Fakhry (2011). Characterization of oils from the fruits leaves and flowers of the bitter orange tree, *Journal of Essential Oil Research*, 23, 45–59.

Erbas, S., H. Baydar (2016). Variation in scent compounds of oil-bearing rose (*Rosa damascena* Mill.) produced by headspace solid phase microextraction, hydrodistillation and solvent extraction, *Records of Natural Products*, 10(5), 555–565.

Farah, A., A. Afifi, M. Fechtal, A. Chhen, B. Satrani, M. Talbi, A. Chaouch (2006). Fractional distillation effect on the chemical composition of Moroccan myrtle (*Myrtus communis L.*) essential oils, *Flavour and Fragrance Journal*, 21(2), 351–354.

Ferhat, M.A., B.Y. Meklati, F. Chemat (2007). Comparison of different isolation methods of essential oil from *Citrus* fruits: Cold-pressing, hydrodistillation and microwave 'dry' distillation, *Flavour and Fragrance Journal*, 22(6), 494–504.

Fernandez, X, F. Chemat (2012). *La chimie des huiles essentielles — Tradition et innovation*, Ed. Vuibert, pp. 59–123.

Fleisher, A. (Florasynth Inc., Teteboro, New Jersey), Z. Fleisher (Crompton & Knowles Corp., Mahwah, New Jersey), Water-Soluble Fractions of the Essential Oils, *Perfumer & Flavorist* (1991), Vol. 16, 37–41.

Gamarra, F.M.C., L.S. Sakanaka, E.B. Tambourgi, F.A. Cabral (2006). Influence on the quality of essential key lime (*Citrus aurantifolia*) oil by distillation process, *Brazilian Journal of Chemical Engineering*, 23(1), 147–151.

Germana, M.A., F. De Pasquale, E. Bazan, E. Palazzolo (1990). Essential oils of the flowers, leaves, and buds of five species of Citrus, *Essenze Derivati Agrumari*, 60(3), 297–312.

González-Mas, M.C., J.L. Rambla, M.P. López-Gresa, M.A. Blázquez, A. Granell (2019). *Volatile Compounds in Citrus Essential Oils: A Comprehensive Review*, *Front Plant Sci.* 2019 Feb 5;10:12.

Grosso, C., J.P. Coelho, F.L.P. Pessoa, J.M.N.A. Fareleira, J.G. Barroso, J.S. Urieta, A.F. Palavra, H. Sovova (2010). Mathematical modelling of supercritical CO_2 extraction of volatile oils from aromatic plants. *Chemical Engineering Science*, 65, 3579–3590.

Hellivan, P.J. (2012). The expanding supercritical fluid CO_2 extract universe, *Perfumer & Flavorist*, 37, 26–34.

Johns Michael, R., E. Johns Julia, V. Rudolph (1992). Steam distillation of tea tree (*Melaleuca alternifolia*) oil, *Journal of the Science of Food and Agriculture*, 58(1), 49–53.

Katiyar, A., D. Singh, B.N. Mishra (2010). Essential oil: Production for health care in current scenario, *Annals of Biological Research*, 1(3), 200–209.

Koedam, A. (1982). The influence of some distillation conditions on essential oil composition, *World Crops: Production, Utilization, Description*, 7(Aromat. Plants), 229–236.

Kumar, R. and Y.C. Tripathi *Training Manual On Extraction Technology of Natural Dye & Aroma Therapy and Cultivation Value Addition of Medicinal Plants* (2nd–4th Nov 2011), organized by Extension Division in collaboration with Chemistry and Non Wood Forest Product Division, Forest Research Institute.

Laurent, J. (2017). Conseils et utilisations des huiles essentielles les plus courantes en officine, Pharm. Thesis, Université Toulouse III - Paul Sabatier.

Lee, Y.W., Y. Lee (2003). Flavor and aroma substances, *Food Science and Technology* (New York, NY, United States), 128, 305–327.

Lutisan, J., J. Cvengros (1995). Mean free path of molecules on molecular distillation, *The Chemical Engineering Journal*, 56, 39–50.

Machale Kishore, W., K. Niranjan, G. Pangarkar Vishwas (1997). Recovery of dissolved essential oils from condensate waters of basil and *Mentha arvensis* distillation, *Journal of Chemical Technology & Biotechnology*, 69(3), 362–366.

Manzan, A.C.C.M., S. Toniolo Fabio, E. Bredow, P. Povh Nanci (2003). Extraction of essential oil and pigments from *Curcuma longa* [L.] by steam distillation and extraction with volatile solvents, *Journal of Agricultural and Food Chemistry*, 51(23), 6802–6807.

Moyler David, A. (1984). Carbon dioxide extracted ingredients for fragrances, *Perfumer & Flavorist*, 9(2), 109–114.

Moyler, D. (1991). Natural ingredients for flavors and fragrances, *Chemistry & Industry*, (1), 11–14.

Nazlina, Z., M.K. Chan, J.G. Angelynn, C.M. Choo (2017). Hydrodistillation of orange essential oil: Kinetic study and chemical composition, *Solid State Science and Technology*, 25(1), 100–107.

Nunez, A.J., J.M.H. Bemelmans, H. Maarse (1984). Isolation methods for the volatile components of grapefruit juice. Distillation and solvent extraction methods, *Chromatographia*, 18(3), 153–158.

Perini, J.F., W.P. Silvestre, F. Agostini, D. Toss, G.F. Pauletti (2017). Fractioning of orange (*Citrus sinensis L.*) essential oil using vacuum fractional distillation, *Separation Science and Technology* (Philadelphia, PA, United States), 52(8), 1397–1403.

Perino-Issartier, S., C. Ginies, G. Cravotto, F. Chemat (2013). A comparison of essential oils obtained from lavandin via different extraction processes: Ultrasound, microwave, turbohydrodistillation, steam and hydrodistillation, *Journal of Chromatography A*, 1305, 41–47.

Perrut, M., V. Perrut (2019). Supercritical Fluid Applications in the Food Industry. *Book: Gases in Agro-Food Processes*, Academic Press, pp. 483–509.

Pino Jorge, A., J. Garcia, A. Martinez Marco (1996). Comparative chemical composition of the volatiles of *Coleus aromaticus* produced by steam distillation, solvent extraction and supercritical carbon dioxide extraction, *Journal of Essential Oil Research*, 8(4), 373–375.

Pino Jorge, A., J. Garcia, A. Martinez Marco (1999). Comparison of solvent extract and supercritical carbon dioxide extract of spearmint leaf, *Journal of Essential Oil Research*, 11(2), 191–193.

Rassem Hesham, H.A., H. Nour Abdurahman, M. Yunus Rosli (2016). Techniques for extraction of essential oils from plants: A review, *Australian Journal of Basic and Applied Sciences*, 10(16), 117–127.

Richter, J., I. Schellenberg (2007). Comparison of different extraction methods for the determination of essential oils and related compounds from aromatic plants and optimization of solid-phase microextraction/gas chromatography, *Analytical and Bioanalytical Chemistry*, 387(6), 2207–2217.

Rincon Melvin, A., R. Maciel Filho, R.W. Maciel Maria (2011). Modeling of molecular distillation parameters: Case study of green coffee oil (*Coffea arabica*), *Journal of Chemistry and Chemical Engineering*, 5(8), 706–710.

Rivera Lagunez, L., G. Vilarem, R. Gomez Solano, M. Estrada Jimenez, J.A. Feijoo Vazquez (2010). Water soluble fractions of caraway (*Carum carvi L.*) essential oil, *Boletin Latinoamericano y del Caribe de Plantas Medicinales y Aromaticas*, 9(6), 495–500.

Rossi, P.C., A.A. Willnecker, J. Berti, A.V. Borgarello, G.N. Mezza, M.C. Pramparo (2011). D-limonene and geranial fractionation from lemon essential oil by molecular distillation, *Latin American Applied Research*, 41(1), 81–85.

Silvestre, W.P., F. Agostini, L.A.R. Muniz, G.F. Pauletti (2016). Fractionating of green mandarin (*Citrus deliciosa Tenore*) essential oil by vacuum fractional distillation, *Journal of Food Engineering*, 178, 90–94.

Suresh, A., S. Velusamy, S. Ayyasamy, M. Rathinasamy (2021). Techniques for essential oil extraction from kaffir lime and its application in health care products — A review, *Flavour and Fragrance Journal*, 36(1), 5–21.

Tovar Laura, P., M.F. Pinto Glaucia, R. Wolf-Maciel Maria, B. Batistella Cesar, R. Maciel-Filho, R. (2011). Short-path-distillation process of lemongrass essential oil: Physicochemical characterization and assessment quality of the distillate and the residue products, *Industrial & Engineering Chemistry Research*, 50(13), 8185–8194.

Xiong, Y., Z. Zhao, L. Zhu, Y. Chen, H. Ji, D. Yang (2013). Removal of three kinds of phtalates from sweet orange oil by molecular distillation, *LWT — Food Science and Technology*, 53(2), 487–491.

Younis, A., A. Riaz, M.A. Khan, A.A. Khan, M.A. Pervez (2008). Extraction and identification of chemical constituents of the essential oil of Rosa species, *Acta Horticulturae*, 766, 485–492.

Zhang, X., C. Xu, M. Zhou (2005). Modeling of falling film molecular distillation apparatus, *Separation Science and Technology*, 40(6), 1371–1386.

Zheljazkov Valtcho, D., L. Cantrell Charles, T. Astatkie, E. Jeliazkova (2013). Distillation time effect on lavender essential oil yield and composition, *Journal of Oleo Science*, 62(4), 195–199.

https://doi.org/10.1142/9781800617780_0006

Chapter 6

Biocatalysis and Biotransformation for Optimised Natural Flavour and Fragrance Ingredients

Mathilde Lecourt* and Sylvain Antoniotti†

Université Cote d'Azur, CNRS, Institut de Chimie de Nice, Nice, France

**mathilde.lecourt@univ-cotedazur.fr*
†*sylvain.antoniotti@univ-cotedazur.fr*

Abstract

Natural flavour and fragrance ingredients, either as individual molecules or complex mixtures, are in high demand from consumers. Different techniques are used industrially to access these ingredients, resulting in different degrees of processing, sometimes taking the resulting products away from naturality. Biocatalysis and biotransformation have been used in various ways to provide improved ingredients, either in terms of extraction or distillation yields, improved sensory properties or safety for mixtures or improved isolation and purification for individual molecules. In this chapter, we provide the reader with examples from the literature and patents where enzymatic or cellular biosolutions have been developed either at the R&D level or at the production stage to access optimised natural flavour and fragrance ingredients.

6.1 Introduction

The use of biocatalysis, either with whole cells or isolated enzymes [1], to access flavour and fragrance ingredients has gained in popularity and has been increasingly applied industrially for several decades [2–4]. The most frequent assets brought by biocatalysis to the field are as follows:

- facilitated access to chiral molecules, which can be of critical importance when enantiodivergent olfactory properties are observed [5];
- preservation of the valuable natural product grade, especially for flavours [6];
- improved sustainability of the processes in terms of safety, energy consumption, limited waste generation and limited petrochemical footprints [7–9] — this is a strong demand from end consumers, particularly for cosmetic products.

Since 2000, novel approaches involving biocatalysis have emerged in order to improve, quantitatively or qualitatively, essential oils and natural extracts. In this regard, pre-treatment of plant material to improve extraction or distillation yields was developed mostly using cellulases, pectinases and glycosidases to facilitate the liberation of volatile molecules from cellular compartments. With a significant paradigm shift, some authors started to work on applying biocatalysis directly to natural complex mixtures such as essential oils and natural extracts in post-treatment procedures, while chemists usually preferred to work on pure substances. These aspects were reviewed in 2014 by our group [10], and more recently, we reviewed the topic of the sustainability and naturality of perfumery biotech ingredients [11]. The field has however gained in scope and complexity recently. Not only have new enzymes been used in pre- and post-treatment approaches, but novel techniques and processes have been discovered and implemented in the industry. This marked the birth of novel classes of ingredients, such as biotech ingredients obtained through metabolic engineering either as *de novo* produced essential oils or pure compounds using recombinant microorganisms from farnesyl diphosphate or even glucose. In this regard, biocatalytic approaches are being developed for the valorisation of waste from the food and natural flavour and fragrance industries (Figure 6.1).

The term "biocatalysis" will be reserved for reactions where an isolated enzyme, i.e. isolated from cells of living organisms, is used. The term "biotransformation" will be used for any type of reaction performed

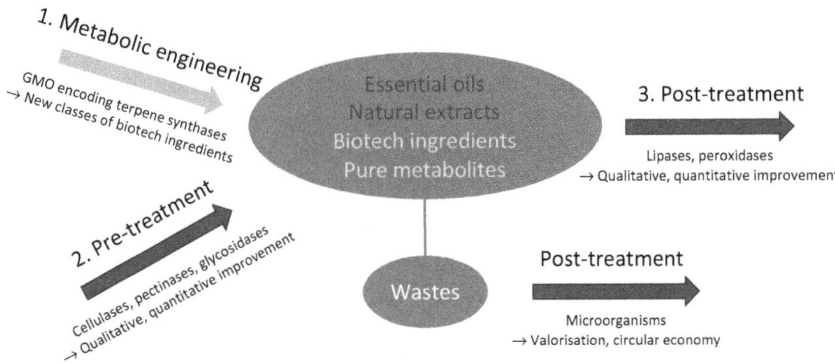

Figure 6.1. Biocatalysis and biotransformation applied to the improvement of natural ingredients.

with living cells (the so-called "whole-cell" processes) either from wild or genetically modified strains. The term "biosynthesis" is applied to describe the chemical synthesis of metabolites *in vivo* by the living organism.

In this chapter, we review the most recent additions to the field since 2010, with a large portion dedicated to new approaches from metabolic engineering and waste valorisation, even though the latter remains in its infancy. With the topic being at the crossroads of various fields of research and industrial activities, the reader may also be interested in more detailed review articles on particular aspects, such as the use of unconventional yeasts for the whole-cell production of flavours and fragrances [12], the use of *Pseudomonas* ssp. for the biotransformation of terpenes [13], the valorisation of orange peel waste [14], industrial biocatalysis [7, 15], biocatalysis in sustainable chemistry [16], catalytic promiscuity of hydrolases [17], redox reactions using isolated enzymes [18], terpenoid cyclases [19], biocatalysis for the synthesis of enantioenriched odorants [4], bio-oxidation of terpenes for flavour industry [20] and perfumery biotech ingredients [11].

6.2 Metabolic Engineering and a New Class of Ingredients

Following the advancements in molecular biology over the past 20 years, it is now possible to select a gene coding for a desired enzyme from its original source (bacteria, fungi, plants, animals, etc.) and transfer it into a recombinant organism, such as *Escherichia coli* or *Aspergillus niger*. If multiple copies are transferred, the host can be used to produce (relatively) large quantities of the enzyme, enabling easier accessibility compared to isolation and purification from the source. The modified organism can also be used as a bioreactor to perform the desired reaction and produce value-added chemicals. Several genes could also be transferred in such a way that a complete or partial biosynthesis could be performed from simple precursors. Finally, the gene can be modified, randomly or selectively, to improve the performance of the enzyme in terms of yield, enantioselectivity, substrate scope, temperature tolerance, etc.

In this section, we review scientific articles and patents featuring heterologous expression of genes coding for enzymes involved in the biosynthesis of odorant molecules in plants. These genetically modified organisms could then be used to produce *de novo* essential oils or individual molecules, and the approach has been recently used to develop new fragrance ingredients, which we can call "biotech ingredients." It is beyond the scope of this chapter to discuss the artificial, synthetic or natural quality of these ingredients; that is a subject addressed by regulatory bodies, not scientists.

6.2.1. *New natural complex mixtures*

Biosynthetic routes to synthesising key constituents of patchouli essential oil have been thoroughly studied, and it was discovered that, in spite of the chemical diversity of sesquiterpene hydrocarbons found in patchouli oil [21], only a handful of enzymes belonging to the family of terpene synthases were involved in the biosynthesis of most of their varieties [22]. In particular, it was found that a single patchoulol synthase cDNA encoded for a multi-functional sesquiterpene synthase. This enzyme could be heterologously expressed from *E. coli*, purified and used in a biocatalytic reaction with terpenyl diphosphates as the substrates (farnesyl, geranyl and neryl). (*E*,*E*)-Farnesyl diphosphate could, for example, through cyclisation, result in (*E*,*E*)-germacradienyl cation, itself leading to guaianyl cation through transannulation [23]. The main sesquiterpenes of patchouli oil, such as (−)-patchoulol, seychellene, patchoulene isomers, and α- and δ-guaiene, were thus formed from this carbenium intermediate (Figure 6.2).

It is striking to observe that starting from a simple linear substrate such as (*E*,*E*)-farnesyl diphosphate, a complex mixture of sesquiterpenes and

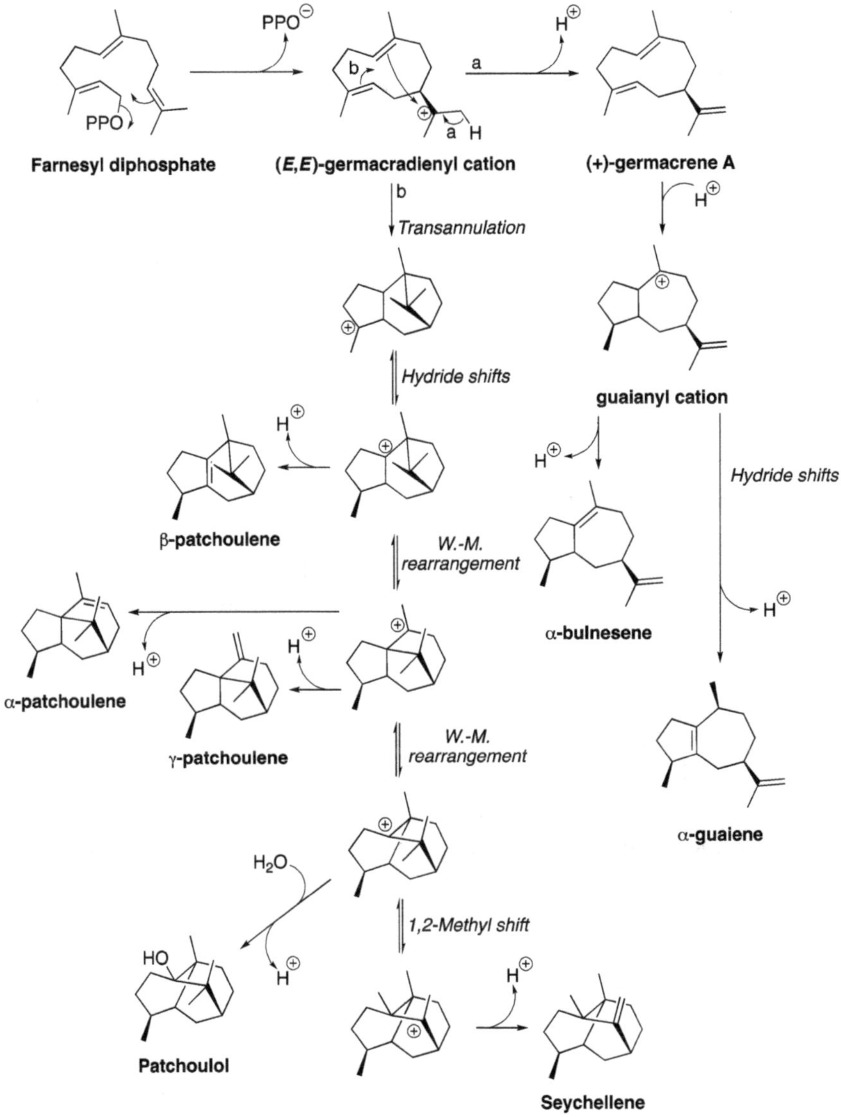

Figure 6.2. Chemical diversity in the sesquiterpene class from a single starting reaction. Hydride shifts could be either 1,2- or 1,3-shifts. W.-M. stands for Wagner–Meerwein.

sesquiterpenoids could be produced with a high similarity with patchouli essential oil. Such an approach allows us to access a novel family of biotech ingredients that have close chemical and sensory features compared with standard patchouli essential oil produced through distillation.

The Swiss company Firmenich & Cie launched in 2014 the ingredient Clearwood®, as a clean equivalent of patchouli essential oil obtained using a sustainable process offering the additional advantage of a better control of supply [24–27]. The naturality of this new ingredient was challenged for two reasons: first, because GMOs were used to assemble the necessary enzymes and reconstruct the biosynthetic route in the bioprocess; second, because of the presence of patchoulol ethyl ether, a peculiar compound never listed among the 130 published components of patchouli essential oil and not previously identified in nature.

In another work by Beutel *et al.*, a recombinant *E. coli* strain expressing a variant of the patchoulol synthase enzyme was used to convert (*E*,*E*)-farnesyl diphosphate into patchoulol and some other sesquiterpenes [28]. The stereospecificity of the enzymatic reaction was examined using (2*Z*,6*E*)- and (2*E*,6*Z*)-farnesyl diphosphates as substrates. Interestingly, the former mostly resulted in α- and γ-gurjunene, and germacrene B, while the latter was mostly converted to an unknown sesquiterpene (of formula $C_{15}H_{24}$ and with a retention index on wax column of 1715) and helminthogermacrene. None of these starting materials could be transformed into patchoulol, the different configurations of C_2–C_3 and C_6–C_7 double bonds, leading to other carbocationic intermediates than the natural isomer. For monoterpene precursors such as geranyl and neryl diphosphates, the formation of limonene, α-terpineol, geraniol and nerol was observed.

Akigalawood® is a captive ingredient produced by the company Givaudan. It also belongs to the class of novel biotech ingredients that have recently appeared in fine fragrances [29]. The launch of Akigalawood was accompanied by some misunderstanding observed on perfume blogs, where some ignorant writers mentioned *akigala* wood as some precious wood from which the ingredient could be extracted. This ingredient is described in a patent as a *rotundone-containing material* [30]. Rotundone is an odorant sesquiterpenoid naturally occurring in natural products and foodstuffs and presenting a spicy-woody olfactory note [31, 32]. The biosynthesis of rotundone from α-guaiene has been studied in *Vitis vinifera* and the genes identified [33].

α-Guaiene and rotundone could be obtained starting with farnesyl diphosphate by using recombinant microorganisms expressing sesquiterpene synthases from *V. vinifera* and *Pogostemon cablin* and oxidases from *Aquilaria spp.*, as mentioned in a patent application from Givaudan [34]. Alternatively, farnesyl diphosphate could also be produced directly from sugar using the microorganism, provided that the MEP pathway was also reconstructed or stimulated. In another series of patents, an enzymatic process to access rotundol from α-guaiene and α-bulnesene was proposed in the presence of laccases [35]. α-Guaiene could be replaced by light fractions of patchouli essential oil, containing various ratios of α-, β-, and δ-guaienes, seychellene and β-, δ-patchoulenes. As laccases are known to oxidise allylic positions, sometimes with the help of a chemical mediator, rotundone is also likely to be formed in these conditions (Figure 6.3).

Ginkgo Bioworks submitted a patent for the formation of related sesquiterpenes and sesquiterpenoids by using chimeric terpene synthases built up through the assembly of nucleic acid sequences from rare or extinct plants [36].

Still in the woody family but related to sandalwood essential oil, the chemoselective hydroxylation of the *pro*-(*Z*) methyl group of α- and β-santalene sesquiterpenes by a monooxygenase belonging the large family of P450, (*Z*)-α and (*Z*)-β-santalols, the odour-impact sesquiterpene alcohols of sandalwood could be formed [37, 38]. These genes were cloned and transferred into recombinant cells, and a fermentation process

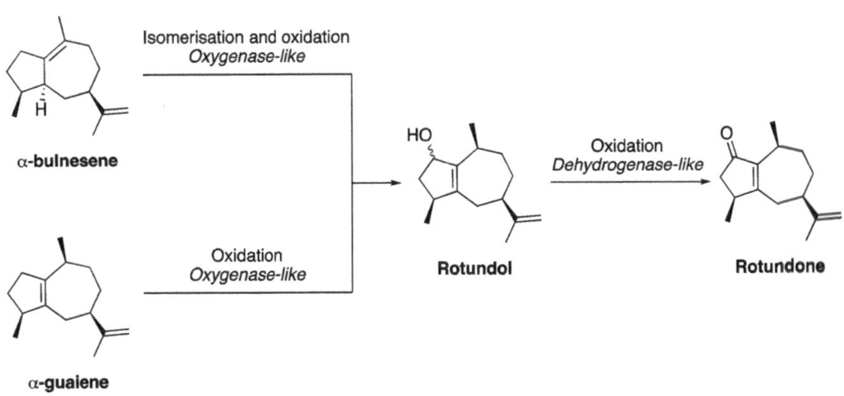

Figure 6.3. Conversion of α-guaiene and α-bulnesene into rotundol and rotundone.

could be implemented industrially to deliver the ingredient Dreamwood®, launched in May 2020 [39].

6.2.2. *Individual molecules produced by recombinant organisms*

Ambrox is a legendary sesquiterpenoid that gives its typical odour to ambergris, one of the six animal raw materials traditionally used in perfumery. It may be described with different tradenames (Ambroxide®, Ambroxan®, Ambrofix®, Ambermox®, Ambermor® and Ambrox® super). Various synthetic routes were developed to meet the global demand while avoiding or limiting ethical, economic and supply issues. The most famous route to ambrox is hemisynthesis from sclareol, a diterpenediol accumulated as a co-product during the manufacture of clary sage essential oil (*Salvia sclarea*). The relatively small market size for clary sage essential oil contrasted with the large demand for ambrox, justifying the development of independent manufacturing routes to sclareol. After successfully cloning two enzymes involved in the biosynthesis of sclareol from geranylgeranyl diphosphate in *S. sclarea*, engineered *E. coli* strains co-expressing geranylgeranyl diphosphate synthase from *Pantoea agglomerans* and both diterpene synthases could be obtained [40, 41]. With this GMO, sclareol could be produced in concentrations of up to 1.5 g/L, with 15% of 13-episclareol stereoisomer and a feed of 30 g/L of glycerol as the carbon source (Figure 6.4) [42, 43].

Engineered *E. coli* strains expressing a mevalonate pathway, a geraniol synthase from *Ocimum basilicum* and an alcohol acyltransferase from *Rosa hybrida* allowed for the formation of geranyl acetate from glucose [44]. To avoid inhibition of cell growth by the product, fermentation was

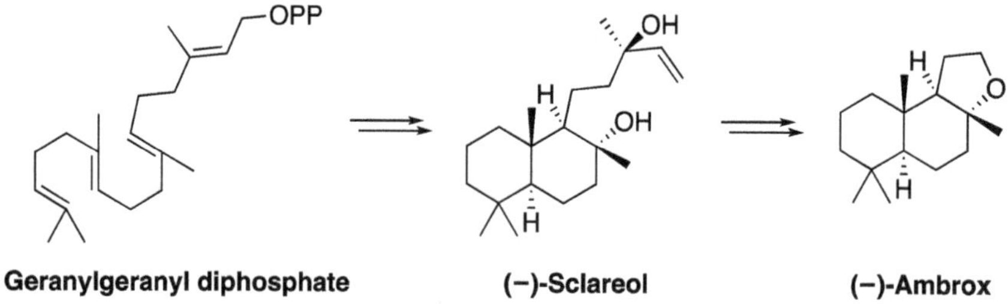

Figure 6.4. Sclareol and ambrox from geranylgeranyl diphosphate.

performed in a biphasic medium consisting of a modified terrific broth (MTB)/dodecane with a glucose feed of 20 g/L, supplemented with acetic acid as the acyl donor (20 mM). In optimised conditions, up to 4.8 mg/L of geranyl acetate could be formed selectively after 95 h.

Vetiver essential oil is obtained by hydrodistillation of the roots of *Chrysopogon zizanioides*, which is found in Haiti, Réunion (a French island), Indonesia, Brazil and Sri Lanka. This essential oil with typical woody-earthy-ambery facets is often described as one of the most complex natural substances used in perfumery from an analytical chemistry viewpoint [45–47]. Most of the odour-impact molecules in vetiver essential oil share a zizaane skeleton, such as khusimone, khusimol, zizaen-3α-ol, ziza-6(13)-en-3-one and 2-epiziza-6(13)-en-3-one (Figure 6.5) [48].

Terpene synthase identified in (+)-zizaene biosynthesis in *Chrysopogon zizanioides* has been transferred into recombinant *E. coli* along with genes coding for enzymes involved in the mevalonate pathway [49]. An engineered whole-cell biotransformation of glucose directly into (+)-zizaene

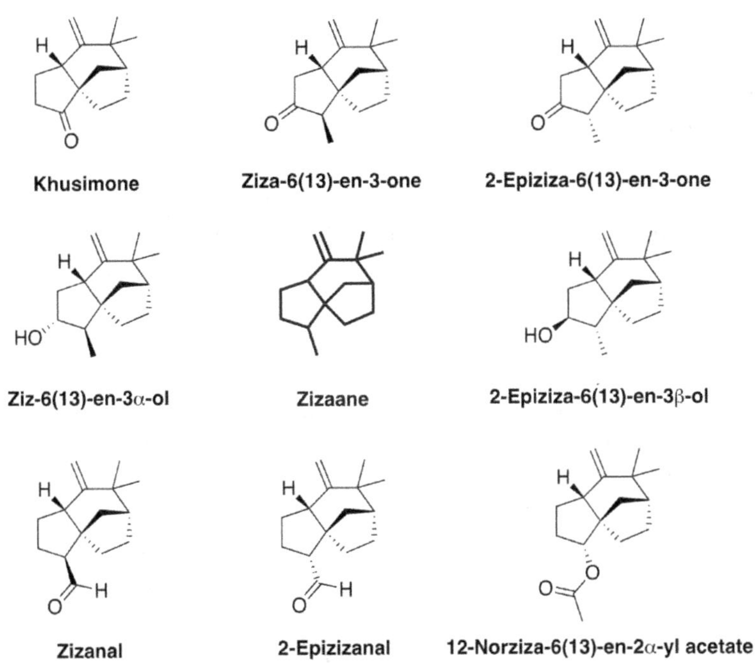

Figure 6.5. Odour-impact molecules featuring a zizaane skeleton found in vetiver essential oil from Ref. [48].

could thus be developed and produced under optimised conditions: 25 mg/L of (+)-zizaene, formed with 10% of β-acoradiene, for a glucose feed of 5 g/L.

6.3 Pre-Treatment of Plant Material Prior to Distillation or Extraction

The very modest yield of distillation in the manufacture of essential oils has always motivated efforts to improve the efficiency of the process. To facilitate the liberation of volatiles from vegetal cells or to hydrolyse glucosylated terpenoids, several enzymes have been used. In another approach, the metabolism is stimulated in order to produce more secondary metabolites, such as terpenoids.

Thus, an elicitation strategy to improve essential oil production using *Rhododendron tomentosum* Harmaja (formerly *Ledum palustre* L.) microshoots was developed based on a pre-treatment by various elicitors, including bacterial lysates from pathogens, to stimulate the secondary metabolism [50]. Dried microshoots were further submitted to small-scale hydrodistillation, and the essential oil obtained was analysed through GC-MS. Unfortunately, only spectra comparison using the NIST 11.0 library was performed for characterisation purposes, and crude GC-MS integration values were used for quantification instead of GC-FID with an internal standard, which is the gold standard for that purpose [51, 52]. Although a similar productivity was observed and the growth parameters of the plant were lower, the chemical compositions of samples submitted to lysates from strains of *Pectobacterium carotovorum* SCRI 1043 (ATCC BAA-672), *Dickeya dadantii* 3937, *E. coli* ATCC 25922, *Enterobacter sakazaki* ZOBR A741 and *Candida albicans* ATCC 90028 were apparently different. In particular, the relatively crude GC-MS proportions of terpenes such as δ-2-carene, sabinene and γ-terpinene seemed to increase, while those of sesquiterpenes and other terpenoids decreased. The authors did not really consider the effect, as notable in their conclusion.

Patchouli essential oil is a highly valuable woody-oriental essential oil, widely used in fine fragrances for decades (*vide supra*). It is obtained through hydrodistillation of leaves from *P. cablin*, mostly harvested in Indonesia [21]. As with many other natural ingredients, its market price has been highly fluctuating over time within a US$60–150/kg range in recent years. With a distillation yield of 2.6% w/w (range: 0.54–5.2%), efforts have been devoted to maximising it through optimisation of fertilisation, harvesting time, drying, fermentation, distillation and microbial pre-treatment of patchouli foliage [53]. The last method was tested with fungal strains of *Aspergillus foetidus* (MTCC 10559), *Penicillium citrinum* (MTCC 6590) and *Trichosporon asteroides* (MTCC 7632). As a result, distillation yields increased by 22–30% after up to eight days of fermentation, the best result being obtained with *T. asteroides*. Unfortunately, chemical analysis of the composition of patchouli essential oil obtained by this means was not offered to the reader.

Pre-treatment of black pepper and cardamom plant material with Lumicellulase®, a commercial mixture of cellulase, β-glucanase, pectinase and xylanase, before hydrodistillation resulted in increases of key aroma compounds in the chemical composition [54]. In particular, β-caryophyllene in black pepper and α-terpenyl acetate in cardamom were reported to increase from 15.0% to 25.6% and from 38.9% to 48.6%, respectively, as determined from raw integration percentages obtained through GC-FID analysis. Such an increase was attributed to the destruction of the cell wall structure by enzymatic pre-treatment, which is supported by scanning electron microscopy (SEM) images.

The effect of enzymatic pre-treatment (pectinase, cellulase and hemicellulase) of plant material from *Sideritis clandestine* in acidic medium on the yield of hydrodistillation was shown to be negligible [55]. The improvement was mostly due to the acidic treatment, as demonstrated by proper control experiments. Some quantitative differences in the chemical composition were also suggested but only assessed by GC-MS (simple quadrupole), although with external and internal calibration using α-pinene and octane, where GC-FID would have been probably more insightful.

For *Thymus capitus* and *Rosmarinus officinalis* L., however, 109% and 20% yield increases, respectively, were measured when using hemicellulase and cellulase [56]. Enzymatic pre-treatment (cellulase and pectinase) of bergamot (*Citrus bergamia*) peels was also shown to improve aroma extraction through pervaporation [57].

6.4 Post-Treatment of Essential Oils and Extracts for Improved Quality

It is possible to modify the chemical composition of essential oils or natural extracts specifically and selectively using enzymes. Such a modification could be motivated by the need for improving the safety of the mixture, the olfactory quality or simply physico-chemical properties.

Essential oil from the Ecuadorian plant *Piper aduncum* L. has thus been treated by endophytic fungal strains, with the rationale that endophytic microorganisms and the host plant have in common a portion of their secondary metabolism and could develop without mutual pathogenicity [58]. The chemical composition of the treated essential oil was analysed through GC-MS; however, no details on the method used for identification and quantification were given. In general, it was observed that monoterpenes and monoterpenoids, such as *trans*-ocimene (presumably β) and piperitone, respectively, underwent hydroxylation and dehydration reactions during the biotransformation procedure. These compounds were also submitted independently to the biocatalyst and presumably afforded (*E*,*E*) and (*Z*,*E*) isomers of 2,6-dimethylocta-1,3,5,7-tetraene (simply called octatetraene isomer 1 and 2 in the paper), 3 isomers of hydroxylated *trans*-ocimene (simply called octatrienol isomers 1, 2 and 3 in the paper) and hydroxylated hydrated *trans*-ocimene (simply called octadiendiol in the paper). Even if no formal nomenclature was used and no determination of the stereochemistry of C–C double bonds and location of alcohol functions was performed, one can guess from these data that unselective allylic oxidation occurred on *trans*-ocimene, leading to three isomers of hydroxylated *trans*-ocimene, which further converted to (*E*,*E*) and (*Z*,*E*) isomers of 2,6-dimethylocta-1,3,5,7-tetraene by dehydration and hydroxylated hydrated *trans*-ocimene by hydration. With regard to piperitone, only a hydroxylation reaction was observed, presumably in the allylic position, and two isomers were obtained in a 3/2 ratio. Unfortunately, no experiment supported the characterisation of the product nor the stereoselectivity of the reaction, particularly since the author referred to (4*R*)-piperitone and (+/−)-piperitone indifferently.

Essential oil from *Eupatorium buniifolium*, a plant from South America, has also been treated with endophytic fungi, which is supposed to be both less sensitive to and less pathogenic to the plant [59]. A detailed chemical analysis of the composition of the modified essential oil was performed through GC-FID for quantification purpose and GC-MS for characterisation using RI, comparison of spectra with commercial databases and enantioselective GC-FID with a permethylated β-cyclodextrin column when necessary. The essential oil being rich in monoterpenes such as limonene, α-(−)-pinene and sabinene, the corresponding hydration and oxidation products were formed, as well as mostly terpinen-4-ol. A study on individual compounds followed, with the description of both types of products, unfortunately without quantitative data on chemical yields.

Various citrus essential oils have been modified enzymatically using a lipase to increase their ester content [60]. Interestingly, the authors used lipase B from *Candida antartica* (CaLB), one of the most commonly used enzymes, which is displayed on the surface of a recombinant yeast strain of *Pichia pastoris*. However, as it is well known that citrus essential oils contain mostly monoterpenes, very few alcohols are likely to react with CaLB, which prefers primary and secondary alcohols and no carboxylic acid. The authors therefore provided both the alcohol (EtOH, hexan-1-ol) and the acid (acetic, butyric, hexanoic) to the system, thus using citrus essential oil as the solvent. The chemical composition of the resulting mixtures contained esters, but none of these was formed from components of the original essential oil used. Sensory analysis was performed and showed an increase in the fruity odour at the expense of the citrus odour, short esters being a well-known contributor to the smell of fruits.

The production of α-terpineol directly from limonene of sweet orange oil (SOO) through biotransformation using fungal strains fermented in cassava waste water was proposed [61]. In this interesting work in terms of sustainability using a co-product (SOO), wastewater and biotransformation, *Penicillium sp.*, *A. niger*, and *Fusarium oxysporum* strains were used. Chemical analysis was carried out using GC-FID with internal calibration (with decane and camphor as standards) and GC-MS by comparing retention indices (RIs) and mass spectra of the samples with those of commercially available standards. Enantioselective GC-FID was performed using a permethylated β-cyclodextrin column. It was observed that a chemo- and enantiospecific hydration reaction of (*R*)-limonene to (*R*)-α-terpineol occurs with a yield of approximately 15%, along with trace amounts of perillyl alcohol formed upon oxidation of limonene (Figure 6.6).

Following this work, SOO was also submitted as the sole source of carbon for biotransformation by *Fusarium oxysporum* fungal strains in order to generate oxygenated products, potentially improving antioxidant

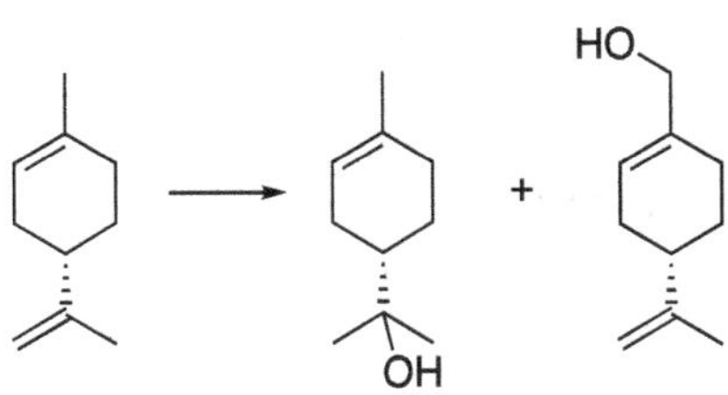

Figure 6.6. Biotransformation of (*R*)-limonene into (*R*)-α-terpineol and perillyl alcohol.

properties [62]. Considering that SOO contains almost exclusively (*R*)-limonene, hydroxylated and hydrated products such as perillyl alcohol and α-terpineol were expected. An analysis of the resulting product was performed through GC-FID for quantification using decane as the internal standard, and GC-MS and products were characterised by comparing their RIs and mass spectra with those of commercial standards. It was shown that small amounts of perillyl alcohol indeed formed upon hydroxylation at the methyl position at C7, together with α-terpineol and carvone, the former by hydration and the latter by hydroxylation, and further oxidation to the carbonyl function. Most of the starting essential oil was lost, and only 20% of unconverted limonene was found in the product, together with less than 1% yield of each product identified. Antioxidant properties *in vitro* of this modified orange essential oil were improved, as demonstrated by DPPH assay, lipid peroxidation inhibition assay and glutathione S-transferase activity assay.

Calamintha nepeta ssp. *glandulosa* is a plant with a minty odour whose essential oil, rich in monoterpenes and monoterpenoids, has been treated with *A. niger* [63]. The analysis of the chemical composition has been performed through GC-MS, but the methodology of characterisation and quantification was not described. Some key products were, however, isolated and analysed through ^{1}H and ^{13}C NMR. The biotransformation delivered both hydration and oxidation products that were only described from a qualitative viewpoint. The major volatile components of *C. glandulosa* essential oil are known to be piperitone oxide, piperitenone oxide and limonene, together with small amounts of terpinen-4-ol, α-terpineol and thymol. As a result, biotransformed products were mostly from limonene, piperitone and piperitenone oxides through simple hydrolysis. Terpinen-4-ol, α-terpineol, carvone, piperitenone, 6-hydroxypiperitenone and 6-hydroxypiperitone were thus found, either formed by hydration and/or oxidation, but no chemical yields were provided.

Ho Sho (*Cinnamon camphora* Nees & Eberm. var. *linaloolifera* Fujita) essential oil has been modified through treatment with a lipase in order to increase its content in linalool esters [64]. Although the most efficient lipase described to date was used — CaLB supported on an organic resin — only modest esterification yields were measured, as this enzyme is reluctant to accommodate tertiary alcohols, such as linalool. Tentative esterification with acetic, propanoic and butyric acid and acetic anhydride was performed and evaluated using GC-FID, but yields up to 5.58% were obtained despite a lengthy optimisation procedure.

As previously mentioned, vetiver essential oil is an important natural raw material for the perfume industry. Different specialities can be found on the market, including light or heart fractions obtained through molecular distillation with some particular olfactory aspects. The crude of fractionated vetiver essential oil, rich in sesquiterpene alcohols, can also be acetylated to yield various qualities of vetiveryl acetate, another woody ingredient with ambery and grapefruit facets and less earthy. Vetiveryl acetate is typically obtained through reaction with acetic anhydride upon heating in a solvent or in the presence of a catalyst. Thanks to the large substrate scope of lipase B from CaLB, which is considered one of the most promiscuous enzymes [65], vetiveryl acetate could be prepared through enzymatic acetylation using natural EtOAc as both the solvent and the acetyl donor [66, 67]. This process presents a high sustainability score since the enzyme used was immobilised and could be recycled more than 12 times with very little decrease in the activity, EtOAc could be recycled upon evaporation, and the reaction was performed at room temperature. With this process, natural vetiveryl acetate was produced for the first time in quantitative yield at the kilogram scale. Comprehensive GCxGC-MS was performed on the product to analyse in detail its chemical composition. It was shown that the enzymatic reaction occurred chemoselectively in favour of primary sesquiterpene alcohols, leaving mostly unconverted secondary alcohols.

Another famous natural extract, oakmoss absolute, presenting earthy and woody notes, is obtained through solvent extraction of the lichen *Evernia prunastri* [68, 69]. This material is composed of monoaromatic compounds and their dimers, linked through ester linkages and called depsides. Unfortunately, some of these monoaromatic compounds are contact allergens, as in the case of atranol and chloroatranol. The European Union (EU) Commission Regulation 2017/1410 of 2 August 2017 had set that cosmetic products containing these substances shall not be made available on the Union market from 23 August 2021 [70]. To solve the problem, a peroxidase-based strategy has been developed to specifically convert atranol and chloroatranol into hydrosoluble compounds, facilitating their removal from oakmoss absolute through simple liquid/liquid extraction [71, 72]. The reaction proceeded in two phases, with a first Dakin oxidation at pH 9 in the presence of H_2O_2, leading to pyrogallol derivatives, followed by a HRP-catalysed oxidation (Figure 6.7). Under these conditions, a cascade starting with dimerisation was proposed to explain the complexity of the product formed, which was subsequently

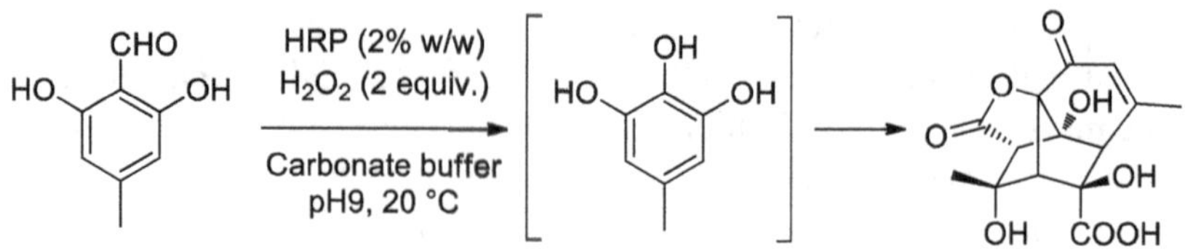

Figure 6.7. Oxidation of atranol to 5-methylpyrogalol, leading eventually to a densely fused dimeric product via a cascade triggered by HRP.

isolated and characterised using mono- and bidimensional ^1H and ^{13}C NMR, MS and HRMS [71]. The modified oakmoss absolute was analysed using both GC-MS and HPLC-MS, and only ppm levels of atranol and chloroatranol remained. Triangular testing was performed with a 56-person panel and showed that the atranol-free absolute was not different at the 99% level of confidence.

6.5 Conclusion and Future Directions

This past decade has witnessed the development of original biotech processes in the fragrance industry thanks to the maturity of molecular biology techniques and the recent advances in the understanding of biosynthetic routes to odorant metabolites in plants. Bringing both novelty and sustainability, the success of these ingredients certainly indicates that the trend will continue to increase in the future. These new approaches can contribute to the supply of scarce or depleting raw materials for which the demand is high.

While the strategy generally relies on the identification of the genes in the plant of interest, cloning and transfer into a recombinant organism, works such as that of Ginkgo Bioworks involving chimeric terpene synthases built by selecting portions of genes found in the DNA of rare or extinct plants open new perspectives [36].

The question of the use of GMOs is however still an issue for some consumers. Consequently, companies tend to avoid processes where GMOs could be involved, regardless of their persistence in final products, which is typically low or null in bioprocesses. R&D phases could still be operated with enzymes obtained through heterologous expression, but enzymes isolated from wild sources should remain available for production phases. Even if the activity of enzymes can be tuned by medium engineering and various types of promiscuity, a gap will remain between conventional biocatalysis and synthetic biology [65, 73–78].

We have seen earlier that cassava wastewater could be used as the reaction medium for the biotransformation of sweet orange oil [61]. Orange peel oil can also be used as raw material for α-terpineol production by the bacterial strain *Penicillium digitatum* (NRRL 1202) [79]. Again, a stereospecific bioconversion of (*R*)-limonene into (*R*)-α-terpineol, the major component of orange peel oil that is a co-product of orange juice manufacture, was observed. With 36,000 tons in 1997 and 70,000 tons in 2009 of limonene produced as a waste or a co-product of the citrus industry, this C_{10} hydrocarbon calls for valorisation. Besides the orange juice industry, it is worth mentioning that most essential oils rich in terpenes designed for use in fragrances are subjected to a deterpenation process, which in fact mostly consists of removing simple monoterpenes such as limonene and pinenes, whose odour is of low interest. Such an example of valorisation of industrial waste and/or co-products of essential oils' manufacture [80, 81], particularly using biocatalytic methods [82], should play an increasing role in the future to satisfy sustainability needs and consumers' desires.

As such, natural grade and biosourced grade could merge at some point in the context of sustainability, with manufacturing processes following the principles of sustainable chemistry, such as waste prevention, resource management, safety and energy frugality, by implementing biocatalysis and biotransformation approaches.

Acknowledgements

This work was supported by the University of Côte d'Azur and CNRS. M.L.'s PhD fellowship is co-funded by IFF-LMR and French Région SUD.

References

[1] K. Faber, W. Kroutil. (2005). New enzymes for biotransformations, *Curr. Opin. Chem. Biol.*, 9, 181–187.

[2] R. Buller, K. Hecht, M.A. Mirata, H.-P. Meyer. (2018). An appreciation of biocatalysis in the swiss manufacturing environment, in *Biocatalysis: An Industrial Perspective* (The Royal Society of Chemistry, Cambridge), pp. 1–43.

[3] S. Serra, C. Fuganti, E. Brenna. (2005). Biocatalytic preparation of natural flavours and fragrances, *Trends Biotechnol.*, 23, 193–198.

[4] E. Brenna, C. Fuganti, F.G. Gatti, S. Serra. (2011). Biocatalytic methods for the synthesis of enantioenriched odor active compounds, *Chem. Rev.*, 111, 4036–4072.

[5] E. Brenna, M. Crotti, F.G. Gatti, F. Parmeggiani, A. Pugliese, S. Santangelo. (2015). Multi-enzymatic cascade procedures for the synthesis of chiral odorous molecules, in *Importance of Chirality to Flavor Compounds* (American Chemical Society, Washington DC), pp. 59–75.

[6] European Flavour and Fragrance Association. (2013). Guidance Document for the Production of Natural Flavouring Substances and (Natural) Flavouring Preparations in the EU, European Flavour and Fragrance Association.

[7] J.-M. Choi, S.-S. Han, H.-S. Kim. (2015). Industrial applications of enzyme biocatalysis: Current status and future aspects, *Biotechnol. Adv.*, 33, 1443–1454.

[8] R.A. Sheldon, F. van Rantwijk. (2004). Biocatalysis for sustainable organic synthesis, *Aust. J. Chem.*, 57, 281–289.

[9] K.P. Dhake, D.D. Thakare, B.M. Bhanage. (2013). Lipase: A potential biocatalyst for the synthesis of valuable flavour and fragrance ester compounds, *Flavour Fragr. J.*, 28, 71–83.

[10] S. Antoniotti. (2014). Tuning of essential oil properties by enzymatic treatment: Towards sustainable processes for the generation of new fragrance ingredients, *Molecules*, 19, 9203–9921.

[11] M. Lecourt, S. Antoniotti. (2020). Chemistry, sustainability and naturality of perfumery biotech ingredients, *ChemSusChem*, 13, 5600–5610.

[12] L. Forti, S. Di Mauro, M.R. Cramarossa, S. Filippucci, B. Turchetti, P. Buzzini. (2015). Non-conventional yeasts whole cells as efficient biocatalysts for the production of flavors and fragrances, *Molecules*, 20, 10377–10398.

[13] G. Molina, M.R. Pimentel, G.M. Pastore. (2013). Pseudomonas: A promising biocatalyst for the bioconversion of terpenes, *Appl. Microbiol. Biotechnol.*, 97, 1851–1864.

[14] I. de la Torre, V. Martin-Dominguez, M.G. Acedos, J. Esteban, V.E. Santos, M. Ladero. (2019). Utilisation/upgrading of orange peel waste from a biological biorefinery perspective, *Appl. Microbiol. Biotechnol.*, 103, 5975–5991.

[15] J. Schrader, M.M.W. Etschmann, D. Sell, J.M. Hilmer, J. Rabenhorst. (2004). Applied biocatalysis for the synthesis of natural flavour compounds — current industrial processes and future prospects, *Biotechnol. Lett.*, 26, 463–472.

[16] R.A. Sheldon, D. Brady. (2019). Broadening the scope of biocatalysis in sustainable organic synthesis, *ChemSusChem*, 12, 2859–2881.

[17] E. Busto, V. Gotor-Fernández, V. Gotor. (2010). Hydrolases: Catalytically promiscuous enzymes for non-conventional reactions in organic synthesis, *Chem. Soc. Rev.*, 39, 4504–4523.

[18] D. Monti, G. Ottolina, G. Carrea, S. Riva. (2011). Redox reactions catalyzed by isolated enzymes, *Chem. Rev.*, 111, 4111–4140.

[19] D.W. Christianson. (2006). Structural biology and chemistry of the terpenoid cyclases, *Chem. Rev.*, 106, 3412–3442.

[20] J.L. Bicas, A.P. Dionísio, G.M. Pastore. (2009). Bio-oxidation of terpenes: An approach for the flavor industry, *Chem. Rev.*, 109, 4518–4531.

[21] T.A. van Beek, D. Joulain. (2018). The essential oil of patchouli, Pogostemon cablin: A review, *Flavour Fragr. J.*, 33, 6–51.

[22] F. Deguerry, L. Pastore, S. Wu, A. Clark, J. Chappell, M. Schalk. (2006). The diverse sesquiterpene profile of patchouli, Pogostemon cablin, is correlated with a limited number of sesquiterpene synthases, *Arch. Biochem. Biophys.*, 454, 123–136.

[23] R. Croteau, S.L. Munck, C.C. Akoh, H.J. Fisk, D.M. Satterwhite. (1987). Biosynthesis of the sesquiterpene patchoulol from farnesyl pyrophosphate in leaf extracts of Pogostemon cablin (patchouli): Mechanistic considerations, *Arch. Biochem. Biophys.*, 256, 56–68.

[24] M. Schalk. Sesquiterpene synthases from patchouli WO2005052163 (A2).

[25] M. Schalk. Sesquiterpene synthases from patchouli WO2005056803 (A1).

[26] M. Schalk, A. Clark. Sesquiterpene syntheses and methods of use WO2004031376 (A2).

[27] M. Schalk, A. Clark. Sesquiterpene synthases and methods of use EP3109219 (A1).

[28] T. Frister, S. Hartwig, S. Alemdar, K. Schnatz, L. Thöns, T. Scheper, S. Beutel. (2015). Characterisation of a recombinant patchoulol synthase variant for biocatalytic production of terpenes, *Appl. Biochem. Biotech.*, 176, 2185–2201.

[29] Anon. https://www.givaudan.com/file/44351/download, downloaded 6 April 2020.

[30] R. Clery, S. Dossenbach, D. Lelievre, B. Schilling. (2018). Perfume compositions (Givaudan SA), US 20180187123 (A1).

[31] L. Caputi, S. Carlin, I. Ghiglieno, M. Stefanini, L. Valenti, U. Vrhovsek, F. Mattivi. (2011). Relationship of changes in rotundone content during grape ripening and winemaking to manipulation of the 'peppery' character of wine, *J. Agric. Food Chem.*, 59, 5565–5571.

[32] C. Wood, T.E. Siebert, M. Parker, D.L. Capone, G.M. Elsey, A.P. Pollnitz, M. Eggers, M. Meier, T. Vössing, S. Widder, G. Krammer, M.A. Sefton,

M.J. Herderich. (2008). From wine to pepper: Rotundone, an obscure sesquiterpene, is a potent spicy aroma compound, *J. Agric. Food Chem.*, 56, 3738–3744.

[33] H. Takase, K. Sasaki, H. Shinmori, A. Shinohara, C. Mochizuki, H. Kobayashi, G. Ikoma, H. Saito, H. Matsuo, S. Suzuki, R. Takata. (2015). Cytochrome P450 CYP71BE5 in grapevine (Vitis vinifera) catalyzes the formation of the spicy aroma compound (−)-rotundone, *J. Exp. Bot.*, 67, 787–798.

[34] A. Goeke, J. Charpentier, B. Schilling, F. Schröder. Production of guaiene and rotundone WO2019110299 (A1).

[35] T. Granier, B. Schilling, E. Locher. 1-hydroxy-octahydroazulenes as fragrances EP2402301(A1).

[36] C. Ridley, J. Wang, S. Marr. Chimeric terpene synthases WO2019161141 (A1).

[37] L. Daviet, L. Rocci, M. Schalk. (2015). Method for producing fragrant alcohols (Firmenich & Cie [CH]), WO2015040197 (A1).

[38] M.L. Diaz-Chavez, J. Moniodis, L.L. Madilao, S. Jancsik, C.I. Keeling, E.L. Barbour, E.L. Ghisalberti, J.A. Plummer, C.G. Jones, J. Bohlmann. (2013). Biosynthesis of sandalwood oil: Santalum album CYP76F cytochromes P450 produce Santalols and Bergamotol, *PLOS One*, 8, e75053.

[39] Anon. https://www.firmenich.com/en_INT/company/news/Dreamwood-Offers-Olfactive-Warmth-Cosmetic-Benefits-and-Positive-Impact.html, downloaded 26 May 2020.

[40] P. Zerbe, J. Bohlmann. (2015). Enzymes for synthetic biology of ambroxide-related diterpenoid fragrance compounds, in J. Schrader, J. Bohlmann (eds.) *Biotechnology of Isoprenoids* (Springer International Publishing, Cham), pp. 427–447.

[41] M. Schalk, L. Pastore, M.A. Mirata, S. Khim, M. Schouwey, F. Deguerry, V. Pineda, L. Rocci, L. Daviet. (2012). Toward a biosynthetic route to sclareol and amber odorants, *J. Am. Chem. Soc.*, 134, 18900–18903.

[42] C. Ignea, F.A. Trikka, A.K. Nikolaidis, P. Georgantea, E. Ioannou, S. Loupassaki, P. Kefalas, A.K. Kanellis, V. Roussis, A.M. Makris, S.C. Kampranis. (2015). Efficient diterpene production in yeast by engineering Erg20p into a geranylgeranyl diphosphate synthase, *Metab. Eng.*, 27, 65–75.

[43] F.A. Trikka, A. Nikolaidis, A. Athanasakoglou, A. Andreadelli, C. Ignea, K. Kotta, A. Argiriou, S.C. Kampranis, A.M. Makris. (2015). Iterative carotenogenic screens identify combinations of yeast gene deletions that enhance sclareol production, *Microb. Cell Fact.*, 14, 60.

[44] M.G. Chacón, A. Marriott, E.G. Kendrick, M.Q. Styles, D.J. Leak. (2019). Esterification of geraniol as a strategy for increasing product titre and specificity in engineered Escherichia coli, *Microb. Cell Fact.*, 18, 105.

[45] P. Weyerstahl, H. Marschall, U. Splittgerber, D. Wolf, H. Surburg. (2000). Constituents of Haitian vetiver oil, *Flavour Fragr. J.*, 15, 395–412.

[46] J.-J. Filippi, E. Belhassen, N. Baldovini, H. Brevard, U.J. Meierhenrich. (2013). Qualitative and quantitative analysis of vetiver essential oils by comprehensive two-dimensional gas chromatography and comprehensive two-dimensional gas chromatography/mass spectrometry, *J. Chromatogr. A*, 1288, 127–148.

[47] L. Paillat, C. Perichet, J.-P. Pierrat, S. Lavoine, J.-J. Filippi, U. Meierhenrich, X. Fernandez. (2012). Purification of vetiver alcohols and esters for quantitative high-performance thin-layer chromatography determination in Haitian vetiver essential oils and vetiver acetates, *J. Chromatogr. A*, 1241, 103–111.

[48] E. Belhassen, N. Baldovini, H. Brevard, U.J. Meierhenrich, J.-J. Filippi. (2014). Unravelling the Scent of Vetiver: Identification of Character-Impact Compounds, *Chem. Biodiversity*, 11, 1821–1842.

[49] F. Aguilar, T. Scheper, S. Beutel. (2019). Modulating the precursor and terpene synthase supply for the whole-cell biocatalytic production of the sesquiterpene (+)-zizaene in a pathway engineered *E. coli*, *Genes*, 10, 478.

[50] A. Jesionek, A. Kokotkiewicz, A. Krolicka, B. Zabiegala, M. Luczkiewicz. (2018). Elicitation strategies for the improvement of essential oil content in Rhododendron tomentosum (*Ledum palustre*) bioreactor-grown microshoots, *Ind. Crops Prod.*, 123, 461–469.

[51] C. Bicchi, A. Chaintreau, D. Joulain. (2018). Identification of flavour and fragrance constituents, *Flavour Fragr. J.*, 33, 201–202.

[52] T. Cachet, H. Brevard, A. Chaintreau, J. Demyttenaere, L. French, K. Gassenmeier, D. Joulain, T. Koenig, H. Leijs, P. Liddle, G. Loesing, M. Marchant, P. Merle, K. Saito, C. Schippa, F. Sekiya, T. Smith. (2016). IOFI recommended practice for the use of predicted relative-response factors for the rapid quantification of volatile flavouring compounds by GC-FID, *Flavour Fragr. J.*, 31, 191–194.

[53] N. Khare, D. Khokhar, S. Patel, N.K. Mishra. (2017). Study on oil recovery after incubation of patchouli foliage, *Trends Biosci.*, 10, 4799–4803.

[54] J. Chandran, K.P.P. Amma, N. Menon, J. Purushothaman, P. Nisha. (2012). Effect of enzyme assisted extraction on quality and yield of volatile oil from black pepper and cardamom, *Food Sci. Biotechnol.*, 21, 1611–1617.

[55] V.D. Dimaki, G. Iatrou, F.N. Lamari. (2017). Effect of acidic and enzymatic pretreatment on the analysis of mountain tea (*Sideritis* spp.) volatiles via distillation and ultrasound-assisted extraction, *J. Chrom. A*, 1524, 290–297.

[56] K. Hosni, I. Hassen, H. Chaâbane, M. Jemli, S. Dallali, H. Sebei, H. Casabianca (2013). Enzyme-assisted extraction of essential oils from thyme (*Thymus capitatus* L.) and rosemary (*Rosmarinus officinalis* L.): Impact on yield, chemical composition and antimicrobial activity, *Ind. Crops Prod.*, 47, 291–299.

[57] F. Galiano, A. Mecchia, M. Castro-Muñoz, A. Tagarelli, R. Lavecchia, A. Cassano, A. Figoli. (2019). Enzyme-mediated extraction of limonene, linalool and linalyl acetate from bergamot peel oil by pervaporation, *J. Membrane Sci. Res.*, 5, 187–193.

[58] L. Scalvenzi. (2014). New frontiers of essential oils research: Biotransformation of the phytocomplex and its pure compounds by endophytic fungi, *Acta Hortic.*, 1030, pp. 125–132.

[59] F.M. Cecati, C. Magallanes-Noguera, C.E. Tonn, C.E. Ardanaz, M. Kurina-Sanz. (2018). Ecofriendly chemical diversification of Eupatorium buniifolium essential oil by endophytic fungi, *Process Biochem.*, 64, 93–102.

[60] K. Zhang, Y. Lin, Z.-J. Diao, W.-H. Zhang, S.-P. Zheng, S.-L. Liang, S.-Y. Han. (2017). Enzymatic process enhances the flavour profile and increases the proportion of esters in citrus essential oils, *Chem. Biodiv.*, 14, e1700187.

[61] M.R. Maróstica, G.M. Pastore. (2007). Production of R-(+)-α-terpineol by the biotransformation of limonene from orange essential oil, using cassava waste water as medium, *Food Chem.*, 101, 345–350.

[62] M.R.M. Junior, T.A.A.R. e Silva, G.C. Franchi, A. Nowill, G.M. Pastore, S. Hyslop. (2009). Antioxidant potential of aroma compounds obtained by limonene biotransformation of orange essential oil, *Food Chem.*, 116, 8–12.

[63] M. Novakovic, D. Bukvicki, V. Vajs, V. Tesevic, S. Milosavljevic, P. Marin, Y. Asakawa. (2018). Microbial transformation of Calamintha Glandulosa essential oil by Aspergillus niger, *Nat. Prod. Commun.*, 13, 1934578X1801300425.

[64] N. Paroul, A. Biasi, A.C. Rovani, C. Prigol, R. Dallago, H. Treichel, R.L. Cansian, J.V. Oliveira, D. de Oliveira. (2010). Enzymatic production of linalool esters in organic and solvent-free system, *Bioprocess Biosyst. Eng.*, 33, 583–589.

[65] K. Hult, P. Berglund. (2007). Enzyme promiscuity: Mechanism and applications, *Trends Biotechnol.*, 25, 231–238.

[66] I. Notar Francesco, J.-J. Filippi, S. Antoniotti. (2017). Sustainable manufacture of a valuable fragrance ingredient: Lipase-catalyzed acylation of Vetiver essential oil and chemoselectivity between sesquiterpene alcohols, *ChemPlusChem*, 82, 407–415.

[67] S. Antoniotti, J.-J. Filippi, I.N. Francesco, J. Ramilijaona. (2015). Biotechnological manufacture of vetiveryl esters (Centre Nat de la Rech Scient (CNRS) [FR]; Université de Nice-Sophia Antipolis [FR]), WO2016193208 (A1).

[68] D. Joulain, R. Tabacchi. (2009). Lichen extracts as raw materials in perfumery. Part 1: Oakmoss, *Flavour Fragr. J.*, 24, 49–61.

[69] D. Joulain, R. Tabacchi. (2009). Lichen extracts as raw materials in perfumery. Part 2: Treemoss, *Flavour Fragr. J.*, 24, 105–116.

[70] Anon. https://eur-lex.europa.eu/eli/reg/2017/1410/oj/fra, downloaded 7 April 2020.

[71] H. Bouges, K. Calabro, O.P. Thomas, S. Antoniotti. (2018). Unusual polycyclic fused product by oxidative enzymatic dimerisation of 5-methylpyrogallol catalysed by horseradish peroxidase/H_2O_2, *Molecules*, 23, 2619.

[72] H. Bouges, A. Monchot, S. Antoniotti. Biotechnological process for the manufacture of atranols-free oakmoss absolute WO2019175171 (A1).

[73] P.J. O'Brien, D. Herschlag. (1999). Catalytic promiscuity and the evolution of new enzymatic activities, *Chem. Biol.*, 6, R91–R105.

[74] S.D. Copley. (2003). Enzymes with extra talents: Moonlighting functions and catalytic promiscuity, *Curr. Opin. Chem. Biol.*, 7, 265–272.

[75] U.T. Bornscheuer, R.J. Kazlauskas. (2004). Catalytic promiscuity in biocatalysis: Using old enzymes to form new bonds and follow new pathways, *Angew. Chem. Int. Ed.*, 43, 6032–6040.

[76] R.J. Kazlauskas. (2005). Enhancing catalytic promiscuity for biocatalysis, *Curr. Opin. Chem. Biol.*, 9, 195–201.

[77] M.S. Humble, P. Berglund. (2011). Biocatalytic promiscuity, *Eur. J. Org. Chem.*, 2011, 3391–3401.

[78] G. Carrea, S. Riva. (2008). Medium engineering, in *Asymmetric Organic Synthesis With Enzymes* (Wiley-VCH Verlag GmbH & Co. KGaA, Weinheim), pp. 1–20.

[79] A.Z.M. Badee, S.A. Helmy, N.F.S. Morsy. (2011). Utilisation of orange peel in the production of α-terpineol by *Penicillium digitatum* (NRRL 1202), *Food Chem.*, 126, 849–854.

[80] A. Navarrete, M. Herrero, A. Martín, M.J. Cocero, E. Ibáñez. (2011). Valorization of solid wastes from essential oil industry, *J. Food Eng.*, 104, 196–201.

[81] R. Sánchez-Vioque, M. Polissiou, K. Astraka, M.D.L. Mozos-Pascual, P. Tarantilis, D. Herraiz-Peñalver, O. Santana-Méridas. (2013). Polyphenol composition and antioxidant and metal chelating activities of the solid residues from the essential oil industry, *Ind. Crops Prod.*, 49, 150–159.

[82] M. Rusanova, K. Rusanov, V. Butterweck, I. Atanassov. (2019). Exploring the capacity of endophytic fungi isolated from medicinal plants for fermentation and phenolics biotransformation of rose oil distillation wastewater, *Biotechnol. Biotechnol. Equip.*, 33, 651–663.

Bibliography

Bauer, K., D. Garbe, H. Surburg. (2001). *Common Fragrance and Flavor Materials: Preparation, Properties and Uses*, 4th ed. Wiley-VCH Verlag GmbH & Co. KGaA.

Faber, K. (2004). *Biotransformations in Organic Chemistry*, 5th ed. (Springer-Verlag, Berlin).

Ohloff, G., W. Pickenhagen, P. Kraft. (2011). *Scent and Chemistry* (Verlag Helvetica Chimica Acta, Zürich).

Sell, C.S. (2003). *A Fragrant Introduction to Terpenoid Chemistry* (The Royal Society of Chemistry, Cambridge).

https://doi.org/10.1142/9781800617780_0007

Chapter 7

Rose Ketones: Past and Present History of This Major Class of Compounds for the F&F Industry*

Fabrice Robvieux

dsm-firmenich SA, Satigny, Switzerland

fabrice.robvieux@firmenich.com

Abstract

The damascone family holds a prominent place among the different olfactive tonalities. Despite a relatively simple structure, damascones have proven to be a great playground for chemists. This interest has not weakened in the past 60 years, both in academic laboratories and in the various companies active in perfumery. With the increasing importance of green and sustainable chemistry, damascones remain a very vivid and active subject of research for the entire industry. In this chapter, different strategies for the synthesis of rose ketones with their advantages and disadvantages are presented.

7.1 Introduction

We all live in an environment that is filled with molecules that can be smelled or tasted. The sense of smell is widespread throughout the animal

*Dedicated to Dr. Charles Fehr on the occasion of his 78th birthday for his major contributions to the synthesis of rose ketones.

kingdom. Volatile compounds are used by animals and plants to communicate, protect themselves, find food, mating partners, etc. It should be pointed out that plants also use chemical substances in various ways. The range of molecules is quite large; therefore, the olfactory system needs to be flexible and efficient [1].

As part of the living world, humans also possess a sense of smell and taste. The importance of smell is clearly illustrated by the large number of genes coding for olfactory receptors: 350–400 genes, i.e. 1.5% of the total genome (~25,000 genes), is devoted to olfactive receptor genes [2].

Smell, as it is directly connected to our emotions, has a very powerful impact on behaviour. This is readily illustrated with the famous example of Proust's madeleine cake. In this well-known story, the narrator remembers his childhood when he consumes a cake [3]. This illustrates how a unique odour can trigger distinct memories from our childhood or from emotional moments — positive or negative — later in life.

Beyond primary functions, important divergences appeared. Odorous materials moved from vital functions to more secular uses. For example, in ancient Egypt, perfumes were used for offerings to the gods. At first, products were extracted from nature as oils or ointments. Then, with the development of organic chemistry and finding inspiration from mother nature's creations, chemists were able to synthesise new synthetic materials. New, unknown or new combinations of smells were accessible, and the perfumery landscape profoundly changed. The higher accessibility of perfumery ingredients combined with lower prices allowed perfumery market growth over the last decades.

Molecules have been classified in large families based on their smell. The most famous classes are musk, green notes, sandalwood and floral notes, as well as fruity notes, etc. [4]. Among all the floral notes, the smell of rose holds a unique place not only for its symbolic role in most cultures but also for the multiple aspects of its smell. In this chapter, the discovery and the tremendous efforts directed towards the isolation, structure elucidations and synthesis of their key components, the rose ketones, are presented.

7.2 Discovery and General Information

The history of rose can be traced back before Roman times. In the thirteenth century, *Rose damascena* (the name is a reference to the city of Damas, Syria) was brought back to Europe [5]. It was not until the early

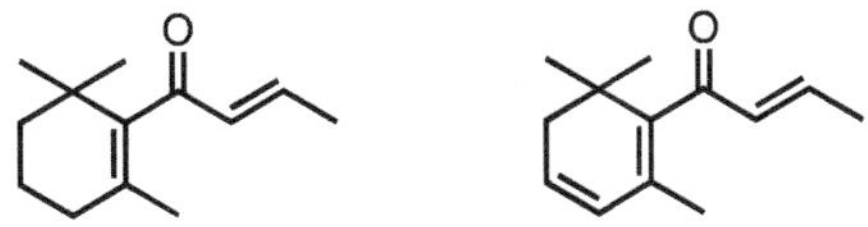

1
β-Damascone

2
β-Damascenone

Figure 7.1. The two main compounds of the damascone family.

1960s, driven by progress in analytical tools, that analytical chemists could tackle the exact composition of the rose oil. This epic history has been described in two publications published 20 years after the real work had been done! [6]. During that time, more than 300 compounds have been identified, among which two molecules stand out (Figure 7.1).

The structure of β-damascenone **2** has already been identified as a component in tobacco and raspberry [7].

Structures containing 13 carbons (called apocarotenoids) are believed to be formed through the degradation of carotenoid-like molecules, which is outside the scope of this review. But they probably originate from the degradation of carotenoid-like molecules [8]. Recently, a discussion on the advantages and disadvantages between chemical and biotechnological synthesis of C_{13}-apocarotenoids has been published [9].

As these compounds can only be found in trace amounts in nature, chemists from academia and also the industry quickly embarked on research programmes towards their total synthesis. Damascone and damascenone can exist as several olefin regioisomers. All have been synthesised, but only four isomers of damascone and β-damascenone are currently used in the fragrance and flavour industry. It should be noted that δ-damascone **4** and γ-damascone **5** have not been found in nature (Figure 7.2).

The different isomers exhibit not only different notes around the rose smell. They also display two orders of magnitude difference in their threshold values, as shown in (Figure 7.2) [10].

The different synthesis methods that will be presented in this chapter span over 60 years of research. As some approaches could be applied to several isomers, all isomers will be discussed at the same time. The order is neither related to the importance in perfumery nor to a chronological order but more by class of reactions. Industrially relevant reactions and

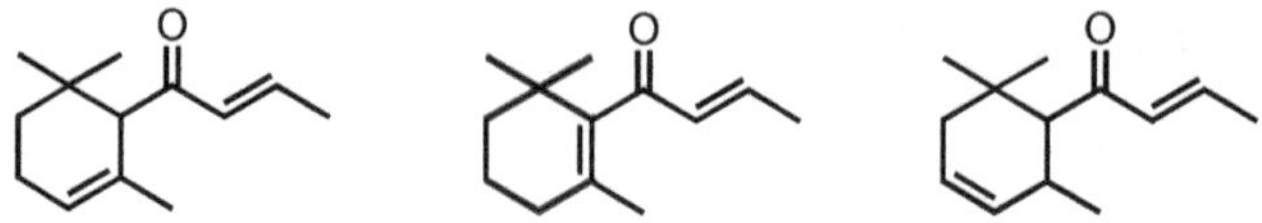

3 α-Damascone $1.4*10^{-4}$ μg/dm³ | **1** β-Damascone $1.9*10^{-4}$ μg/dm³ | **4** δ-Damascone $2.1*10^{-5}$ μg/dm³

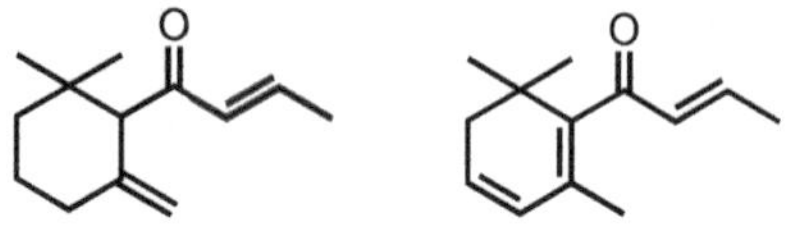

5 γ-Damascone $4.5*10^{-3}$ μg/dm³ | **2** β-Damascenone $1.3*10^{-5}$ μg/dm³

Figure 7.2. Threshold value for different damascones (indicated below each molecule).

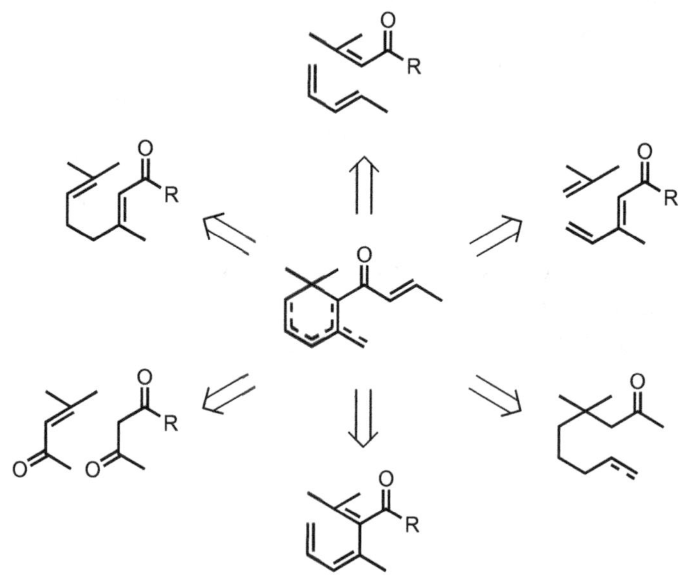

Scheme 7.1. Retrosynthetic approaches based on cyclisation reactions.

methodologies will be emphasised, and the balancing of interests will be discussed.

In the first half of this chapter, various cyclisation reactions are considered (Scheme 7.1).

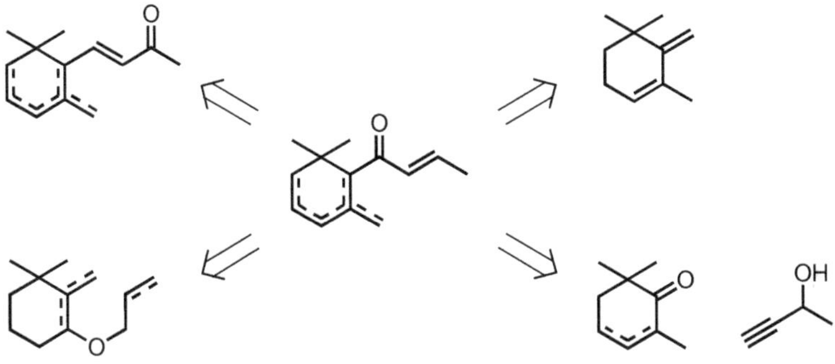

Scheme 7.2. Miscellaneous retrosynthetic approaches.

Then, other synthetic disconnections based on functional rearrangements, nucleophilic addition are highlighted (Scheme 7.2).

So, let's start our journey through the history of rose ketones spanning the past 60 years.

7.3 Cyclisation Reactions

The α-damascone was the first damascone to be synthesised. From a historical perspective β-damascone was first synthesised but mostly as an intermediate *en route* to the more prestigious β-damascenone.

7.3.1 *Pseudodamascone route*

In 1975, K.-H. Schulte-Elte published an interesting approach. It is related to the synthesis of another important class of molecules called ionones [11]. The synthesis starts from dehydrolinalool **6** (Scheme 7.3).

The reaction takes advantage of the Saucy–Marbet reaction [12]. The product is obtained as a mixture of *cis* and *trans* enones that could not be separated by distillation. This synthesis presents several appealing aspects from an industrial point of view. The starting material is cheap and readily available. All steps can be scaled up and are cost-effective. But unfortunately, some off-notes present in the final product completely undermine this route. Despite a lot of hard work, no cost-efficient purification leading to an olfactory-accepted material could be found. During the investigation, an impurity responsible for a strong green off-note was identified and went on to become the very successful captive Dynascone® [13].

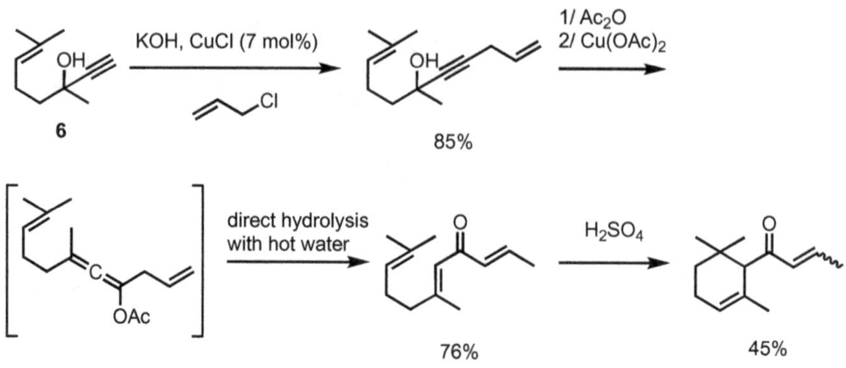

Scheme 7.3. Pseudodamascenone route.

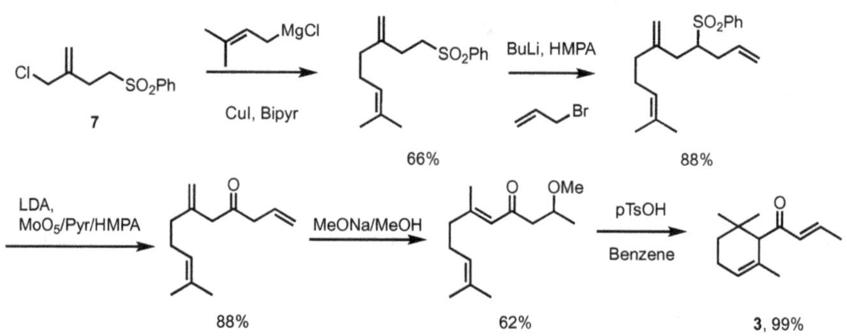

Scheme 7.4. Sulphone as a ketone surrogate.

7.3.2 *Sulphone-based route*

In 1984, Prof. J. Otera published an interesting extension of the chemistry developed in his lab [14]. The strategy took advantage of the transformation of a sulphone into a ketone using oxidative conditions (Scheme 7.4).

7 is a versatile synthon and can be used towards several interesting intermediates. α-Damascone **3** can be synthesised, but many steps are clearly not compatible with the challenges of an industrial synthesis, especially the reaction conditions, which they used for the oxidative degradation of the sulphone. Vedej's reagent (MoO_5/Pyridine/HMPA) has to be used, which is clearly a major obstacle for scale-up.

Sulphone is also a versatile functional group. In the late 1970s, Prof. Torii used two of the main properties of this functional group to generate an alkene: firstly, as an activating group that allows alkylation in the

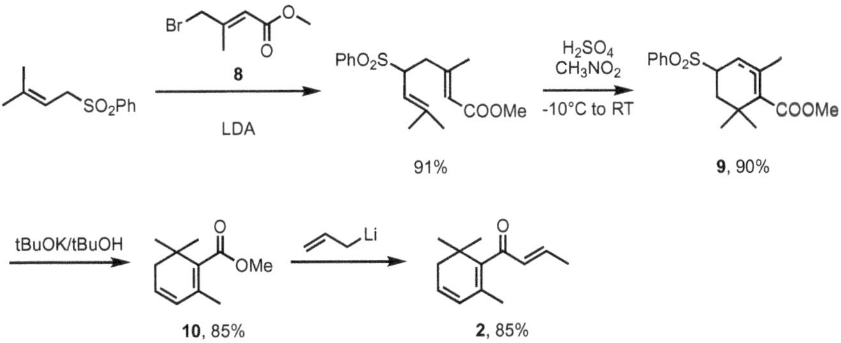

Scheme 7.5. Sulphone as leaving group.

Scheme 7.6. Wittig olefination approach.

α position, and secondly, as a very good leaving group in their synthesis of β-damascenone **2** [15] (Scheme 7.5).

The precursor for the cyclisation is obtained by alkylating the prenyl-sulphone with an activated methallyl bromide **8**. The acid-catalysed cyclisation takes place with concentrated sulphuric acid diluted in nitromethane at low temperatures. The ratio between conjugated/unconjugated isomers is 95 to 5. Final elimination was achieved by treating **9** with *t*BuOK, yielding methyl safranate **10** in excellent yield. Then, the mixture of isomers was converted to the desired product. More details of this final step will be discussed later in this chapter.

7.3.3 *Wittig reaction*

A very different synthesis was reported by Prof. Büchi towards β-damascenone **2** [16]. The central idea was to build one of the double bonds via a Wittig reaction followed by an electrocyclisation (Scheme 7.6).

The sequence starts with the addition of methyl Grignard halide (2 equiv.) on ethyl diacetylacetate, followed by an acetic anhydride quench. A critical point is the synthesis of the phosphonium salt **11**.

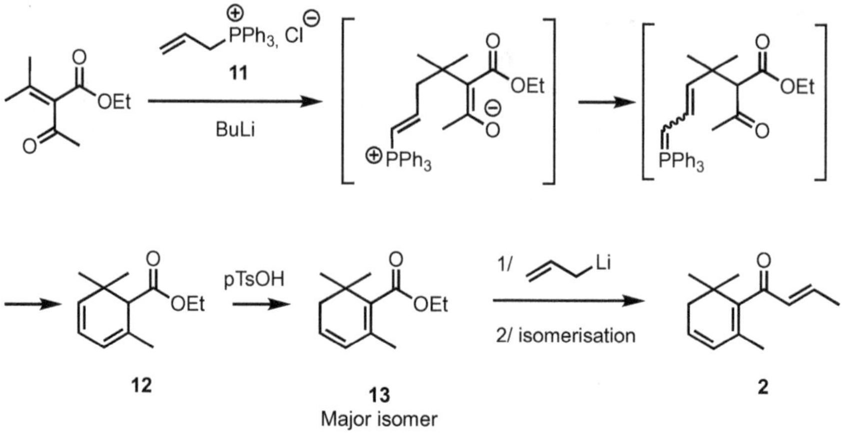

Scheme 7.7. Mechanism of the tandem Michael addition/Wittig olefination reactions.

Depending on the synthesis used, an isomerisation reaction occurs to give propenyltriphenylphosphonium chloride as the major product. The nature of the counter anion also plays a key role. Chloride is critical in this reaction. Having solved these problems, the key reaction gives none of the expected triene but instead forms directly the desired ethyl α-safranate **12** (Scheme 7.7).

This result can be rationalised if Michael addition occurs first, followed by an intramolecular Wittig reaction. The ethyl α-safranate **12** could be isomerised to the ethyl β-safranate **13** with *p*-TsOH. Büchi also found that allyl lithium was able to carry out the final step with a high yield [17].

7.3.4 *Methylheptenone route*

In the early 1980s, Toyotama published a patent on the synthesis of γ-damascone **5**, starting from methyl heptenone **13** (Scheme 7.8) [18].

In the patent description, yields are not given. The Lewis acid-mediated cyclisation uses a typical catalyst for this kind of reaction, $SnCl_4$. No comment is given for the selectivity of the Wittig reaction, but steric hindrance of the *gem*-dimethyl can be evoked. The final aldol step uses typical reaction conditions that will be described in more detail later. Despite the cheap starting materials, the use of stoichiometric $SnCl_4$ meant that this process could generate a large amount of toxic waste stream. Absence of yields makes any further comment difficult.

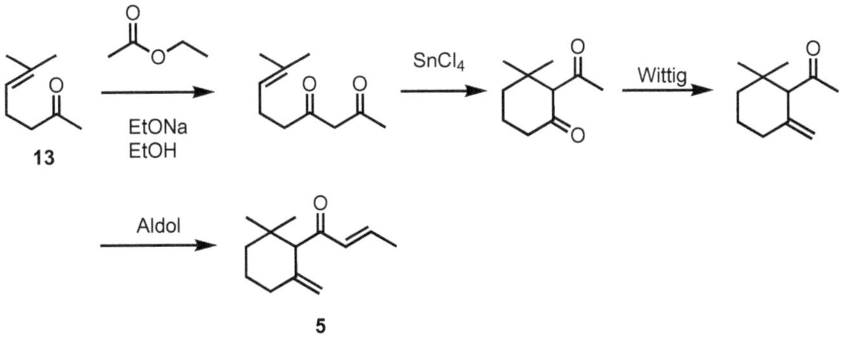

Scheme 7.8. Lewis acid-mediated cyclisation.

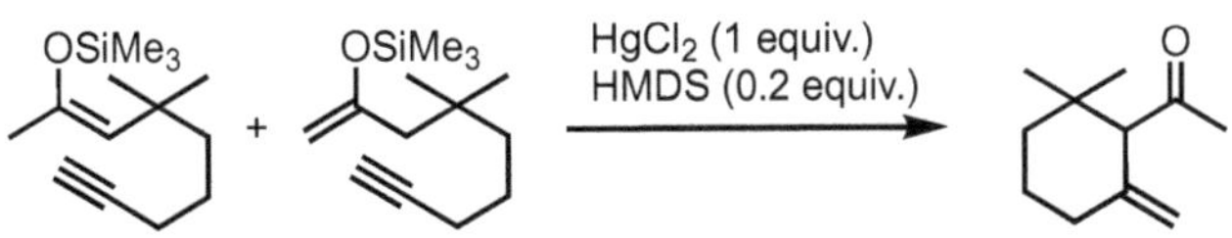

14, 7/3 **15**, 79%

Scheme 7.9. Silylenol ether addition to alkyne approach.

7.3.5 *Mercury-catalysed cyclisation*

The same intermediate can be synthesised by another route. Silyl enol ethers **14** can cyclise onto a terminal alkyne to give cyclic compounds [19]. The cyclisation is promoted by a stoichiometric amount of mercury and a catalytic amount of hexamethyldisilazane (HMDS) (Scheme 7.9).

The use of a mercury salt poses insurmountable problems when run on a scale. The disconnection is appealing, especially with the recent renaissance of gold catalysis [20]. However, synthesis of the required ketone is not obvious. It is also worth mentioning that the conditions used during the quench are particularly critical to ensure the alkene remains in the γ position.

7.3.6 *Zinc-catalysed cyclisation*

BASF patented the cyclisation of an allylic chlorine derivative on an internal alkyne **17** with $ZnCl_2$ (Scheme 7.10) [21].

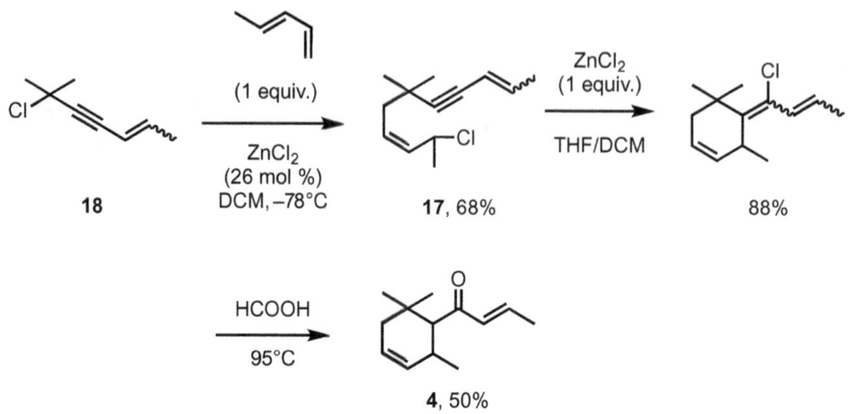

Scheme 7.10. Zinc-catalysed cyclisation reaction.

The starting chloride **18** compound is easily accessible from two simple building blocks. However, both cryogenic conditions and stoichiometric amounts of $ZnCl_2$ during the cyclisation step make it difficult to achieve an economically viable process. The final yield also has a detrimental effect on the overall yield of this sequence.

7.3.7 *Cyclisation to methyl cyclogeraniate*

Damascones and damascenone are closely related molecules and can be traced back to common intermediates. Until the 1980s, damascones were synthesised by different routes. After several years of intense research, chemists at Firmenich made an important breakthrough. A unified synthesis was designed with methyl cyclogeraniate as the key intermediate. The latter is synthesised starting from the cheap methylheptenone **13** (Scheme 7.11).

A Horner–Wadsworth–Emmons reaction installs the α-β-unsaturated ester. The cyclisation is well known and gives a mixture of α-**18** and β-**19** in favour of the former isomer [22]. Isomers present in the mixture can easily be separated by distillation. The cyclisation is done either with Brönsted acids or with heterogeneous acidic supports, which are industrialisable and cost-efficient. Then, the challenging part of the synthesis arises. As already mentioned, Büchi had shown that allyllithium could add to the ester to give the desired compound; however, from an industrial point of view, this is not viable due to the expensive and cryogenic

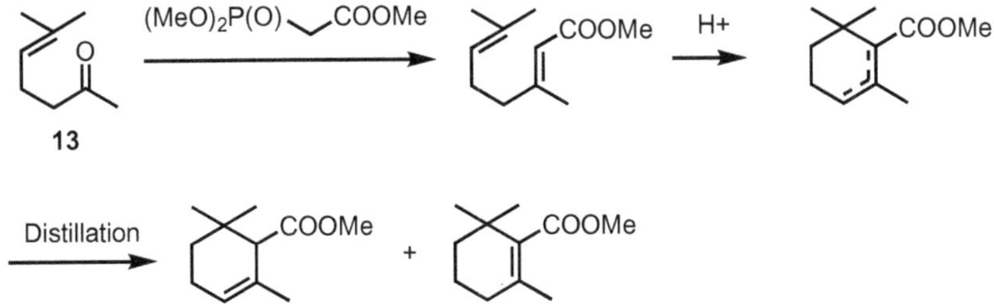

13

18 19

Scheme 7.11. Cyclogeraniate synthesis.

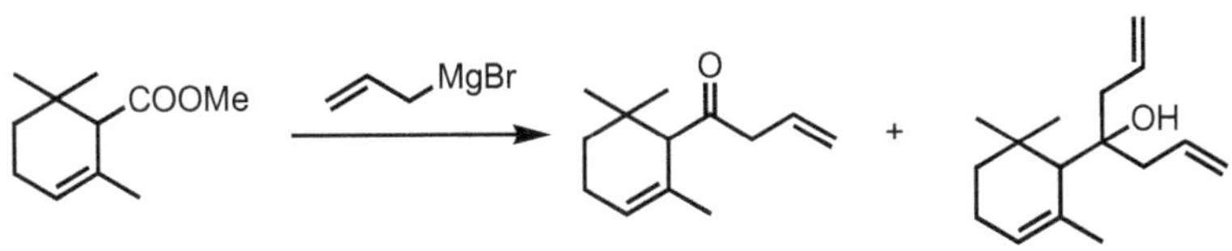

18 20 21

Scheme 7.12. Double-allyl Grignard addition.

conditions required. The industrial alternative is an allyl Grignard reagent, but then selectivity issues arise. A double addition product is obtained as the major product (Scheme 7.12).

It should be pointed out that depending on the quench method, the side-chain deconjugated product is obtained first and could be isomerised either by acid or basic treatment. Allyl Grignard reagent addition is unselective and has been recently reviewed [23]. To illustrate this amazing reactivity in a competitive experiment, allylic Grignard reagents still add to an excess of ketone or aldehyde even in the presence of excess water [24].

At the same time, part of the research was devoted to the recycling of the double addition adduct **21**. This retro-ene reaction was known in vapour phase thermolysis [25] (Scheme 7.13).

Due to its simplicity, this reaction was scaled up. Again, in the case of α-damascone **3**, one of the downsides was the formation of the by-products **22** and **23**, which were difficult to remove and gave an unpleasant smell to the final product (Scheme 7.14).

The alcohol **21**, being triply homoallylic, can adopt a conformation where the retro-ene with the internal double can proceed, giving the cyclohexene **22**. Beside thermal rearrangements, different methods are known

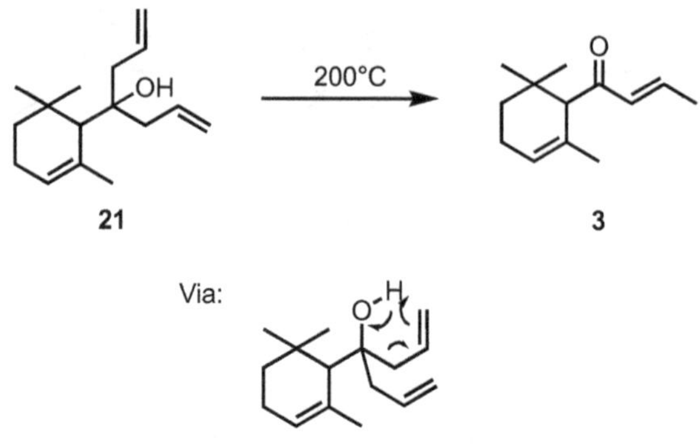

Scheme 7.13. Retro-ene reaction for the damascone synthesis.

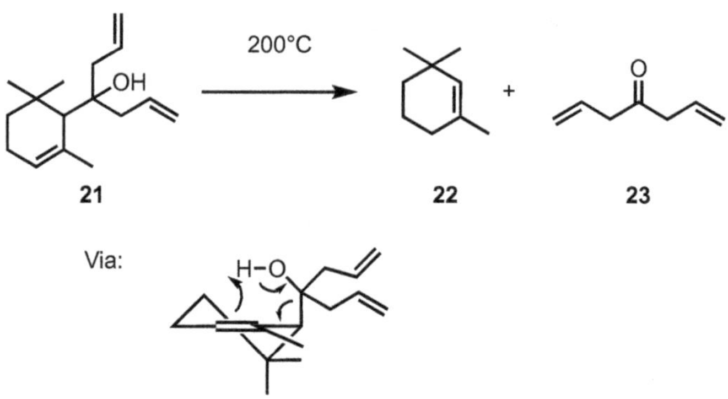

Scheme 7.14. Unselective retro-ene reaction for by-product formation.

in the literature to achieve this transformation of alcohol **21** into damascone. Fragmentation was observed as a side reaction during the thermal rearrangement of homoallylic tertiary alkoxides (with Li, MgCl and MgBr as counter-anions) and was in competition with the anionic oxy-Cope rearrangement (with Na and K as counter-anions) of alkoxide [26].

As a consequence, homoallylic alcohol **21** was treated with KH in HMPA at RT. A mixture of conjugated and deconjugated damascones was obtained. However, *t*BuOK in DMF at 70°C could also be used for the same reaction [27]. Despite these milder reaction conditions, the same by-products could be observed for the rearrangement of the precursor of α-damascone **3**.

Büchi showed that allyllithium is a very efficient and selective reactant. Clearly, the reactivity between the allyllithium reagents and Grignard reagent was very different. The former was able to act both as a nucleophile and a base. To allow this reaction to go to completion, an excess of the expensive reagent was needed. Therefore, known methods that favour enolate formation for the Grignard reagent were explored [28]. Despite extensive studies, no satisfactory experimental conditions were found. Large-scale syntheses could not be improved at that time.

To solve this issue, a more elegant approach was later found by C. Fehr [29]. If the addition of the allyl Grignard reagent was done in the presence of a stoichiometric amount of a strong base, i.e. LDA, the only product obtained in the reaction was the mono-allylated ketone (Scheme 7.15).

After the first addition of allyl Grignard reagent, the intermediate ketone was swiftly deprotonated by LDA, giving the enolate, while the ketone remained "protected", thus avoiding double addition until the final quench.

The yields and selectivity (mono vs. double addition) were impressive. Selectivity for the α-damascone **3** and β-damascenone **2** was 98/2 and 96/4, respectively. However, for β-damascone **1**, slight erosion was observed with 90/10. The advantage of using a mixed organolithium and magnesium reagent is clearly demonstrated. Syntheses of rose ketones had become truly unified. The strategic importance of this methodology resulted in worldwide patent applications [30].

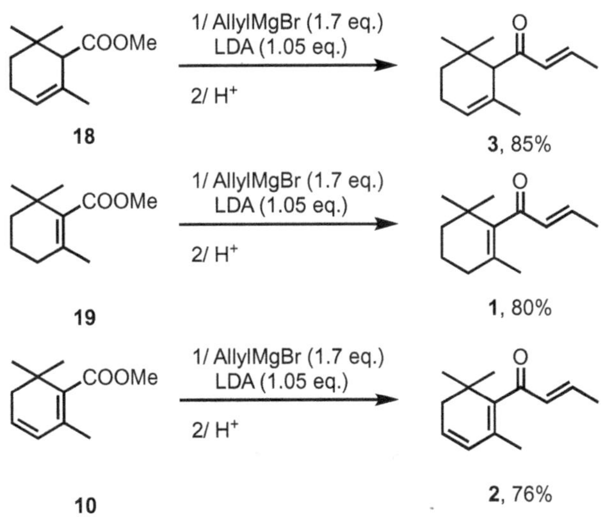

Scheme 7.15. Scope of the allyl Grignard/LDA methodology.

This methodology was also applied to the γ-damascone **5** by simply changing the order of reactants. The key concept was to kinetically deprotonate in γ position the methyl β-cyclogeraniate **19** with a strong base [31]. The ester enolate bearing an *exo* olefin still remained susceptible to a nucleophilic attack by the allyl Grignard reagent [32] (Scheme 7.16).

The first step was a critical factor for this process and was patented [33]. This trick allowed Firmenich to commercialise pure γ-damascone **5**.

It should be emphasised that, in addition to organolithium and Grignard reagents, hydride reagents can be added. Vitride® (Sodium bis(2-methoxyethoxy)aluminium hydride) was the reagent of choice. This new approach allowed the direct synthesis of aldehydes from esters. Due to the importance of aldehydes in the fragrance chemistry and the ease of implementation of this process, a patent was also filed [34].

7.3.8 *Cyclisation to cyclogeranic acid*

In the previous approach, a keto-enolate was thought to be an intermediate. One could argue that the same enolate can stem from the addition of allyl Grignard reagent to the ketene **24** (Figure 7.3).

Obviously, this methodology can only work if the ketene can be generated and be stable enough to allow the nucleophilic addition. Ketenes are in our collective psyche unstable compounds that dimerise very quickly. However, if they are substituted with bulky groups, they are relatively

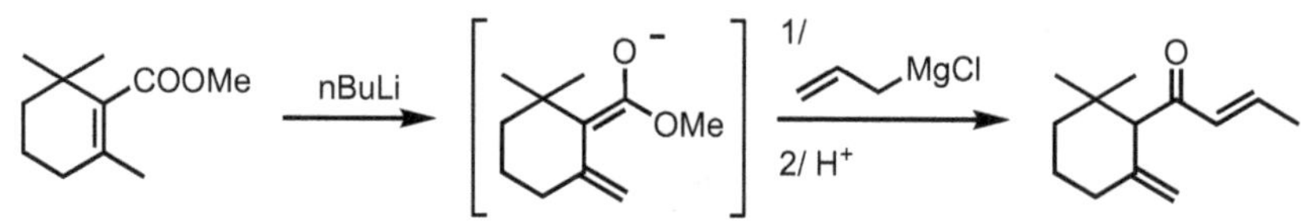

19

76%

Scheme 7.16. γ-Damascone synthesis.

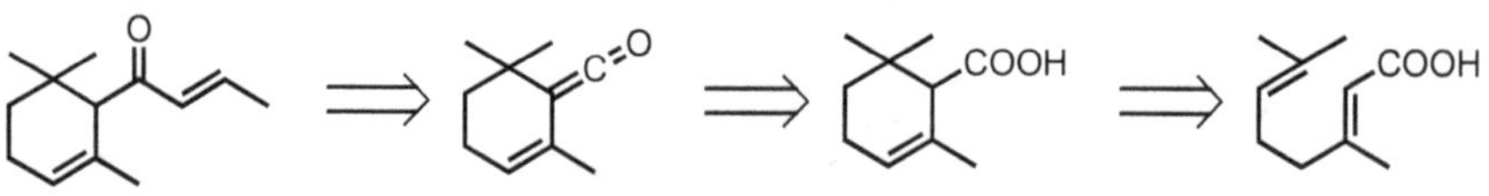

3 24 25 26

Figure 7.3. Ketene-based retrosynthetic approach.

stable. In some instances, ketenes can even be persistent [35]. Two different easy and cheap syntheses of ketenes are possible (Scheme 7.17).

With proper R and R', ketene can be generated either via the mixed anhydride and pyrolysis under vacuum [36] or via the acyl chloride followed by base treatment [37]. In this chapter, the synthesis and distillation of **24** are described. Furthermore, the addition of a nucleophile, particularly a Grignard reagent or alkyllithium, to ketenes was known [38]. A more extensive study was started with the aim of developing an industrial process. Both routes were evaluated, and after some refinements, the acyl chloride pathway was selected and run on a scale [39] (Scheme 7.18).

The cyclisation of geranic acid **26** is well known and selectively affords the α-cyclogeranic acid **25** [40]. The ketene **24** is generated in

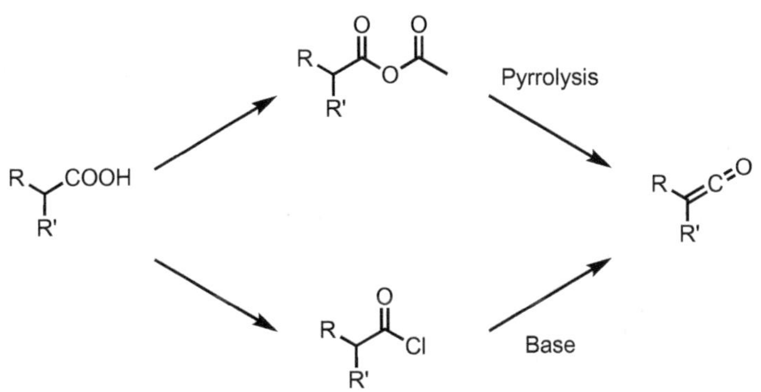

Scheme 7.17. General route towards ketene.

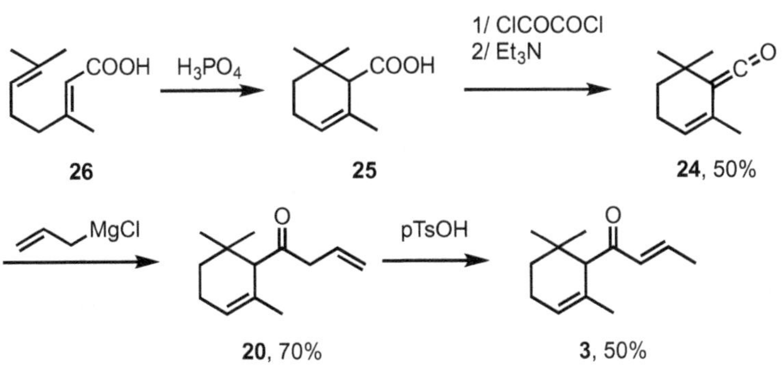

Scheme 7.18. α-damascone synthesis based on the ketene approach.

two steps in a one-pot reaction. The allyl Grignard reagent addition worked smoothly and gave the product as a mixture of non-conjugated product **20** and conjugated product **3** (80/20). The last step afforded the α-damascone **3** in a moderate yield. The process has been further optimised to make it industrially and economically viable.

The ketene route generates an enolate which can be trapped by different electrophiles (TMSCl or Ac_2O, for instance). This possibility gave the opportunity to further elaborate the molecule. For example, a simple proton can be the electrophile. This paved the way towards enantioselective protonation [41].

However, before further discussion, it should be emphasised that important findings were made independently by Tidwell and Seebach. They have shown that stereospecific formation of enolates is possible by the addition of organolithium reagents on a ketene [42].

Of the pieces together, the synthesis of enantio-enriched α-damascone **30** appeared feasible. After tremendous effort, *N*-methyl ephedrine **27** proved to be the best chiral proton source. The first development was made using a stoichiometric amount of this reactant [43] (Scheme 7.19).

Careful and thorough work has been done, and a very subtle influence of the counter-anion on the enantioselectivity of the protonation was found. Going beyond first limitations and by a careful understanding of the reaction mechanism, C. Fehr and co-workers were able to develop a catalytic version [44]. The trick is that the generated ephedrine-alkoxide **28** is able to deprotonate the intermediate ketone **29** to regenerate the ephedrine-alcohol **27**. This process could be called an auto-catalytic system (Scheme 7.20).

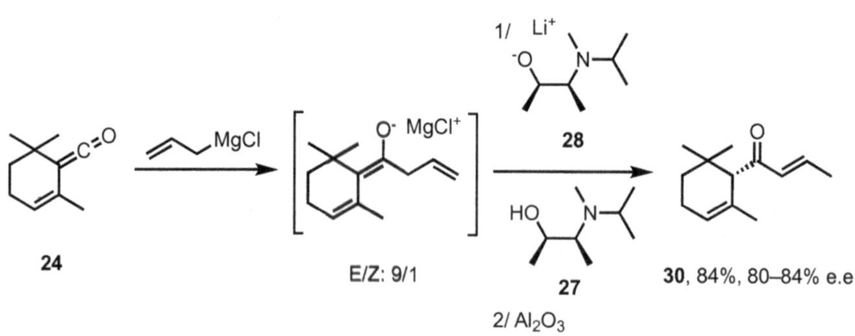

Scheme 7.19. Enantioselective protonation for α-damascone synthesis.

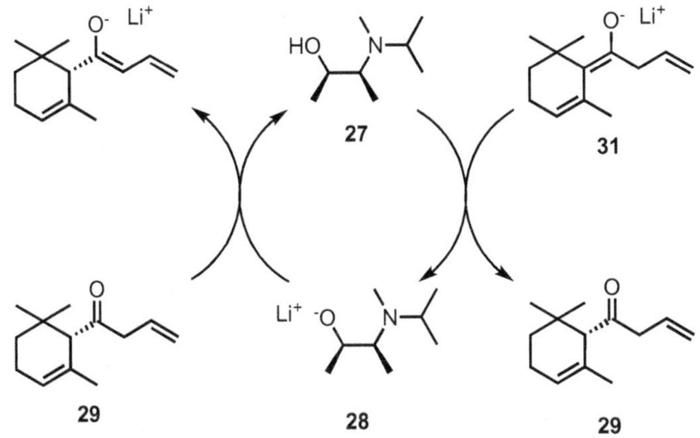

Scheme 7.20. Catalytic enantioselective protonation mechanism.

In the best conditions, 0.3 equivalents of the ephedrine derivative at −60°C afford, after acidic quench, ketones **29** and **30** in an 86% yield and a 93% e.e. A lower amount of **27** gives lower conversion, and the reaction rate decreases dramatically. In order to get a better understanding of the exact mechanism, several control experiments were tested. Ketone **29** does not protonate **31** enantioselectively without **27**.

Extension of this chemistry was accomplished with the use of aryl thiol ester instead of the ester [45]. The enantioenriched aryl thiol ester could be further elaborated into different interesting compounds like cyclogeranic acid and cyclocitral. The potential of this technology was so high that it was protected by a patent [46]. A similar approach has been developed in academia with the use of a dithioester [47]. But from an economic standpoint, this last approach is not industrially relevant.

The final development of this chemistry was made with the use of the enol instead of the enolate. Despite the common perception that enols are unstable, they can be isolated if sterically hindered [48]. C. Fehr and co-workers have shown that the enol of (*R*/*S*)-α-damascone could be protonated with **28** and gave (*S*)-α-damascone in an 86% yield and a 93% e.e. [49].

Beside these sophisticated approaches, classical resolution with chiral phenylethylamine has been used to access enantioenriched cyclogeranic acid [50]. Potentially enantioenriched damascones are accessible by this approach.

7.3.9 *Diels–Alder reaction*

The six-membered ring can be constructed through a Diels–Alder reaction. Cookson *et al.* reported the synthesis of α-damascone in 1973 [51]. They used an inverse electron-demand Diels–Alder catalysed with $AlCl_3$ (Scheme 7.21).

The first two steps of the synthesis are rather laborious, giving respectively a 30% yield of an *E*/*Z* mixture and the Diels–Alder adduct in less than 20% yield. The final aldolisation/elimination steps were based on the work of Zimmerman [52]. As an aside, the aldol reaction has been subject to numerous studies. The use of Grignard reagents seems to be essential. A possible explanation is that magnesium can form either a complex or coordinated intermediate, bringing the reactants together. The final product can also be stabilised, bringing the reactants together and ensuring high selectivity and reactivity (Figure 7.4).

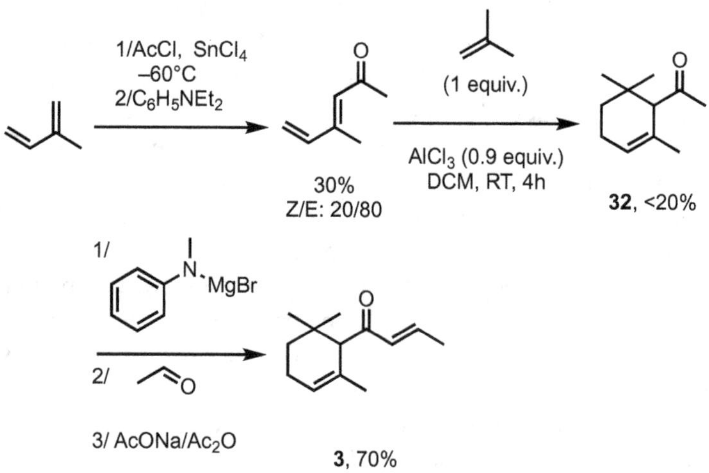

Scheme 7.21. Inverse electron-demanding Diels–Alder.

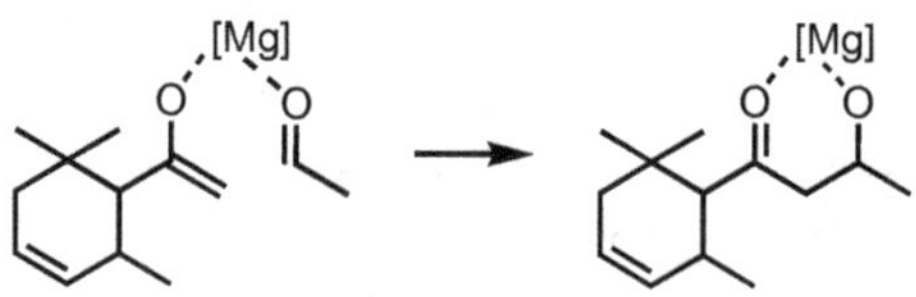

Figure 7.4. Plausible Mg chelating effect during the aldol reaction.

Nowadays, this aldol reaction is still used on a scale for the final aldol step in the industrial synthesis of damascone [55b].

The same group extensively explored all possible combination of reactants and was the first to propose a rapid synthesis of *β*-damascenone **2** and *δ*-damascone **4** [53] (Scheme 7.22).

The starting material was easily accessible from mesityl oxide **33**, and the Diels–Alder step gives a moderate yield. The second alkene was introduced via the elimination of the bromide under basic conditions at elevated temperatures. The final aldol step occurs smoothly, as already described. Alternatively, if **34** is treated with a stoichiometric amount of zinc, debromination occurs and *δ*-damascone **4** is obtained. Despite this novel synthesis, the debromination step is challenging for scale-up. A few years later, another straightforward synthesis of *β*-damascenone **2** and *δ*-damascone **4** was published [54]. Among the damascones, *δ*-damascone **4** is the one possessing both the substitution pattern and the double-bond position that are most suited for a Diels–Alder reaction. In this chapter, mesityl oxide **33** and penta-1,3-diene **36** undergo a Diels–Alder reaction catalysed by 50 mol% $AlCl_3$ (Scheme 7.23).

From an experimental point of view, the diene **36** is added to a mixture of mesityl oxide **33** and $AlCl_3$ in DCM at reflux. A fivefold excess of the diene was needed, as polymerisation was a major competing pathway. It should be emphasised that the diene is a mixture of *cis*/*trans* isomers in

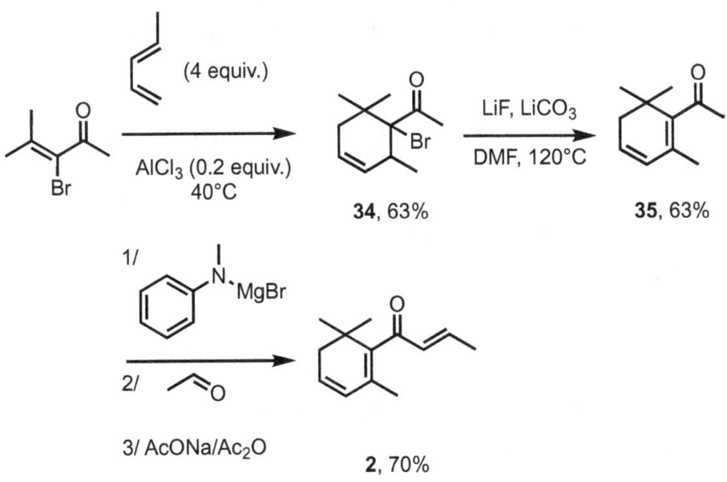

Scheme 7.22. *β*-damascenone synthesis via a Diels–Alder reaction.

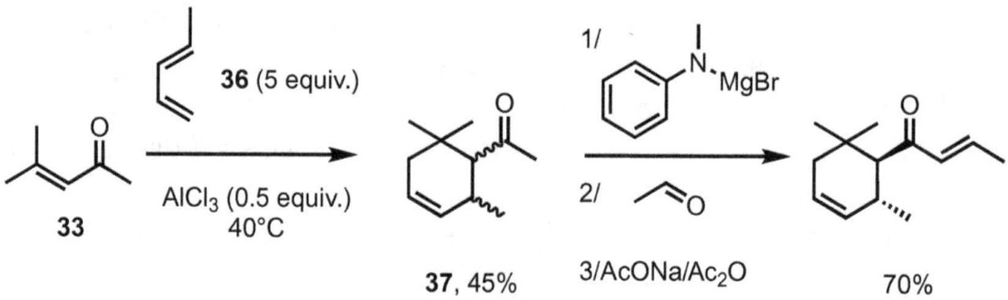

Scheme 7.23. δ-damascone synthesis via a Diels–Alder reaction.

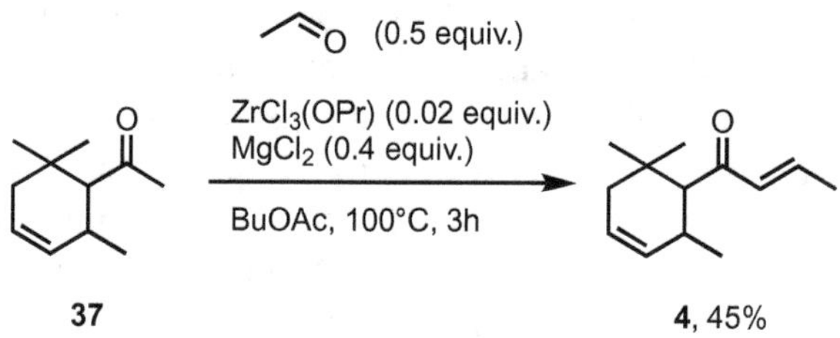

Scheme 7.24. Catalytic aldol reaction for lateral chain synthesis.

a ratio of 30/70, and only the *trans* reacted in the Diels–Alder reaction. It was the first time that a β-disubstituted dienophile was used in a Diels–Alder reaction, and a remarkable yield of 45% was obtained. The Diels–Alder adduct **37** was obtained as a mixture of *cis*/*trans* in a ratio of 87/13. However, the *trans* isomer is desired, and epimerisation occurs during the aldol step. IFF was the first to file several patents on this approach [55]. This pathway is most probably the current industrial synthesis of δ-damascone **4**. Recently, Procter and Gamble has patented variations of the epimerisation/aldol steps, showing that the interest around this molecule is still high [56].

As shown previously, this Diels–Alder approach is selective and feasible on a scale; however, the aldol step suffers from the use of a stoichiometric amount of a strong base.

In 2012, Firmenich showed that the aldol step could be promoted catalytically [57]. A mixture of $ZrCl_3(OPr)$ and $MgCl_2$ could be used with acetaldehyde, giving reasonable yields of δ-damascone **4** (Scheme 7.24).

This new aldol reaction represented not only a major advance towards δ-damascone in terms of cost but also as part of the green chemistry principles [58].

As we have seen previously, the use of β-disubstituted dienophile made the Diels–Alder step difficult. One possible trick is to use electron-withdrawing groups to further enhance the electrophilicity of the dienophile. The Meldrum's acid derivative **38** was used, and later on, isopropylidene malonate was also found to be a versatile reactant [59] (Scheme 7.25).

The isopropylidene dienophile **38** was obtained in a 54% yield by condensing Meldrum's acid with acetone in the presence of pyridine. The Diels–Alder reaction worked well, given the formation of two contiguous quaternary centres [60]. The allyllithium addition gave a mixture of the expected product and some δ-damascone **4**. Two further steps were needed to convert **39** to δ-damascone **4**.

Finally, a relatively long synthesis of the γ-damascone **5** was described by Snowden. This route was based on a Diels–Alder reaction with activated diene to create the substituted cyclohexenone. Several steps were required to give the desired compound [27c].

7.3.10 *Acetoacetate cyclisation*

Another possible disconnection of the cyclohexene moiety is the condensation of an acetoacetate derivative to an enone (Figure 7.5).

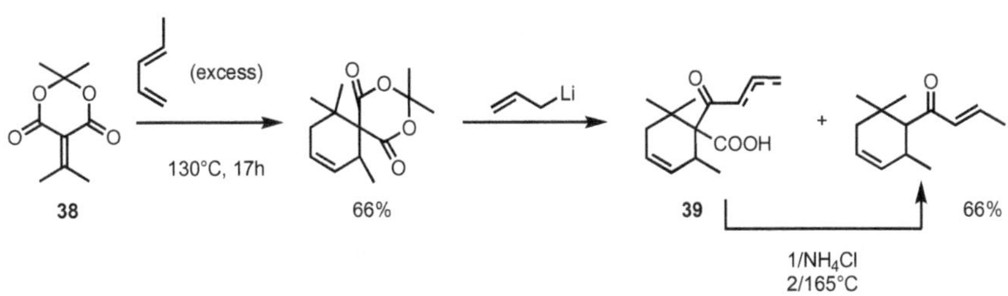

Scheme 7.25. Meldrum's acid-based damascone synthesis.

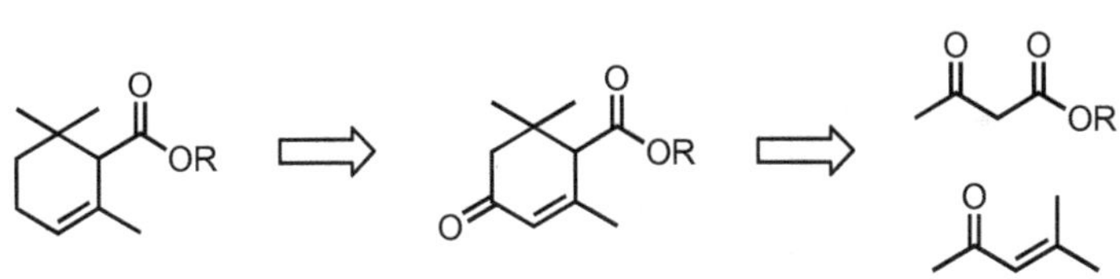

Figure 7.5. Michael addition-based synthesis.

Product **40** was described as a potential intermediate for the synthesis of xanthophyll and zeanthin [61]. This approach was identified and later used for the synthesis of α-damascone **3** (Scheme 7.26) [62].

The major drawback of this synthesis is the poor selectivity observed during the first step, in addition to a low yield. The mixture of regioisomers obtained can only be separated after extensive column chromatography. In order to get rid of the ketone, the thioketal is introduced. The elaboration of the lateral chain is quite long with an overall yield of 50%. Finally, Raney nickel selectively reduced the thioacetal.

A similar approach has been published in the early 1970s by Hoffmann-La Roche [63]. They targeted carotenoid molecules that shared the same trimethyl cyclohexanone core as the damascones (Scheme 7.27).

On the one hand, this synthesis suffered from selectivity issues and poor yields, but, on the other hand, the starting materials were cheap and available. It was claimed that selective hydrolysis and decarboxylation of **41** to give isophorone **42** was observed and possible. It was tried prior to purification and gave easier access to **40**. Later, these results were disputed and a specific thioketalisation of **40** has been developed [64]. **40** has been further elaborated in various ways that could be potentially used to give damascones

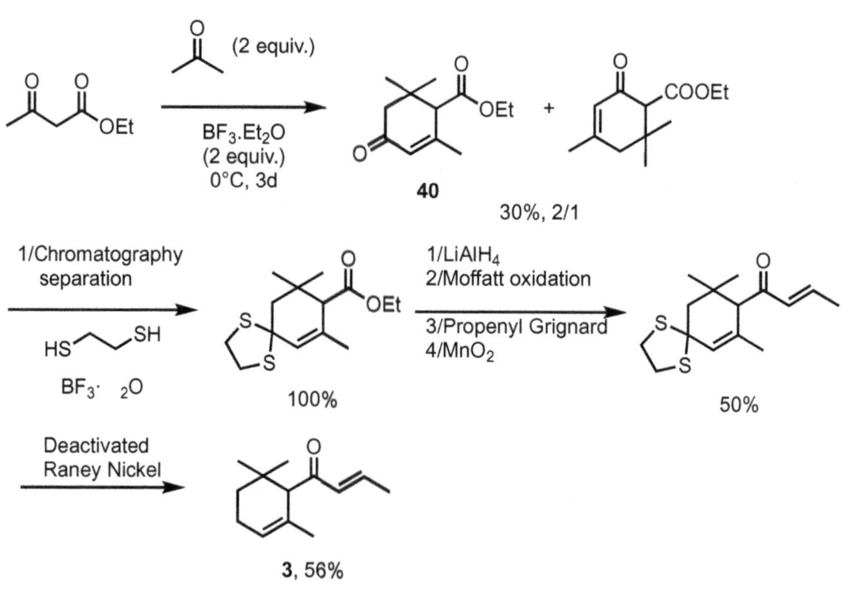

Scheme 7.26. Advanced isophorone derivative synthesis for α-damascone synthesis.

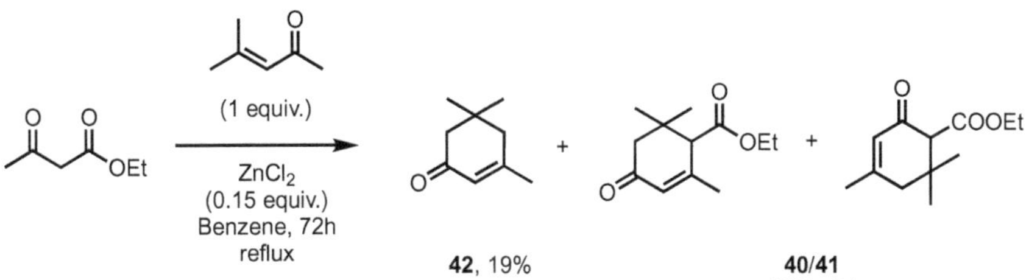

42, 19% **40/41** 40%, 4/1

Scheme 7.27. Zinc catalysed addition of acetoacetate to mesityl oxide.

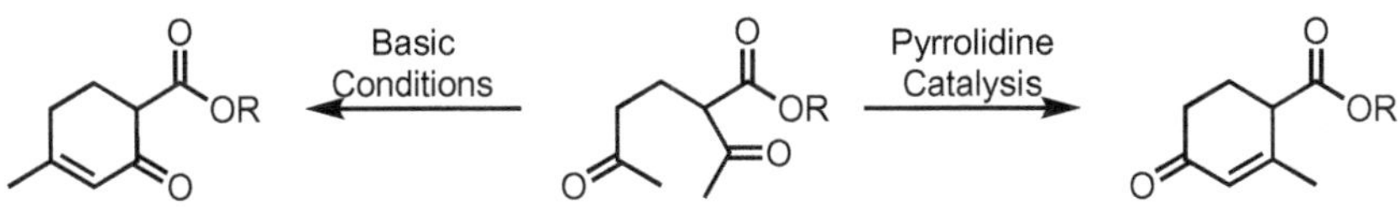

43 Major Product Observed

44 Major Product Observed

Figure 7.6. Regioselectivity for the aldol reaction.

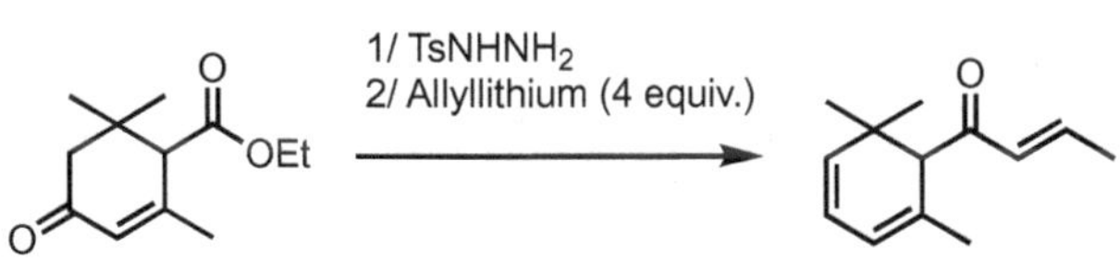

40 **45**, 70%

Scheme 7.28. Shapiro reaction applied to the synthesis of α-damascenone.

or damascenone. This methodology was disclosed in a Firmenich patent as a potential entry to the damascone/damascenone series [*25b*]. Both published methodologies (either BF_3 or $ZnCl_2$) or recent improvements from Firmenich [65] suffered from low yields and selectivities.

The major issue of this approach is the regioselectivity of the dioxo-ester cyclisation. It has been reported that under basic conditions, product **43** is favoured [66]. It was later demonstrated that the use of pyrrolidine as a catalyst gave a complete reversal of selectivity for **44** [67] (Figure 7.6).

In 1991, a Korean group published syntheses of the three damascenone isomers based on a common intermediate [68]. Hydrazone synthesis followed by a Shapiro reaction afforded the α-damascenone **45** (Scheme 7.28).

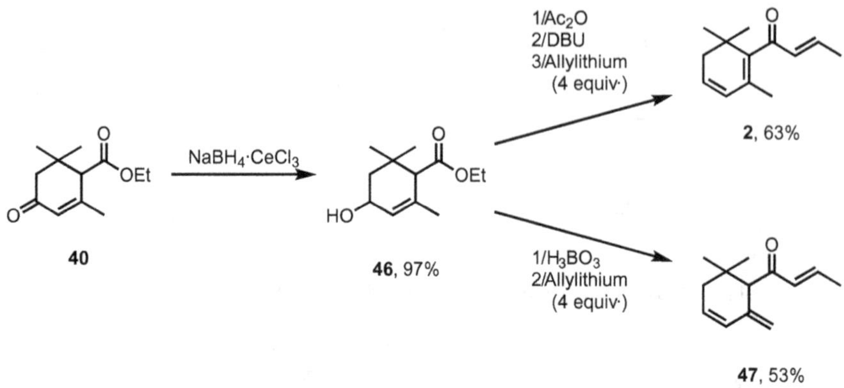

Scheme 7.29. From a common intermediate towards the damascenone family.

An excess of allyllithium was found to be the best solution to promote the Shapiro reaction and the allyl addition to the ester. Simultaneous and selective reduction of the ketone moiety to a hydroxyl ester **46** was achieved using the Luche conditions. From this common precursor, β- and even the unnatural γ-damascenone **47** could be accessed (Scheme 7.29).

The β-damascenone **2** was also obtained after elimination under basic conditions of the acetate derivative, followed by allyllithium addition. The acidic elimination of the alcohol followed by allyllithium addition afforded the γ-damascenone **47** isomer not yet found in nature.

The construction of the cyclohexene ring nucleus from acyclic precursors has resulted in many successful approaches. Several of these are currently used in the industrial production of damascones and damascenone.

7.4 Rearrangement Reactions

7.4.1 *Wharton rearrangement route*

By comparing the functional group position in the damascone and ionone families of molecules, a [1,3] shift of the oxygen was envisaged by a Wharton transposition [69]. In the early 1970s, Ohloff was able to synthesise the α-damascone **3** using this synthetic route [70] (Scheme 7.30).

As (+)-α-ionone can be obtained by resolution of the racemic material, access to enantioenriched α-damascone was possible [71]. The [1,3]

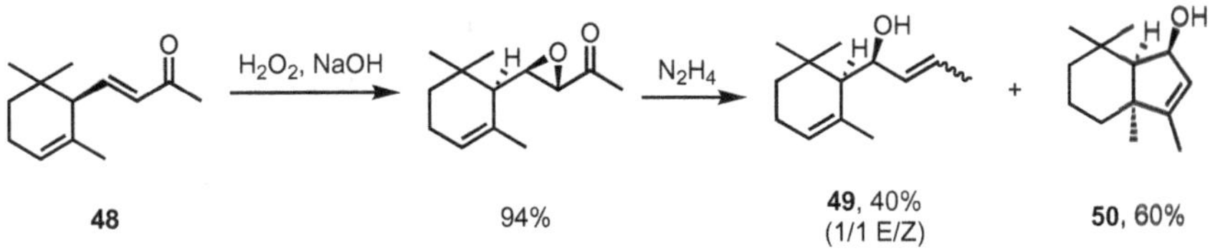

Scheme 7.30. From α-ionone to α-damascone.

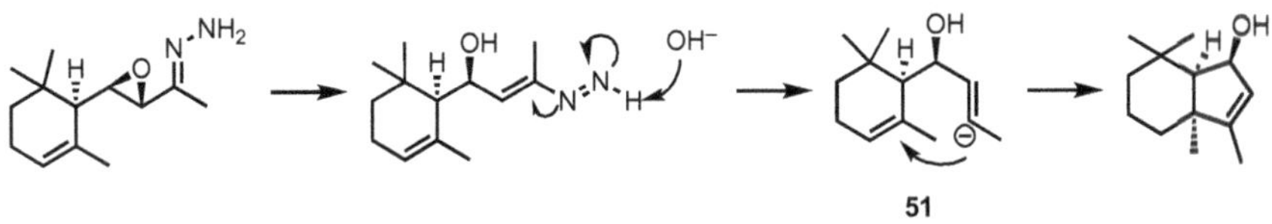

Scheme 7.31. First hypothesis for the formation of the side-product.

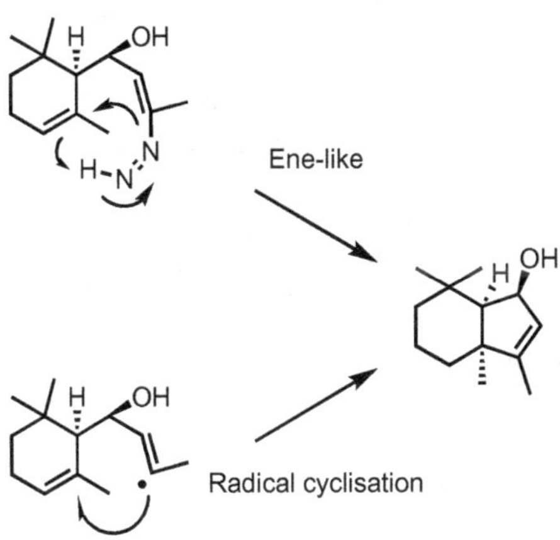

Scheme 7.32. Other possible mechanisms for the formation of the side-product.

rearrangement worked well, but an unexpected product **50** was formed predominantly. A possible mechanism was proposed at that time (Scheme 7.31).

During the fragmentation process, a vinyl carbanion **51** is generated and adds to a trisubstituted olefins. Several years later, Stork *et al.* published a paper, where they argued that the vinyl carbanion **51** could not survive in the alcoholic reaction medium [72]. They proposed two new possible reaction mechanisms (Scheme 7.32).

The first mechanistic proposal was based on an ene-like mechanism, and the second was based on a radical generated, followed by a 5-exo-dig cyclisation.

From an industrial perspective, even though the starting material is available in a racemic form, safety issues were a primary concern. The use of hydrazine necessitates the use of specialised equipment on a scale. A better selectivity towards the desired product seemed difficult to achieve. In addition, a final oxidation step was required to give the desired damascone.

This rearrangement was also applied to the β- and γ-damascone and was patented [73].

7.4.2 *Meisenheimer rearrangement route*

While exploring other rearrangements, Rautenstrauch used the Meisenheimer rearrangement as the key step in the synthesis of α-damascone **3** [74]. α-Ionol was again the starting material to test this reaction (Scheme 7.33).

Whereas the yield of the rearrangement is very good, the selectivity was poor. In fact, the Meisenheimer rearrangement was in competition with the elimination pathway giving a mixture of **52** and **53**. Apart from the number of steps required to synthesise the amine oxide and the selectivity issues, product **52** required further steps to give α-damascone **3**, during which several by-products were formed. Sadly, this synthesis, far from being industrialisable, was never further developed.

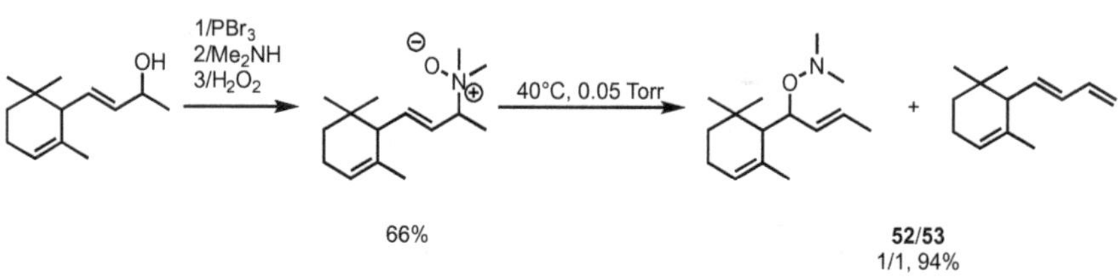

Scheme 7.33. From α-ionol to an α-damascone precursor.

7.4.3 [*1,3*] *Transposition reaction*

Transforming β-ionone **53** into β-damascone **1** via a new transposition was disclosed by Büchi *et al.* in 1972 [75,76]. This transposition proceeds via an isoxazole **54** that is further reduced (Scheme 7.34).

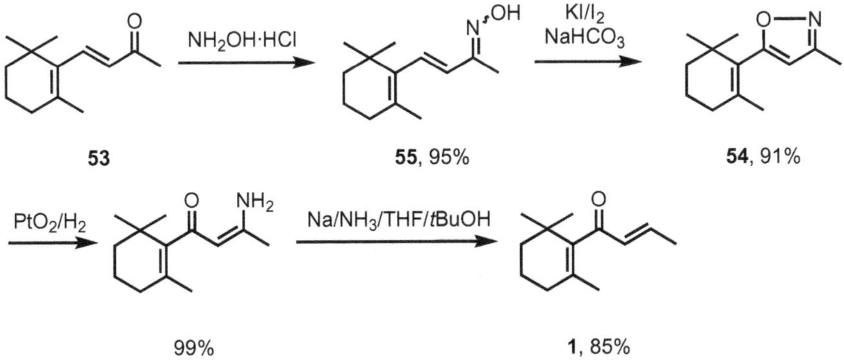

Scheme 7.34. From β-ionone to β-damascone.

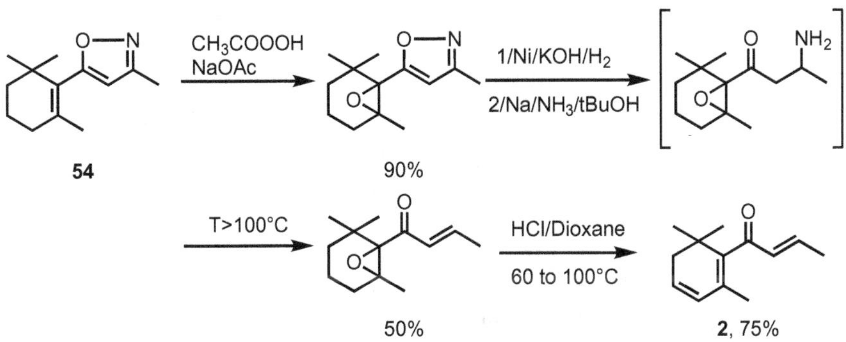

Scheme 7.35. From β-ionone to β-damascenone.

The sequence started with the synthesis of the oxime **55**, followed by an oxidative cyclisation with either a stoichiometric or catalytic amount of iodine. All attempts to employ a cheaper and environmentally friendly halonium source (Cl^+ or Br^+) were unsuccessful.

To convert this synthesis to a more practical one, the use of iodine had to be avoided. Chemists from Firmenich developed an electrochemical reduction, thereby avoiding the use of iodine which proved problematic on a large scale. The synthesis of β-damascone **1** was scaled up using this approach; however, in parallel, a more cost-efficient synthesis of β-damascone **1** starting from cyclohexanone was developed [77].

This strategy was also applied to α-damascone **3** with success. The most valuable β-damascenone **2** was accessible starting from β-ionone **53** (Scheme 7.35).

7.4.4 [2,3] *Wittig rearrangement*

By looking closely at the relative position of the ketone and the alkene in the damascones, the [2,3] Wittig rearrangement is a powerful method to install these two motifs. The first example of this rearrangement was described in the early 1970s [78] and pure γ-damascone **5** was obtained from a relatively simple starting material (Scheme 7.36).

The starting alcohol **56** was synthesised in four steps from 2,2-dimethylcyclohexanone, following a known procedure [79]. Phosphorus tribromide was used to convert **56** into bromide **57**. The key [2,3] Wittig rearrangement gave the expected product with a yield of 63%. To complete the synthesis, a reduction and an oxidation step had to be performed. Depending on the order, the final isomeric mixture is different. As drawn in Scheme 7.34, if the oxidation is run first followed by semi-hydrogenation of the alkyne, a 4-to-1 mixture in favour of *cis* is obtained. Whereas if $LiAlH_4$ is used to reduce the olefin in **58**, only the *trans* product is obtained. This synthesis suffers from several drawbacks in order to be industrialised. However, it shows how powerful Wittig rearrangement is to get an exo-double bond. An expensive alternative to the chromium oxidation was published by Mukaiyama in collaboration with Hasegawa Co Ltd. The generated magnesium alkoxide of the γ-damascol is oxidised with 1,1′-azodicarbonyldipiperidine [80].

More recently, Mori *et al.* published a related synthesis [81] of (S)-α-damascone **30**. Based on the same retrosynthetic analysis, two pathways

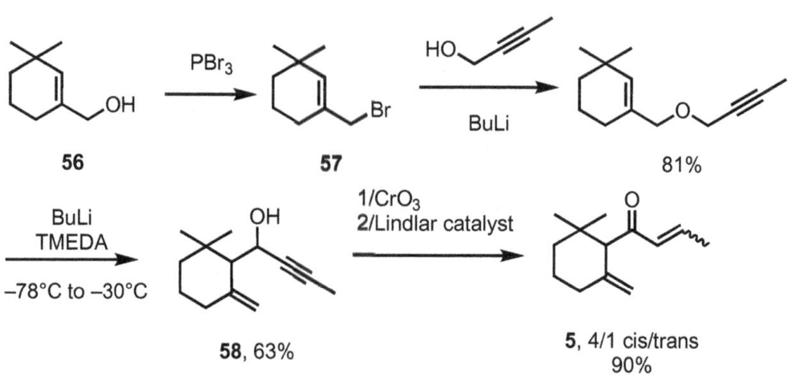

Scheme 7.36. Synthesis of γ-damascone based on a [2,3] Wittig rearrangement.

were investigated. The direct [2,3] Wittig rearrangement was performed on **59** (Scheme 7.37).

Although this critical step worked well to give a reasonable yield of the desired alcohol, it was contaminated with trace amounts of unidentified and inseparable by-products.

To overcome this problem, Still reported the [2,3] Wittig rearrangement via lithium/tin exchange, which gave a much cleaner reaction [82] (Scheme 7.38).

The synthesis is straightforward, and two different approaches to the enantioselective synthesis of allylic alcohol **60** were investigated. Although Corey's asymmetric reduction (CBS reduction) [83] gave the alcohol **60** in an excellent e.e. (96% e.e.), the enzymatic resolution of acetate with pig liver esterase was selected as a more practical method [84]. The use of tin made this synthesis undesirable on a scale.

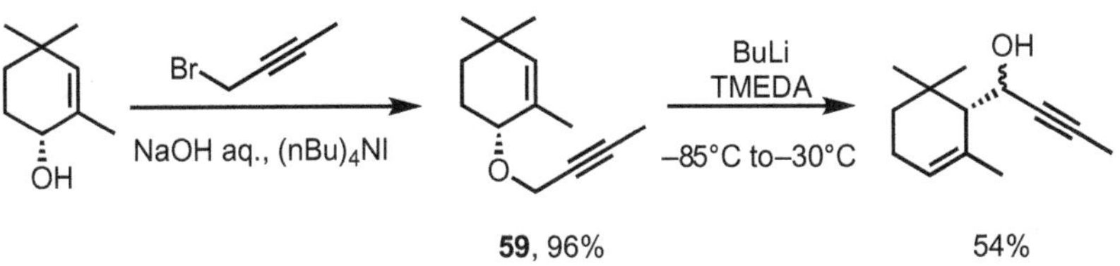

Scheme 7.37. Synthesis of α-damascone based on a [2,3] Wittig rearrangement.

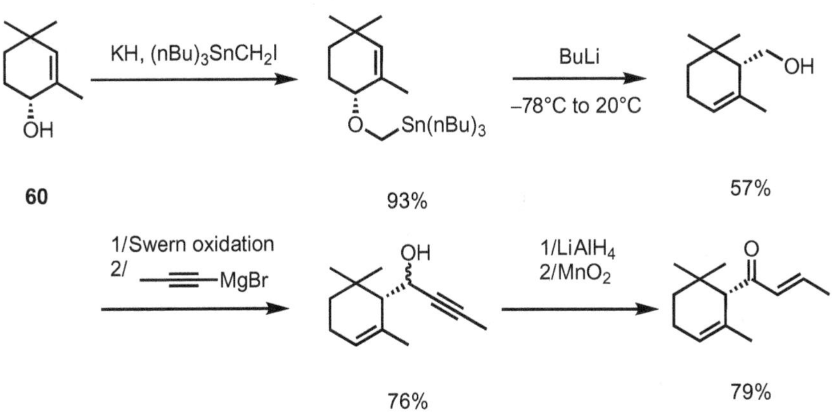

Scheme 7.38. Synthesis of α-damascone based on a lithium–tin exchange [2,3] Wittig rearrangement.

7.4.5 [2,3] *Sigmatropic rearrangement*

Stella published the synthesis of both γ-damascone and β-damascenone employing a related [2,3] Sigmatropic rearrangement of cyano-ammonium ylides [85]. Dialkylaminonitriles have long been known as acyl anion equivalents [86]. Monoalkylation of commercially available N,N-dimethylaminoacetonitrile **61** with allyl bromide gave nitrile **62** in excellent yield (Scheme 7.39).

Then alkylation with bromide **57** followed by rearrangement gave γ-damascone **5** in a moderate-range yield after seven days at RT (Scheme 7.40).

The hydrolysis of the aminonitrile **63** to give the ketone occurred during the purification step over silica gel [87]. At the same time, the double bond isomerised into the conjugated position [88].

The same laboratory had previously published a synthesis of γ-cyclocitral **64** based on the same methodology [89]. Later, Hasegawa Co. Ltd. combined the γ-cyclocitral **64** synthesis, the addition of allyl Grignard reagent and oxidation to synthesise γ-damascone **5** [90] (Scheme 7.41).

In this case, the hydrolysis is done with an aqueous solution of oxalic acid, which offers a practical alternative to silica gel.

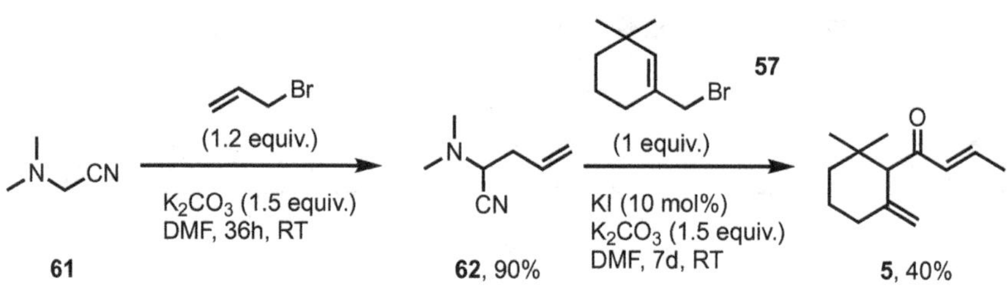

Scheme 7.39. Synthesis of γ-damascone based on a [2,3] sigmatropic rearrangement.

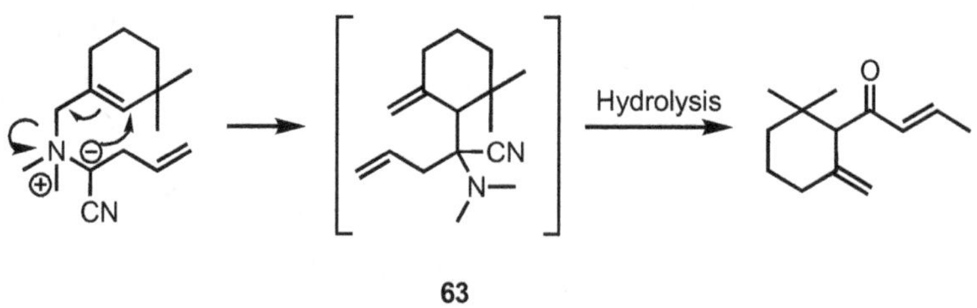

Scheme 7.40. Plausible mechanism of the [2,3] sigmatropic rearrangement.

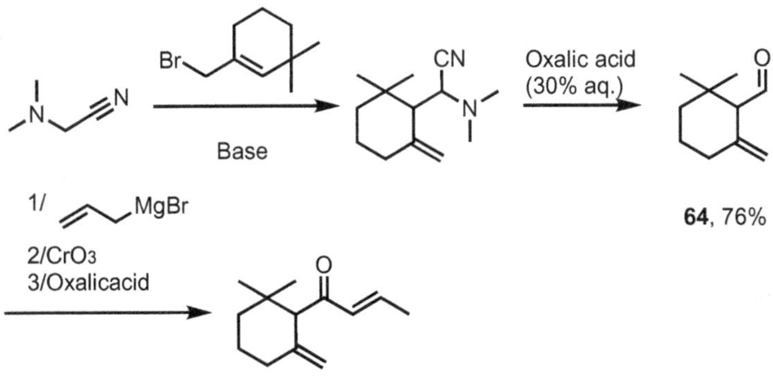

64, 76%

5, 78%

Scheme 7.41. Synthesis of γ-cyclocitral *en route* towards γ-damascone.

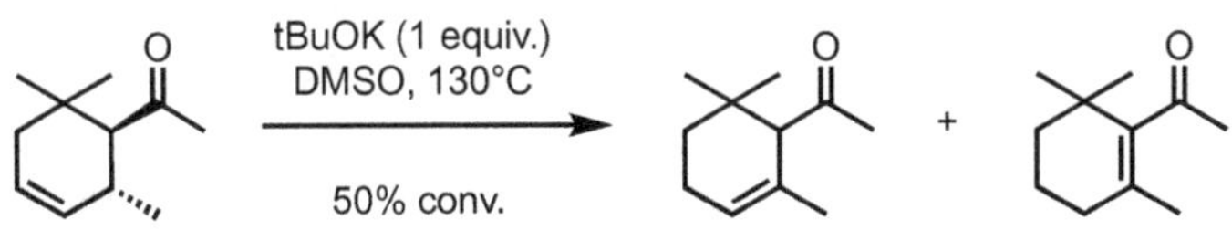

37 **32**, 32% yield **66**, 14% yield

Scheme 7.42. Alkene isomerisation on a common precursor of damascones.

Rearrangement reactions represent a powerful access to damascone skeletons. Several other retrosynthetic approaches have been explored to synthesise the rose ketones.

7.5 Rearrangement Reactions

δ-Damascone **4** is the cheapest and most accessible damascone isomer, and significant effort has been dedicated to transforming this isomer into the more appreciated α- or β-isomers. Takasago patented the use of metal alkoxides like *t*BuOK [91]. A mixture of α- and β-Damascone is obtained. The ratio claimed is within a 65–99% range for the former and 1–35% range for the latter. The same approach was used with the **37**, which is a potential precursor of the δ-damascone **4**. Of course, in order to maximise the full potential of this methodology, a higher selectivity is needed to be achieved. (Scheme 7.42).

7.6 Propargylic Addition and Allyl Grignard Reagent

7.6.1 *β-Cyclocitral route*

The first synthesis used on an industrial scale was the scale-up of laboratory synthesis starting from *β*-cyclocitral **66**, available from BASF via an oxidative degradation of *β*-ionone **53**. It should be noted that this oxidative degradation is unsuccessful with *α*-ionone **48** [92] (Scheme 7.43).

Despite the efficient access to *β*-damascenone **2**, this route suffered from several major drawbacks. Oxidation with chromium or manganese generated toxic waste. Then, bromination/debromination steps produced large amounts of waste and caused some safety issues upon scale-up. Trace amounts of brominated by-products were always present and were difficult to remove. The scarcity of *β*-cyclocitral **66** jeopardised the economic and technical viability of this process [7(a), 93]. Recently, Nippon Zeon patented a similar approach using 1-propynylmagnesium bromide. Several different end-game syntheses were also tested [94].

To circumvent these issues, a new synthesis was developed, implemented on a scale and allowed rapid growth in production volume [10].

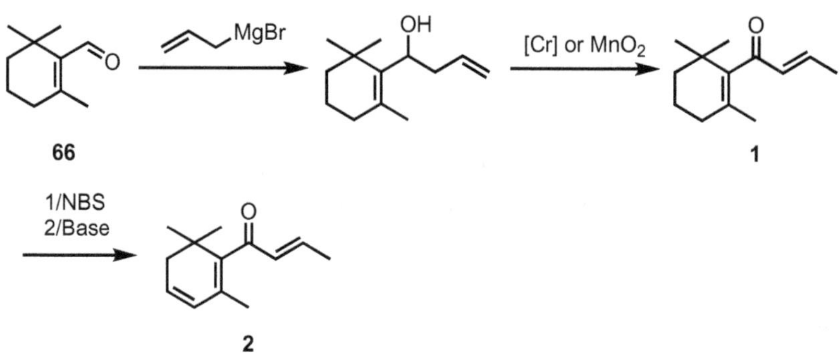

Scheme 7.43. From *β*-cyclocitral towards *β*-damascenone.

7.6.2 *Trimethyl cyclohexanone route*

The availability of 2,2,6-trimethylcyclohexan-1-one **67** prompted chemists to look at new synthesis of the damascone family. The route was straightforward and avoided previous difficulties [95] (Scheme 7.44).

The addition of the 1-butyn-2-ol gives the diol in good yields. Acidic treatment afforded directly the *β*-damascone **1** [96].

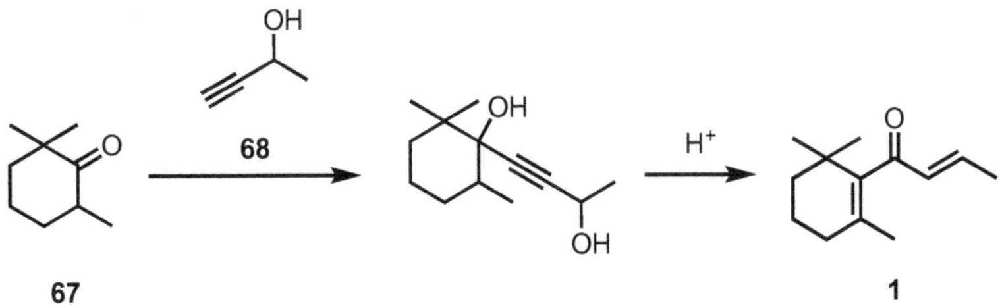

Scheme 7.44. *β*-damascone from a trisubstituted cyclohexanone.

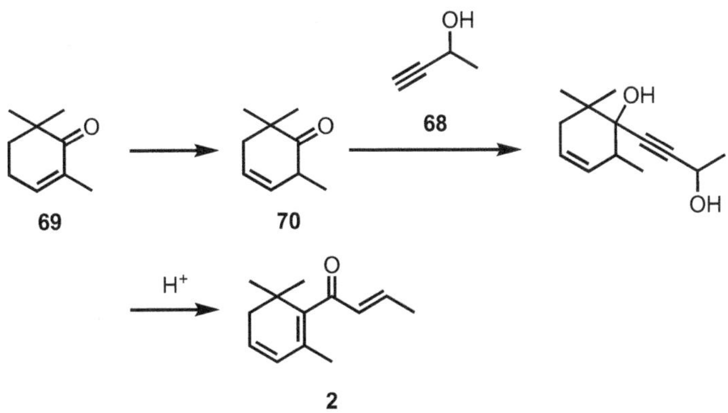

Scheme 7.45. *β*-damascenone from a trisubstituted cyclohexenone.

Several years later, IFF patented a related synthesis using acetylene addition, followed by a Rupe rearrangement with formic acid, and the resulting methyl ketone **65** was transformed to *β*-damascone **1** by aldol reaction [97].

The synthesis of the starting trimethyl ketone was available via several routes: the condensation of acrolein with 2-methylpentan-3-one [98a], poly-methylation of cyclohexanone [98b] or photooxidation of the enamine [98c–e].

7.6.3 *Trimethyl cyclohexenone route*

In the 1980s, BASF also supplied the 2,6,6-trimethylcyclohex-2-en-1-one **69**. By extending the same chemistry, a more efficient synthesis of *β*-damascenone **2** was developed at Firmenich [99] (Scheme 7.45).

The formation of the deconjugated enone **70** was well known [100]. Addition of the propargylic alcohol **68** gave the diol, which rearranged under acidic conditions to give the β-damascenone **2**. Firmenich was obliged to develop its own synthesis of the starting enone **69**, as BASF stopped producing it. Oxophorone (2,6,6-trimethylcyclohex-2-ene-1,4-dione) was used as the starting point of the synthesis, and this new quality was commercialised in 1982 after the final purification step was solved [101]. A similar approach was patented around the same period by PFW Company [102].

As shown previously, the β-damascone **1** serves as an advanced precursor for the most desirable β-damascenone **2** via a net oxidation reaction. Several strategies have already been presented (NBS or epoxidation). Among the methodologies available within Firmenich, photooxygenation was used for the production of several key ingredients (Scheme 7.46) [103].

The starting material of the desired epoxide could be prepared from β-cyclocitral **66** via allyl Grignard reagent addition. The photochemical reaction gave a moderate 60% yield. The major by-product resulted from the abstraction of the exomethyl proton, which led to a fragmentation reaction, giving **72** and **73** in a 40% yield (Figure 7.7).

The epoxide **71** was treated with either HCl acid or acidic clay and gave the desired β-damascenone **2** in good yields [104].

Similarly, epoxidation and rearrangement of α-methylcyclogeraniate **18** gave a concise synthesis of methyl safranate **10**, which could also be transformed to β-damascenone **2** (Scheme 7.47).

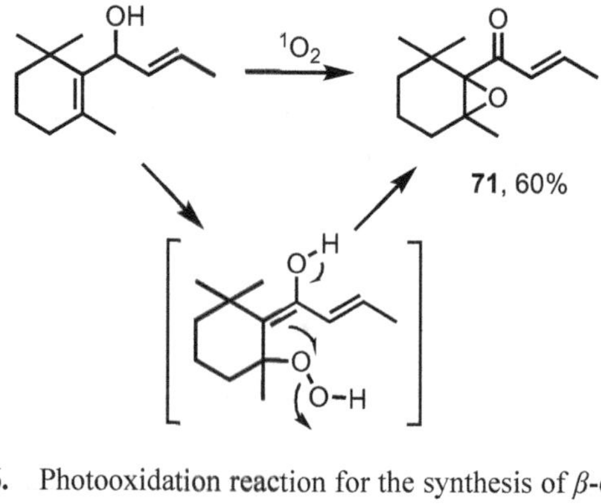

Scheme 7.46. Photooxidation reaction for the synthesis of β-damascenone.

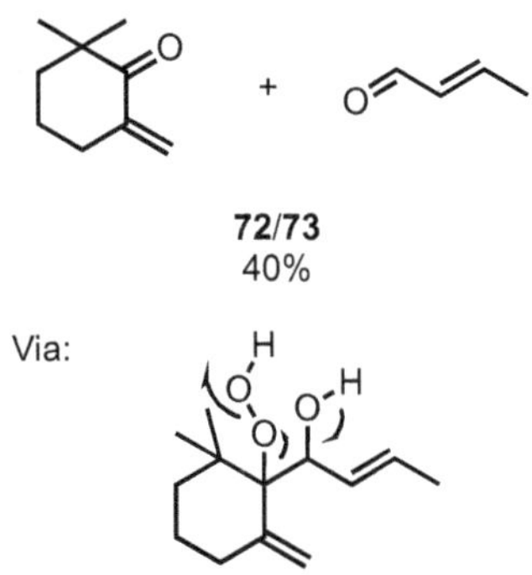

Figure 7.7. Undesired fragmentation reaction during the photooxidation reaction.

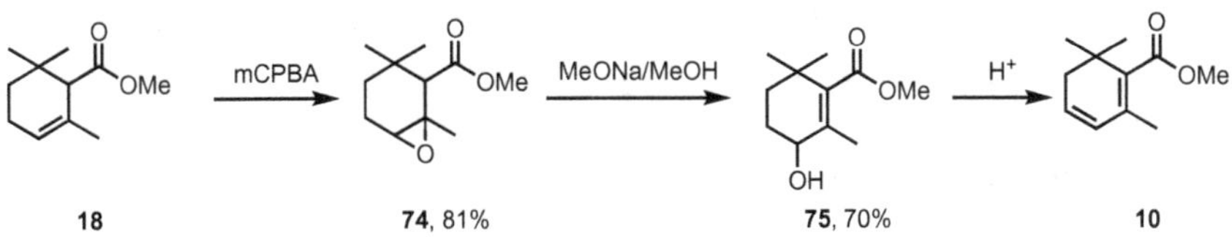

Scheme 7.47. Methyl safranate synthesis from α-methylcyclogeraniate.

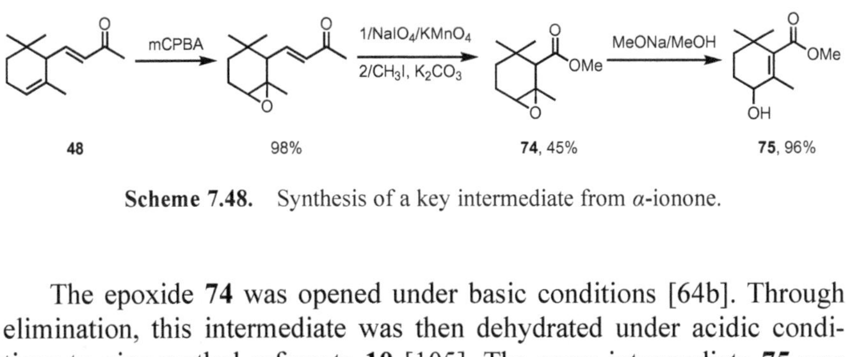

Scheme 7.48. Synthesis of a key intermediate from α-ionone.

The epoxide **74** was opened under basic conditions [64b]. Through elimination, this intermediate was then dehydrated under acidic conditions to give methyl safranate **10** [105]. The same intermediate **75** was obtained via a degradation of α-ionone **48** (Scheme 7.48) [106].

The epoxidation with *m*-CPBA (meta-chloro perbenzoic acid) was highly selective for the more electron-rich olefin. Larger-scale epoxidation used peracetic acid [64b]. Then, oxidative cleavage gave the epoxy ester in moderate yields. Elimination under basic conditions afforded **75**. This approach from α-ionone **48** used a cheap starting material but suffered from the use of a huge excess of $NaIO_4$.

7.7 Miscellaneous Approaches

Apart from these syntheses described so far, whether implemented on a scale or not, several novel approaches to the damascenones have been published more recently.

7.7.1 *Γ-Pyronene-based route*

The first synthesis covered starts from γ-pyronene **76**: a readily available and potentially sustainable starting material [107]. This family of natural compounds possesses the right core skeleton and only needs a slight redox adjustment (Scheme 7.49).

Previous work [108] had shown that epoxide **77** could be opened with $MgBr_2$. It was thought that the allyl Grignard reagent served as both a Lewis acid to rearrange the epoxide to aldehyde and then as a nucleophile to give alcohol **78** in very good yields. The described chromium oxidation and olefin isomerisation on alumina were not suited for large-scale manufacture.

7.7.2 *Regio-selective enolate acylation*

Another effort to synthesise γ-damascone involved the regioselective synthesis of dione, which could eventually be converted to the desired compound (Scheme 7.50) [109].

76 **77**, 70% (After purification) **78**, 91% **3**, 90%

Scheme 7.49. Pyronene as a potential starting material α-damascone synthesis.

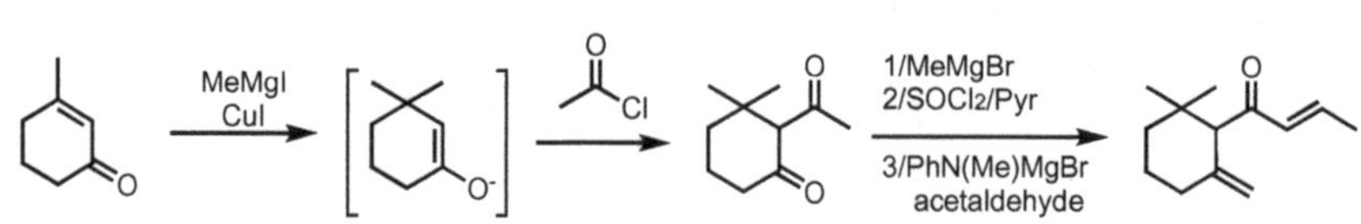

79 **80**, 62% **5**, 57%

Scheme 7.50. γ-damascone from 3-methylcyclohex-2-en-1-one.

Catalytic cuprate addition to the enone **79** followed by regioselective enolate acylation gave dione **80** in reasonable yields. The selective addition of the methyl Grignard reagent to the less hindered ketone of the dione afforded keto-alcohol, which upon dehydration gave a moderate yield of **15**, the precursor of γ-damascone **5**.

7.7.3 *Boron-mediated synthesis*

In 1986, a short and boron-based synthesis of α- and γ-damascones was disclosed [110]. Addition of methyl Grignard reagent to the same enone **79** followed by dehydration gave the required starting material **81** (Scheme 7.51).

Highly selective allylic lithiation of **82**, followed by quenching with $ClBEt_2$, afforded the allylic borane **83**. A reaction of **83** with crotonaldehyde **84** gave γ-damascol **85** in very good yields. Oxidation gave the corresponding γ-damascone **5**. The same strategy could be applied to yield α-damascone **3**. Despite the high selectivity, this synthesis suffered from major drawbacks if applied on an industrial scale, such as deprotonation at low temperatures and stoichiometric borylation.

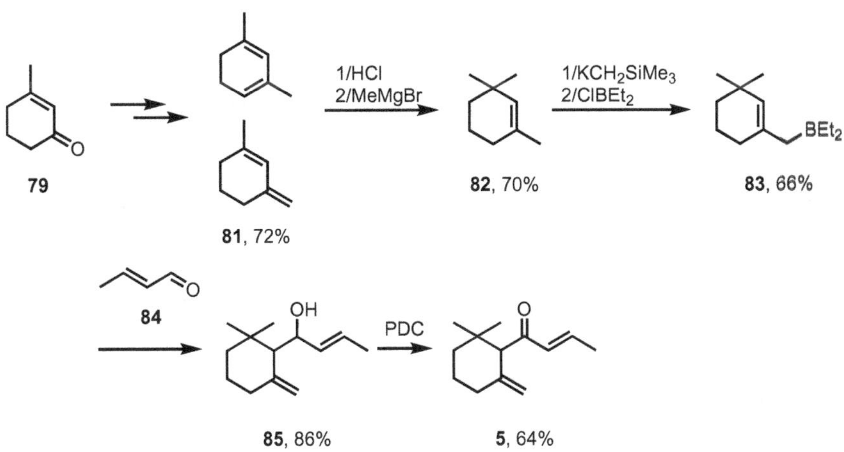

Scheme 7.51. γ-damascone via a boron allylation reaction.

7.7.4 *Glutaconic acid-based route*

A route based on glutaconic acid towards β-damascenone **2** was reported by Cornforth and co-workers (Scheme 7.52) [111].

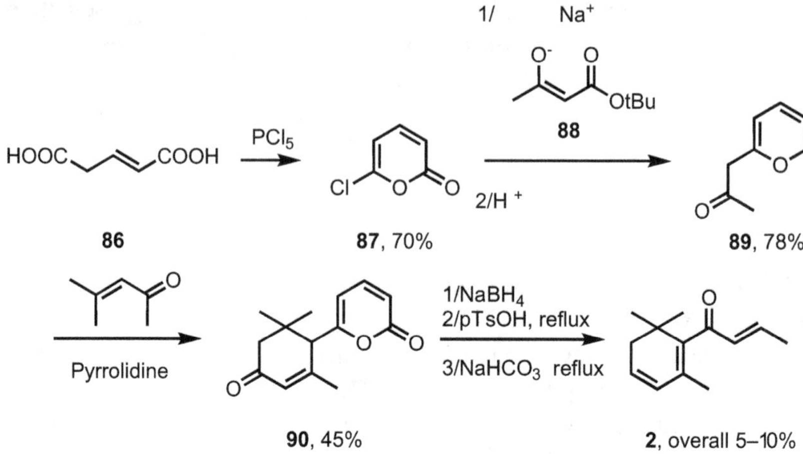

Scheme 7.52. β-damascenone synthesis starting from glutaconic acid.

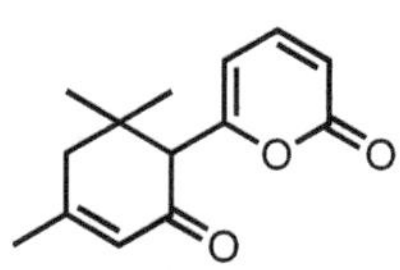

91

Figure 7.8. Product derived from opposite selectivity.

The starting material **86** could be used as an *E*/*Z* mixture, as only the *E* isomer reacted to give the chloropyranone **87**. Conjugated addition followed by the elimination of the chlorine by the anion of acetoacetate **88** followed by decarboxylation lead to **89**. The authors found that the use of pyrrolidine gave **90** selectively. It should also be mentioned that when *t*BuOK was used as a base, the other isomer **91** was obtained selectively (Figure 7.8) [67].

Having established the core structure, selective reduction followed by elimination and hydrolysis steps gave β-damascenone **2** in a poor overall yield.

7.8 Conclusion

Many ingenious approaches to the synthesis of β-damascenone and each of the damascone isomers have been investigated, some of which are used for the industrial production of this important class of molecules. This chapter has discussed some of the most interesting approaches from published patents and academic literature. It should be mentioned that even in approaches that are used for the industrial production of this interesting class of molecules, significant improvements and optimisation have been done, which will continue to remain unpublished. Several unnatural synthetic analogues have also been synthesised and found application in perfumery [112].

Acknowledgements

The author would like to thank Dr. A. Birkbeck, Dr. C. Chapuis, Dr. F. Santoro, Dr. C. Fehr and Dr. G. Oddon for their help in correcting and refining the manuscript.

References

[1] (a) S. Firestein. (2001). How the Olfactory System Makes Sense of Scents, *Nature*, 413, 211–218. (b) P. Mombaerts. (2004). Genes and Ligands for Odorant, Vomeronasal and Taste Receptors, *Nature Rev. Neurosci.*, 5, 263–278. (c) C.G. Galizia, P.-M. Lledo. (2013). Neurosciences, Chapter 13 "Olfaction," 1st Edition (Springer Spektrum, Berlin), pp. 253–284.

[2] The Nobel Prize in Physiology and Medicine. (2004). Awarded to Richard Axel and Linda B. Buck "For their Discoveries of Odorant Receptors and the Organization of the Olfactory System." https://www.nobelprize.org/nobel_prizes/medicine/laureates/2004/.

[3] E.T. Troscianko. Cognitive Realism and Memory in Proust's Madeleine Episode, *Memory Studies*, 6, 437–456. https://ora.ox.ac.uk/objects/uuid:abcad420-01fa-48a0-b298-09ee493ae3c4.

[4] (a) G. Fráter, J.A. Bajgrowicz, P. Kraft. (1998). Fragrance Chemistry, *Tetrahedron*, 54, 7633–7707. (b) P. Kraft, J.A. Bajgrowicz, C. Denis, G. Fráter. (2000). Odds and trends: Recent development in the chemistry of odorants, *Angew. Chem. Int. Ed.*, 39, 2980–3010. (c) M. Gautschi, J.A. Bajgrowicz, P. Kraft. (2001). Fragrance chemistry — milestones and perspectives, *Chimia*, 55, 379–387. (d) N. Armanino, J. Charpentier, F. Flachsmann, A. Goeke, M. Liniger, P. Kraft. (2020). What's hot, what's not: The trends of the past 20 years in the chemistry of odorants, *Angew. Chem. Int. Ed.*, 59, 16310–16344.

[5] (a) J. Garnero. (1982). La Rose et Son Parfum, *Parfums, Cosmét. Arômes*, 47, 31–46. (b) B. Schaefer. (2013). Natural Products in the Chemical

Industry, in B. Schaefer (ed.) Chapter 3 "Flavours and Fragrances" (Springer), pp. 52–53.

[6] (a) G. Ohloff, E. Demole. (1987). Importance of the odoriferous principle of bulgarian rose oil in flavor and fragrance chemistry, *J. Chromatogr. A*, 406, 181–183. (b) E.sz. Kovátz. (1987). Composition of Essential Oils. Part 7. Bulgarian oil of Rose (*Rosa damascena* Mill.), *J. Chromatogr. A*, 406, 185–222.

[7] (a) E. Demole, P. Enggist, U. Säuberli, M. Stoll, E.sz. Kovátz. (1970). Structure et Synthèse de la Damascénone (triméthyl-2,6,6-trans-crotonyl-1-cylohexadiène-1,3), Constituant Odorant de l'Essence de Rose Bulgare (*Rosa damascena* Mill.), *Helv. Chim. Acta*, 53, 541–551. (b) E. Demole, D. Berthet. (1971). Identification de la Damascénone et de la β-Damascone dans le Tabac Burley, *Helv. Chim. Acta*, 54, 681–682. (c) E. Demole, D. Berthet. (1972). A chemical study of burley tobacco flavor (*Nicotiana tabacum* L.) I. volatile to medium-volatile constituents (b.p. ≤84°/0.001 Torr), *Helv. Chim. Acta*, 55, 1866–1882. (d) M. Winter, P. Enggist. (1971). Recherches sur les Arômes de Framboise, *Helv. Chim. Acta*, 54, 1891–1898.

[8] (a) S. Isoe, S. Katsumura, T. Sakan. (1973). The synthesis of damascenone and β-damascone and the possible mechanism of their formation from carotenoids, *Helv. Chim. Acta*, 56, 1514–1516. (b) M.A. Sefton, G.K. Skouroumounis, R.A. Massywestropp, P.J. Williams. (1989). Norisoprenoids in vitis vinifera white wine grapes and the identification of a precursor of damascenone in these fruits, *Aust. J. Chem.*, 42, 2071–2084. (c) G.K. Skouroumounis, M.A. Sefton. (2000). Acid-catalyszed hydrolysis of alcohols and their β-D-glucopyranosides, *J. Agric. Food Chem.*, 48, 2033–2039.

[9] V.F. Cataldo, J. López, M. Cárcamo, E. Agosin. (2016). Chemical vs. biotechnological synthesis of C13-apocarotenoids: Current methods, applications and perspectives, *Appl. Microbiol. Biotechnol.*, 100, 5703–5718.

[10] A. Williams. (2002). Rose ketones: Celebrating 30 years of success, *Perfumer & Flavorist*, 27, 18–31.

[11] (a) K.H. Schulte-Elte, H. Strickler, F. Gautschi, W. Pickenhagen, M. Gadola, J. Limacher, B.L. Müller, F. Wuffli, G. Ohloff. (1975). Synthesen und Cyclisierungsreaktionen der Pseudodamascone. Ein neuer Zugang zu β-Damasconverbindungen, *Liebigs Ann. Chem.*, 484–508. (b) E. Kovátz, E. Demole, G. Ohloff, M. Stoll. (1971). Procédé pour la préparation de composés cyclooléfiniques carbonylés, (Firmenich SA) CH-513097. (c) Y. Nakatani, K. Kubota (née Soeda), R. Tahara, Y. Shigematsu. (1974). Nouvelle Synthèse de l'α-Damascone, *Agric. Biol. Chem.*, 38, 1351–1354.

[12] G. Saucy, R. Marbet. (1967). Über eine Neuartige Synthese von β-ketoallenen durch Reaktion von Tertiären Acetylencarbinolen mit Vinyläthern eine Ergiebige Methode zur Darstellung des Pseudojonons und Verwandter Verbindungen, *Helv. Chim. Acta*, 50, 1158–1167.

[13] A.F. Morris, F. Näf, R.L. Snowden. (1991). Dynascone: The ultimate captive, *Perfumer & Flavorist*, 16, 33–35.

[14] T. Mandai, K. Mizobuchi, M. Kawada, J. Otera. (1984). New synthetic methods for α-geraniol, boll weevil pheromone, and β-damascone employing 2-(Hydroxymethyl)-4-8-phenylthio)-1-butene as a building block, *J. Org. Chem.*, 49, 3403–3406.

[15] S. Torii, K. Uneyama, H. Ichimura. (1979). Synthesis of β-damascenone from prenyl phenyl sulfone. A (C5 + C5 + C3) Procedure, *J. Org. Chem.*, 44, 2292–2294.

[16] (a) G. Büchi, H. Wüst. (1971). Synthetic studies on damascenones, *Helv. Chim. Acta*, 54, 1767–1776. (b) G. Büchi, H. Wüst. (1975). Procédé pour la Préparation de Cétones Insaturées Cycloaliphatiques, (Firmenich SA) CH-562186.

[17] J.J. Eisch, A.M. Jacobs. (1963). A convenient preparation of allyllithium, *J. Org. Chem.*, 28, 2145–2146.

[18] S. Shiyoujirou, I. Nobuhiko. (1980). Preparation of α-Damascone, (Toyotama) JPS-55105639.

[19] M.-A. Boaventura, J. Drouin. (1987). Cyclisation de Composés Acétyléniques Carbonylés via leur Ether d'Enol Silylé: II. Synthèse et Réactivité de Quelques Ethers d'Enols Dérivés d'Aldéhydes ou de Cétones Acétyléniques Terminaux, *Bull. Soc. Chim. Fr.*, 1015–1026.

[20] D. Hack, M. Blümel, P. Chauhan, A.R. Philipps, D. Enders. (2015). Catalytic conia-ene and related reactions, *Chem. Soc. Rev.*, 44, 6059–6093.

[21] (a) E. Bäuml, H. Mayr. (1984). 2,6,6-Trialkyl-1-(1'-haloalk-2'-en-1'-ylidene)cyclohex-3-enes, their preparation and their use, (BASF AG) DE-3402993. (b) W. Heilmann, A. Rahman, E. Bäuml, H. Mayr. (1988). Carbocationic cyclizations and rearrangements in the damascone series, *Tetrahedron*, 44, 6047–6054.

[22] (a) M. Kurbanov, A.V. Semenovskii, V.A. Smit. (1973). Cyclization of isoprenoid compounds, *Bull. Acad. Sci. USSR, Div. Chem. Sci.*, 22, 370–373. (b) R.J. Armstrong, L. Weiler. (1983). Synthesis of (+/–)-trixagol by an electrophilic cyclization of an allylsilane, *Can. J. Chem.*, 61, 2530–2539. (c) P. Loeliger. (1974). Preparation of Cyclohexadiene Compound, (Hoffmann-La Roche) US-3962311.

[23] N. Bartoloa, J.A. Reada, E.M. Valentína, K.A. Woerpel. (2017). Reactions of allylmagnesium halides with carbonyl compounds: Reactivity, structure, and mechanism, *Synthesis*, 49, 3237–3246.

[24] G. Osztrovszky, T. Holm, R. Madsen. (2010). Ultrafast grignard addition reactions in the presence of water, *Org. Biomol. Chem.*, 8, 3402–3404.

[25] (a) A. Viola, E.J. Ioro. (1970). Vapor phase thermolyses of 3-hydroxy-1,5-hexadienes. V. Preparation of allyl vinyl ketone, *J. Org. Chem.*, 35, 856–858. (b) K.H. Schulte-Elte. (1975). Procédé pour la Préparation de cétones alicycliques Insaturées, (Firmenich SA) CH-563951.

[26] (a) R.A. Benkeser, M.P. Siklosi, E.S. Modzen. (1978). Reversible grignard and organolithium reactions, *J. Am. Chem. Soc.*, 100, 2134–2139. (b) D.A. Evans, D.J. Baillargeon, J.V. Nelson. (1978). A general approach to the synthesis of 1,6-dicarbonyl substrates. New applications of base-accelerated oxy-cope rearrangements, *J. Am. Chem. Soc.*, 100, 2242–2244.

[27] (a) R.L. Snowden, B.L. Müller, K.H. Schulte-Elte. (1982). Fragmentation of homoallylic alkoxides. Synthesis of propenyl and 2-methylpropenyl ketones from carboxylic esters, *Tetrahedron Lett.*, 23, 335–338. (b) R.L. Snowden, S.M. Linder, B.L. Müller, K.H. Schulte-Elte. (1987). α-cleavage of bis(homoallylic) potassium alkoxides. Two-step preparation of propenyl ketones from carboxylic esters. Synthesis of ar-turmerone, α-damascone, β-damascone, and β-damascenone, *Helv. Chim. Acta*, 70, 1858–1878. (c) K.H. Schulte-Elte, R.L. Snowden, B.L. Müller. (1982). Process for the preparation of polyunsaturated cycloaliphatic ketones, (Firmenich SA) US-4479011. (d) R.L. Snowden, S.M. Linder. (1988). α-cleavage of bis(homoallylic) potassium alkoxides. Synthesis of β-damascone, *Helv. Chim. Acta*, 71, 1587–1597.

[28] (a) F. Huet, G. Emptoz, A. Jubier. (1973). Formation d'Enolates Cétoniques par Action d'Organomagnésiens Mixtes dans l'Hexamethylphosphorotriamide sur des Esters, *Tetrahedron*, 29, 479–485. (b) I. Kikkawa, T. Yorifuji. (1980). Convenient synthesis of alkyl ketones from carboxylic esters, *Synthesis*, 11, 877–880.

[29] (a) C. Fehr, J. Galindo. (1986). Synthesis of (E)-1-propenyl ketones from carboxylic esters and carboxamides by use of mixed organolithium-magnesium reagents, *Helv. Chim. Acta*, 69, 228–235. (b) C. Fehr, J. Galindo, R. Perret. (1987). General synthesis of ketones from carboxylic esters and carboxamides by use of mixed organolithium-magnesium reagents: Syntheses of artemisia ketone, *Helv. Chim. Acta*, 70, 1745–1752.

[30] (a) C. Fehr. (1983). Procédé pour la Préparation de Composés Cétoniques, (Firmenich SA) EP-093840. (b) C. Fehr, J. Galindo. (1988). Procédé pour la Préparation de Cétones Cycloaliphatiques, (Firmenich SA) EP-260472.

[31] J.G. Bauman, R.C. Hawley, H. Rapoport. (1985). An efficient synthesis of aklavinone and related 11-deoxyanthracyclinones, *J. Org. Chem.*, 50, 1569–1573.

[32] C. Fehr, J. Galindo. (1988). Reaction of ester enolates with nucleophiles. Stereocontrolled formation of ketone and aldehyde enolates, *J. Org. Chem.*, 53, 1828–1830.

[33] C. Fehr, J. Galindo. (1990). Esters alicycliques et leur utilisation à titre d'ingrédients parfumants, (Firmenich SA) EP-0378825.

[34] C. Fehr, J. Galindo. (1988). Procédé pour la Préparation d'Aldéhydes Cycloaliphatiques, (Firmenich SA) EP-0262426.

[35] A.D. Allen, A. Stevenson, T.T. Tidwell. (1989). Hydration reactivity of persistent conjugated ketenes, *J. Org. Chem.*, 54, 2843–2848.

[36] (a) H.W. Moore, D.S. Wilbur. (1980). Cyanoketenes. Mechanism of tert-butylcyanoketene cycloaddition to aldo-and ketoketenes, *J. Org. Chem.*, 45, 4483–4491. (b) P.W. Raynolds, J.A. DeLoach. (1984). Ketenes. 18. Evidence for an intermediate in the reaction of ketenes with silyl enol ethers, *J. Am. Chem. Soc.*, 106, 4566–4570.

[37] J.D. Wuest, A.M. Madonik, D.C. Gordon. (1977). Vinylketenes. Synthesis of (+)-actinidine, *J. Org. Chem.*, 42, 2111–2113.

[38] (a) J.-C. Combret. (1964). Action des Magnésiens vinyliques sur les cétènes, *C. R. Acad. Sci. Paris*, 259, 1416–1417. (b) J.-E. Dubois, B.L. Zhang, C. Lioin. (1981). Synthèse en Série Encombrée. Préparation de l'Acide Ditertiobutyl-Methylacetique tBu2MeCCOOH et de Quelques Composés Cétoniques Dérivés de cette Structure, *Tetrahedron*, 37, 4189–4194. (c) D. Lenoir, H.R. Seiklay, T.T. Tidwell. (1982). Highly crowed enolates from reaction of di-tert-butylketene with alkyllithium reagents, *Tetrahedron Lett.*, 23, 4987–4990.

[39] (a) F. Naef, R. Decorzant. (1986). Grignard and hydride addition to a ketene intermediate: A novel access to β-damascone and β-cyclocitral, *Tetrahedron Lett.*, 27, 3245–3250. (b) F. Naef, R. Decorzant. (1987). Procédé pour la Préparation de Cétones Cycloaliphatiques, (Firmenich SA) EP-0224897.

[40] (a) L. Ruzicka, H. Schinz. (1940). Veilchenriechstoffe IX). Über die Syntheses der Kernmathylierten Jononhomologen 1,1,3,6-Tetramethyl-2-(buten-2-ylon-2)-cyclohexen-(2 bzw.3), *Helv. Chim. Acta*, 23, 959–974. (b) G. Stork, A.W. Burgstahler. (1955). The stereochemistry of polyene cyclization, *J. Am. Chem. Soc.*, 77, 5068–5077. (c) P. Fankhauser, P. Hanselmann, B. Jackson. (1988). Geranic acid derivatives, (Firmenich SA and Lonza LTD) WO-0014080.

[41] C. Fehr. (1996). Enantioselective protonation of enolates and enols, *Angew. Chem. Int. Ed.*, 35, 2566–2587.

[42] (a) L.M. Baigrie, H.R. Seiklay, T.T. Tidwell. (1985). Stereospecific formation of enolates from reaction of unsymmetrical ketenes and organolithium reagents, *J. Am. Chem. Soc.*, 107, 5391–5396. (b) R. Häner, T. Laube, D. Seebach. (1985). Regio- and diastereoselective preparation of aldols from α-branched ketone enolates generated from BHT ester enolates and

organolithium reagents-in situ generation and trapping of ketenes from ester enolates, *J. Am. Chem. Soc.*, 107, 5396–5403.

[43] C. Fehr, J. Galindo. (1988). Synthesis of (R)-(+)- and (S)-(-)-β-Damascone by Tandem Grignard Reaction-Enantioselective Protonation: Evidence for the Intermediacy of a Chiral Complex, *J. Am. Chem. Soc.*, 110, 6909–6911.

[44] C. Fehr, J. Galindo. (1994). Catalytic Enantioselective Protonation of Enolates, *Angew. Chem. Int. Ed.*, 33, 1888–1889.

[45] (a) C. Fehr, I. Stempf, J. Galindo. (1993). Highly Enantioselective Protonation of Thiol Ester Enolates, *Angew. Chem. Int. Ed.*, 32, 1042–1044. (b) C. Fehr, I. Stempf, J. Galindo. (1993). Enantioselective Addition of Aromatic Thiols to a Ketene, *Angew. Chem. Int. Ed.*, 32, 1044–1046.

[46] C. Fehr, J. Galindo. (1994). Procédé pour la Préparation d'Esters et Thioesters Optiquement Actifs, (Firmenich SA) EP-593917.

[47] (a) P. Gosselin. (1986). Nouvelle Application de la Réaction des Organomagnésiens sur les Dithioesters: Synthèse de la β-Damascone, *Tetrahedron Lett.*, 27, 5495–5496. (b) P. Gosselin. (1987). New Terpenic Synthons: Methyl α- and β-Dithiocyclogeranates, *Synthesis*, 3, 267–270.

[48] A.J. Kresge. (1996). Reactive Intermediates: Carboxylic Acid Enols and Other Unstable Species, *Chem. Soc. Rev.*, 25, 275–280.

[49] C. Fehr. (2007). Catalytic Enantioselective Tautomerization of Isolated Enols, *Angew. Chem. Int. Ed.*, 46, 7119–7121.

[50] C. Fehr, J. Galindo. (1995). Syntheses of the enantiomers of α-cyclogeranic acid, β-cyclocitral, and β-damascone: Enantioselective protonation of enolates, *Helv. Chim. Acta*, 78, 539–552.

[51] R.C . Cookson, R.M. Tuddenham. (1973). A Lewis acid catalyzed diels-alder reaction with inverse electron demand; An alternative synthesis of α-damascone, *J. Chem. Soc., Chem. Commun.*, 19, 742–742.

[52] A.T. Nielsen, C. Gibbons, C.A. Zimmerman. (1951). Condensation of aldehydes with ketones. Methylanilinomagnesium bromide as a condensing agent, *J. Am. Chem. Soc.*, 73, 4696–4701.

[53] K.S. Ayyar , R.C. Cookson, D.A. Kagi. (1973). Simple synthesis of β-damascenone and related compounds, *J. Chem. Soc., Chem. Commun.*, 5, 161–161.

[54] (a) K.S. Ayyar, R.C. Cookson, D.A. Kagi. (1975). Synthesis of β-Damascone [trans-1-1(2,6,6-Trimethylcyclohex-3-enyl)but-2-en-1-one] and β-Damascenone [trans-1-(2,6,6-Trimethylcyclohexa-1,3-dienyl)but-2-en-1-one], *J. Chem. Soc., Perkin Trans.* 1, 1727–1736. (b) D.A. Kagi, K.S. Ayyar, R.C. Cookson, R.M. Tuddenham. (1974). Catalyzed diels-alder reactions in the synthesis of perfumery materials, *Soap, Perfum.*

Cosmet., 47, 29–34. (c) D.H.R. Barton, R.C. Cookson. (1956). The Principle of Conformation Analysis, *Q. Rev. Chem. Soc.*, 10, 44–82.

[55] (a) B.D. Mookherjee, R.W. Trenkle, R.A. Wilson, F.L. Schmitt, M.H. Vock, E.J. Granda, J. Vinals, J. Kiwala, W.L. Schreiber. (1979). Trans, trans-*β*-Damascone, ainsi que les Mélanges Contenant des Proportions Importantes de ce Composé, les Procédés de Préparation de celui-ci et les Usages Organoleptiques de ce Dernier, (IFF INC.) FR-2422616. (b) B.D. Mookherjee, R.W. Trenkle, R.A. Wilson, F.L. Schmitt, M.H. Vock, E.J. Granda, J. Vinals, J. Kiwala, W.L. Schreiber. (1977). Composition of Matter Comprising more than 50% Trans, trans-*β*-Damascone for Augmenting or Enhancing the Aroma and Taste of a Smoking Tobacco, (IFF INC.) US-42111242. (c) B.D. Mookherjee, R.W. Trenkle, R.A. Wilson, F.L. Schmitt, M.H. Vock, E.J. Granda, J. Vinals, J. Kiwala, W.L. Schreiber. (1979). Use of Trans, trans-*β*-Damascone and Mixtures containing 80% of Trans, trans-*β*-Damascone in Augmenting or Enhancing the Aroma and Taste of a Detergent, (IFF INC.) US-4198309.

[56] (a) Y.G. Aouad, D.M. Bourgeois, G.S. Miracle, J. Wevers, H.Q.N. Gunaratne, K.R. Seddon, E. Boros, S.P. Katdore, T. Belhocine. (2007). Process for Epimerising Cyclohexenyl Ketones and its Application in Aldol Condensation Process, (The Procter & Gamble Company) US-20080021246. (b) D.M. Bourgeois, G.S. Miracle, P.J. Porter, E. Boros, K.R. Seddon, H.Q.N. Gunaratne. (2009). Process for Epimerizing Cyclohexenal Ketones with Subsequent Aldol Condensation to Produce Fragrance Compounds, (The Procter & Gamble Company) WO-2010080504.

[57] D. Jacoby. (2001). Catalytic System for Aldol reactions, (Firmenich SA) WO-2002094755.

[58] R.A. Sheldon. (2018). Metrics of green chemistry and sustainability: Past, present, and future, *ACS Sustainable Chem. Eng.*, 6, 32–48.

[59] (a) W.G. Dauben, A.P. Kozikowski, W.T. Zimmerman. (1975). Meldrum's acid derivative as a useful dienophilic component: Synthesis of *β*-damascone, *Tetrahedron Lett.*, 16, 515–517. (b) C. Chapuis. (1991). Dimethyl isopropylidene malonate, the "forgotten" dienophile, *Tetrahedron Lett.*, 32, 355–358.

[60] M. Büscheb, S. Dorich, S. Hanessian, D. Tao, K.B. Schenthal, L.E. Overman. (2016). Synthetic strategies toward natural products containing contiguous stereogenic quaternary carbon atoms, *Angew. Chem. Int. Ed.*, 55, 4156–4186.

[61] H. Rubinstein. (1962). Intermediate in the synthesis of xanthophyll and zeaxanthin, *J. Org. Chem.*, 27, 3886–3887.

[62] H.J. Liu, H.-K. Hung, G.L. Mhehe. (1976). A total synthesis of *β*-damascone, *Tetrahedron Lett.*, 17, 4129–4132.

[63] J.D. Surmatis, A. Walser, J. Gibas, R.A. Thommen. (1970). Study on the condensation of mesityl oxide with acetoacetic esters, *J. Org. Chem.*, 35, 1053–1056.

[64] (a) H.-J. Liu, H.-K. Hung, G.L. Mhehe, L.D. Weinberg. (1978). Synthesis of trans- and cis-β-Damascone, *Can. J. Chem.*, 56, 1368–1371. (b) O.D.A. Dailey Jr. (1987). New synthetic route to (+/–)-strigol, *J. Org. Chem.*, 52, 1984–1989.

[65] D. Jacoby. (2002). A Process for the Preparation of Michael-Adducts, (Firmenich SA) EP-1238962.

[66] R.N. Lacey. (1960). 6-acetylcyclohex-2-enones: The condensation of β-diketones with α,β-unsaturated ketones, *J. Chem. Soc.*, 1625–1633.

[67] A.L. Begbie, B.T. Golding. (1972). A new synthesis of ethyl 2-methyl-4-oxocyclohex-2-enecarboxylate (hagemann's ester) and its methyl and t-butyl analogues, *J. Chem. Soc., Perkin Trans. 1*, 602–605.

[68] W.Y. Lee, S.Y. Jang, J.-G. Lee, W.K. Chae. (1991). Novel syntheses of isomers of damascenone from ethyl 2,6,6-trimethyl-4-oxo-2-Cyclohexene-1-carboxylate, *Bull. Korean Chem. Soc.*, 12, 31–35.

[69] C. Dupuy, J.L. Luche. (1989). New developments of the wharton transposition, *Tetrahedron*, 45, 3437–3444.

[70] G. Ohloff, G. Uhde. (1970). Über eine Ungewöhnliche Cyclisationsreaktion bei der Umsetzung von (+)-Epoxy-β-Dihydrojonon mit Hydrazinhydrat, *Helv. Chim. Acta*, 53, 531–541.

[71] R.B. Woodward, T.P. Kohman, G.C. Harris. (1941). A new optically active reagent for carbonyl compounds; The resolution of dl-camphor, *J. Am. Chem. Soc.*, 63, 120–124.

[72] G. Stork, P.G. Williard. (1977). Five- and six-membered-ring formation from olefinic α,β-epoxy ketones and hydrazine, *J. Am. Chem. Soc.*, 99, 7067–7068.

[73] E. Kovátz, G. Ohloff, E. Demole, M. Stoll. (1973). Procédé pour la Préparation de Cétones Insaturées, (Firmenich SA) CH-537352.

[74] V. Rautenstrauch. (1973). Controlled allylic transformations via the meisenheimer rearrangement, *Helv. Chim. Acta*, 56, 2492–2508.

[75] (a) G. Büchi, J.C. Vederas. (1972). Interchange of functionality in conjugated carbonyl compounds through isoxazoles, *J. Am. Chem. Soc.*, 94, 9128–9132. (b) K.H. Schulte-Elte, B.L. Müller, G. Ohloff. (1973). Synthetische Übergänge von der Jonon- in die Damaconreihe, *Helv. Chim. Acta*, 56, 310–320.

[76] G.H. Büchi, J.C. Vederas. (1974). Procédé Nouveau pour la Préparation de Composés Carbonylés Alicycliques, (Firmenich SA) CH-553187.

[77] M. Yan, Y. Kawamata, P.S. Baran. (2017). Synthetic organic electrochemical methods since 2000: On the verge of a renaissance, *Chem. Rev.*, 117, 13230–13319.

[78] K.H. Schulte-Elte, V. Rautenstrauch, G. Ohloff. (1971). Synthesen der Stereoisomeren (+/–)-β-Damascone, *Helv. Chim. Acta*, 54, 1805–1812.

[79] P. Seifert, H. Schinz. (1951). Zur Herstellung α-Ungesättigter, Cyclischer Aldehyde, *Helv. Chim. Acta*, 34, 728–735.

[80] (a) O. Takazawa, K. Saigo, K. Narasaka, T. Mukaiyama. (1977). New synthesis of β-damascone, *Chem. Lett.*, 757–758. (b) M. Mukaiyama, K. Saigou, O. Takazawa. (1978). Preparation of β-damascone, (Hasegawa Co LTD) JPS-5334751.

[81] (a) K. Mori, P. Puapoomchareon. (1991). Preparation of optically pure 2,4,4-trimethyl-2-cyclohexen-1-ol, a new and versatile chiral building block in terpene synthesis, *Liebigs Ann. Chem.*, 1053–1056. (b) K. Mori, M. Amaike, M. Itou. (1993). Synthesis of (S)-β-damascone, *Tetrahedron*, 49, 1871–1878.

[82] W.C. Still, A. Mitra. (1978). A highly stereoselective synthesis of Z-trisubstituted olefins via [2,3]-sigmatropic rearrangement. Preference for a pseudoaxially substituted transition state, *J. Am. Chem. Soc.*, 100, 1927–1928.

[83] E.J. Corey, C.J. Helal. (1998). Reduction of carbonyl compounds with chiral oxazaborolidine catalysts: A new paradigm for enantioselective catalysis and a powerful new synthetic method, *Angew. Chem. Int. Ed.*, 37, 1986–2012.

[84] R.A. Sheldon, P.C. Pereira. (2017). Biocatalysis engineering: The big picture, *Chem. Soc. Rev.*, 46, 2678–2691.

[85] L. Stella, A. Amrollah-Madjdabadi. (1987). Utilisation du N,N-diméthylamino-acétonitrile pour la Synthèse Régiosélective de Diallylcétones Masquées. Application à la Synthèse de la β-Damascone, *Bull. Soc. Chim. Fr.*, 6, 350–354.

[86] J.D. Albright. (1983). Reactions of acyl anion equivalents derived from cyanohydrins, protected cyanohydrins and α-dialkylaminonitriles, *Tetrahedron*, 39, 3207–3233.

[87] G. Stork, A.A. Ozorio, A.Y.W. Leong. (1978). N,N-diethylaminoacetonitrile a generally useful latent acyl carbanion, *Tetrahedron Lett.*, 19, 5175–5178.

[88] L. Stella. (1988). Procédé pour la Préparation de β-Damascone, (Firmenich SA) CH-668419.

[89] L. Stella. (1984). Synthèse Pratique du β-Cyclocitral, *Tetrahedron Lett.*, 25, 3457–3458.

[90] I. Masamichi, I. Minoru, K. Kunio. (1987). Production of 2-Methylene-6,6-Dimethyl or 5,6,6-TrimethylCyclohexylCarbaldehyde, (Hasegawa Co LTD) JPS-6251637.

[91] S. Watanabe, H. Ujihara, T. Yamamoto, T. Hagiwara. (2000). Production Process of Cyclohexenyl Methyl Ketones, (Takasago Int. Corp.) EP-1162190.

[92] N. Müller, W. Hoffmann. (1975). Cyclocitrale aus Iononen, *Synthesis*, 12, 781–781.

[93] W. Schalk, P. Wirth, G.E.M. Dannenberg, V. Schmied-Kowarzik, P. Weinhold. (1968). Ungesättigte Cycloaliphatische Ketone, (Firmenich SA) DE-1807568.

[94] K. Watanabe. (2001). Method for Producing Damascone or Damascenone, (Nippon Zeon Co) JP2001247504.

[95] G. Ohloff, V. Rautenstrauch, K.H. Schulte-Elte. (1973). Modellreaktionen zur Biosynthese von Verbindungen der Damascon-Reihe und ihre Präparative Anwendung, *Helv. Chim. Acta*, 56, 1503–1513.

[96] (a) M. Baumann. (1979). Verfahren zur Herstellung von *β*-Damascon und *β*-Damascenon, (BASF AG) DE-2804597. (b) W. Hoffmann, M. Baumann. (1981). Verfahren zur Herstellung von *β*-Damascon und *β*-Damascenon, (BASF AG) DE-2925176.

[97] R.W. Trenkle, B.D. Mookherjee, J.B. Hall, R. Kasper, M.H. Vock, R. Schreck, E.J. Granda, J.F. Vinals. (1981). Process for Preparing Acyl Trimethyl Cyclohexene Derivatives and Use of Intermediates Therefor in Augmenting or Enhancing the Aroma or Taste of a Consumable Material, (IFF INC.) US-4250332.

[98] (a) M. Baumann, W. Hoffmann, A. Nürrenbach. (1979). Synthesen und einige Umsetzungen von 2,6,6-Trimethyl-2-cyclohexenon, *Liebigs Ann. Chem.*, 1945–1951. (b) M.W.C. Keuly, P. Gradeff. (1962). Nouveau Procédé d'Alcoylation de Cétones et Produits Obtenus, (Société des Usines Chimiques Rhône-Poulenc) Bel-640555. (c) J.E. Huber. (1968). Photooxydation of enamines — A partial synthesis of progesterone, *Tetrahedron Lett.*, 29, 3271–3272. (d) C.S. Foote, J.W.-P. Lin. (1968). Chemistry of singlet oxygen. VI. photoxygenation of enamines: Evidence for an intermediate, *Tetrahedron Lett.*, 29, 3267–3270. (e) V. Van Rheenen. (1969). Copper-catalyzed oxygenation of enamines, *J. Chem. Soc., Chem. Commun.*, 314–315.

[99] (a) P. Fankhauser. (1981). Process for Preparing Novel Acetylenic Compounds useful as Starting Materials for Preparing an Alicyclic Ketone, (Firmenich SA) US-4453013. (b) P. Fankhauser. (1982). Novel Acetylenic Compounds useful as Starting Materials for Preparing an Alicyclic Ketone, (Firmenich SA) US-4324729.

[100] (a) M.S. Kharasch, P.O. Tawney. (1941). Factors determining the course and mechanism of grignard reactions. II. The effect of metallic compounds on the reaction between isophorone and methyl-magnesium bromide, *J. Am. Chem. Soc.*, 63, 2308–2316. (b) O. Isler, H. Lindlar, M. Montavon, R. Rüegg, G. Saucy, P. Zeller. (1956). *Helv. Chim. Acta*, 39, 2041–2053.

[101] (a) J.J. Becker, K.-H. Schulte-Elte, H. Strickler, G. Ohloff. (1977). Procédé pour la Préparation de 2,2,6-Triméthylecyclohex-5-ène-1,4-dione, (Firmenich SA) CH-585168. (b) P. Wirth, G.E.M. Dannenberg, V. Schmied-Kowarzik, P. Weinhold, D. Gudel. (1972). Verfahren zur Herstellung von ungesättigten Ketonen und Alkoholen, (Firmenich SA) DE-2242751. (c) H. Bellut. (1988). Process for preparing 2,6,6-trimethyl cyclohex-2-ene-1,4 dione, (Hüls Chemische Werke AG) DE-3842547. (d) Firmenich SA unpublished results.

[102] L.A. Hulshof. (1980). Process for the Production of 1-Crotonyl-2,6,6-trimethylcyclohexa-1,3-diene and esters of Alkynlcyclohexane Derivatives as Intermediates therefor, (PFW B.V.) EP-25638.

[103] (a) K.-H. Schulte-Elte, B. Müller, G. Ohloff. (1971). 203. Die Farbstoffsensibilisierte Photo-Oxygenierung von *β*-Damascol. Ein Einfaches Verfahren zur Darstellung von *β*-Damascenon, *Helv. Chim. Acta*, 54, 1899–1910. (b) G. Ohloff. (1975). Singlet oxygen: A reagent in organic synthesis, *Pure Appl. Chem.*, 43, 481–502.

[104] (a) K.-H. Schulte-Elte, B. Egger, B. Müller. (1981). Oxygenated Alicyclic Compounds and Process for Preparing Same, (Firmenich SA) US-4369328. (b) E. Kovátz, E. Demole. (1972). Procédé de Préparation de Cétones Insaturées, (Firmenich SA) CH-528470.

[105] (a) C. Chapuis, K.H. Schulte-Elte. (1995). 14. Preparation and absolute configuration of some iris essential oil constituents of the 5-methylsafranic-acid series, *Helv. Chim. Acta*, 78, 165–176. (b) I. Kadas, G. Arvai, G. Horvath. (1998). Improved Synthesis of 7,7-Dimethyl-1,4-Dioxo-2,3,4,5,6,7-Hexahydroinden-2-yl-Acetic and Propionic Acid From Citral, *Org. Prep. Proced. Int.*, 30, 79–85.

[106] D.W. Brooks, E. Kennedy. (1983). Efficient synthesis of methyl 3-oxo-2,6,6-trimethylcyclohex-1-ene-carboxylate, an a-ring intermediate for (+/–)-strigol, *J. Org. Chem.*, 48, 277–278.

[107] (a) B. Boulin, B. Arreguy-San Miguel, B. Delmond. (1998). Le *β*-Pyronène: Synthon d'Accès au Safranal et Précurseur d'Intermédiaires de Synthèse de la Forskoline et du Strigol, *Tetrahedron*, 54, 2753–2762. (b) B. Boulin, M. Taran, B. Arreguy-San Miguel, B. Delmond. (2007). New preparation methods for *β*-damascone, *β*-damascone, and *β*-damascenone using pyronenes, *Synth. Commun.*, 37, 2579–2591. (c) C. Chapuis, R. Brauchli. (1993). Preparation of optically active flowery and woody-like odorant ketones via corey-chaykovsky oxiranylation: Irones and analogues, *Helv. Chim. Acta*, 76, 2070–2088.

[108] F. Marc, B. Soulet, D. Serramedan, B. Delmond. (1994). L'Epoxy-Pyronène: Synthon d'Accès aux Cyclocitrals et aux Ionones et Précurseur de Nouveaux Composés Terpénoïdes, *Tetrahedron*, 50, 3381–3388.

[109] F. Näf, R. Decorzant. (1974). 146. Regiospecific acylation, alkylation, and aldol condensation using magnesium enolates resulting from the conjugate addition of grignard reagents to α, β-unsaturated ketones, *Helv. Chim. Acta*, 57, 1317–1327.

[110] (a) M. Zaidlewicz. (1986). Synthesis with organoboranes. 21. Synthesis of α- and β-damascone, *Tetrahedron Lett.*, 27, 5135–5138. (b) M. Zaidlewicz. (1988). Method of Obtaining Damascone, PL-262333.

[111] (a) J. Cornforth, J.E. Hawes, R. Mallaby. (1992). A stereospecific synthesis of (+/–)-abscisic acid, *Aust. J. Chem.*, 45, 179–195. (b) Personal communication at Firmenich in 1972.

[112] P. Chaumont-Olive, J. Sanchez-Quesada, A.M. Collado Perez, J. Cossy. (2021). Synthetic approaches to the damascone and damascenone isomers, *Tetrahedron*, 82, 131932.

https://doi.org/10.1142/9781800617780_0008

Chapter 8

Lewis Acid-Catalysed Cyclisations in the Field of Fragrances

Raphaël Saget[*,†,‡], Piotr Jaunky[†,§] and Elisabet Duñach[*,||]

[*] *Université Cote d'Azur, CNRS*
Institut de Chimie de Nice
Nice, France

[†] *MANE, Centre de Recherche en Chimie Organique,*
Le Bar-sur-Loup, France

[‡] *raphael.saget@mane.com*
[§] *piotr.jaunky@mane.com*
[||] *dunach@unice.fr*

Abstract

The catalytic activity of Lewis superacids, in particular metal trifluoromethanesulfonates (triflates), is highlighted in several cyclisation reactions of interest in the field of fragrance compounds. Lewis superacids allow for efficient and selective processes under mild conditions, proceeding with low catalyst loading and generally shifting from stoichiometry to catalysis. This chapter reports examples of multiple reaction types, such as Diels–Alder, Friedel–Crafts, Prins, carbonyl-ene and related -ene processes, as well as cyclisations involving unsaturated alcohols, carboxylic acids and thiols.

8.1 Introduction

In the field of fragrance chemistry, the design of new odorant ingredients widens the palette of perfumers. The innovation in synthetic organic chemistry allows us to move from nature-identical compounds to inspiring novel structures. As perfumer Ernest Beau already remarked one century ago, "One has to rely on chemists to find new aroma chemicals, creating new, original notes".

The field of fragrance compounds presents a large structural diversity. Most fragrances, initially extracted from plants, provided the basic constituents of perfumes and perfuming goods. The production of natural compounds and raw materials may nowadays be difficult to ensure. Chemical and biochemical methodologies provide synthetic alternatives as well as the possibility of preparing novel structures and different odours. The main inspiration for new fragrance compounds generally relies on terpene chemistry [1]. The development of new synthetic routes leading to the discovery of novel odorants still constitutes a major challenge [2, 3].

The extremely diverse synthetic approaches to cyclisation reactions of terpenoids and polyprenoid-type moieties have been mostly reported using strong protic acids, under homogeneous or heterogeneous media [4]. These cyclisations generally proceed through cationic-type mechanisms. As an alternative to protic media, classical Lewis acids such as $AlCl_3$, BF_3. Et_2O, $SnCl_4$, $TiCl_4$, $ZnBr_2$ or $MeAlCl_2$ have also been reported to promote cyclisations and C–C bond formation, though generally in stoichiometric or even over-stoichiometric amounts [5].

In recent decades, it has become possible to modulate the acidity of metal cations. Superacids and Lewis superacids have emerged as new species, generally enabling the transition from stoichiometry to catalysis in numerous transformations [6]. A superacid can be defined as a protic compound presenting an acidity superior to pure sulphuric acid, [7] and the metallic salts of these superacids can be considered as Lewis superacids. The catalytic activity of these compounds has been reviewed [8, 9].

Particularly interesting for catalysis are metallic triflates [$M(OSO_2CF_3)_n$ or $M(OTf)_n$] and triflimides [$M(NTf_2)_n$], allowing a highly positive charge on the metal centre, thus enhancing their Lewis acid activity in C–C and C–O bond forming reactions [6].

In this chapter, we present, in a non-exhaustive manner, several types of cyclisation reactions of interest in the field of aroma compounds, mostly occurring under catalytic Lewis acid conditions. In some cases, stoichiometric and protic acid reactions, as well as processes not involving cyclisations, will be referenced for comparison.

8.2 Fragrant Molecules via Lewis Acid-Catalysed Diels–Alder Reactions

The pericyclic reaction was discovered in 1928 by Diels and Alder, [10] who were later awarded the Nobel Prize in chemistry in 1950. The reaction typically involves the cycloaddition of a 1,3-diene with a dienophile, leading to a cyclohexene derivative. The Diels–Alder reaction has found widespread use in fragrance chemistry, and commonly encountered dienes include isoprene, piperylene, 2-methylpenta-1,3-diene, cyclopentadiene and myrcene (Figure 8.1). Although thermal conditions were initially applied, Lewis acid catalysis of the Diels–Alder reaction is now routinely used to allow milder conditions, better yields and higher stereoselectivity [11]. Classical Lewis acids such as $AlCl_3$, $TiCl_4$, $SnCl_4$, $ZnCl_2$, $ZnBr_2$ and $BF_3.OEt_2$ are typically used.

The synthesis of Acetarolle® [12] involves the Diels–Alder reaction of isoprene with 2,2-dimethyl-6-ethylidene cyclohexanone, catalysed by $BF_3 OEt_2$ (Figure 8.2). The intermediate spiroketone was further reduced and acetylated to afford Acetarolle®, which exhibits woody and ambery notes.

Triplal® and Isocyclocitral®, with green notes, are obtained by the thermal Diels–Alder reaction of 2-methylpenta-1,3-diene with acrolein and crotonaldehyde, respectively (Figure 8.3) [13].

Tropicalia®, with its sweet fruity note, is prepared from cyclopentadiene and ethyl acrylate, catalysed by $BF_3 OEt_2$ (Figure 8.4) [14]. The reaction of cyclopentadiene and 3-methylpent-3-ene-2-one with $AlCl_3$ catalysis affords the earthy, camphoraceous odorant Presantone® (Figure 8.4) [15].

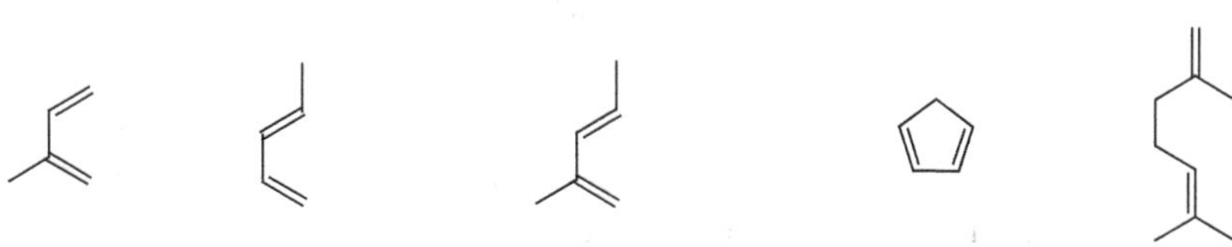

Isoprene *Piperylene* *2-Methylpenta-1,3-diene* *Cyclopentadiene* *Myrcene*

Figure 8.1. Typical 1,3-dienes used for fragrance compounds in Diels–Alder cyclisations.

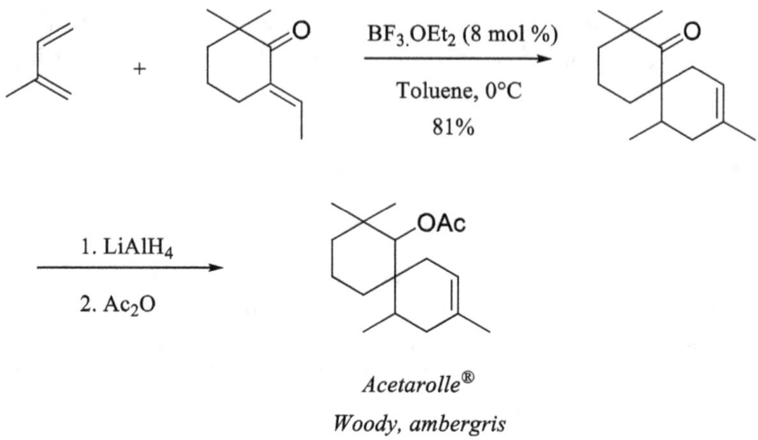

Figure 8.2. Synthesis of Acetarolle®.

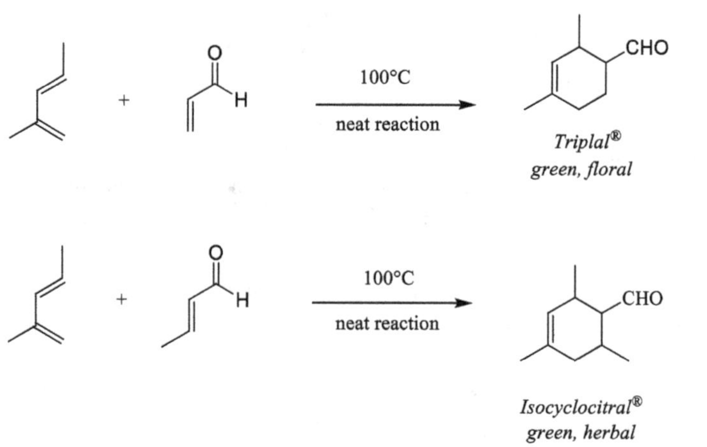

Figure 8.3. Odorants from 2-methylpenta-1,3-diene.

Myrcene, obtained from the pyrolysis of beta-pinene, [16] is one of the most commonly used 1,3-diene for the synthesis of fragrant materials via Diels–Alder reactions. Thus, Lyral®, a benchmark muguet odorant, can be prepared from the thermal reaction of myrcene and acrolein followed by the acid-catalysed hydration of the Diels–Alder adduct known as Myrac aldehyde® (Figure 8.5). Lyral® is best prepared by prior hydration of myrcene to myrcenol, followed by the thermal or Lewis acid ($ZnCl_2$, $SnCl_4$)-catalysed Diels–Alder reaction with acrolein [17].

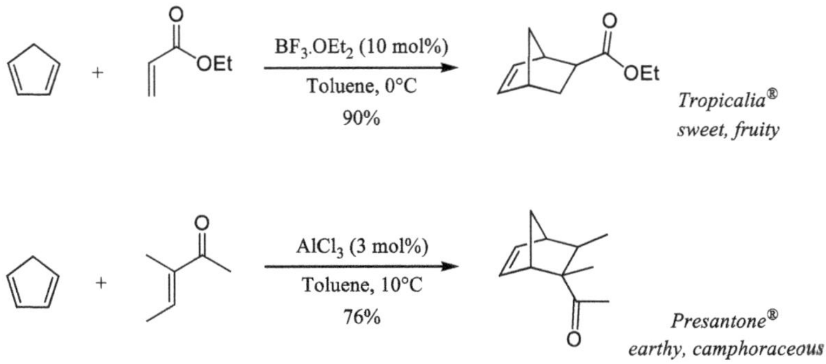

Figure 8.4. Synthesis of Tropicalia® and Presantone®.

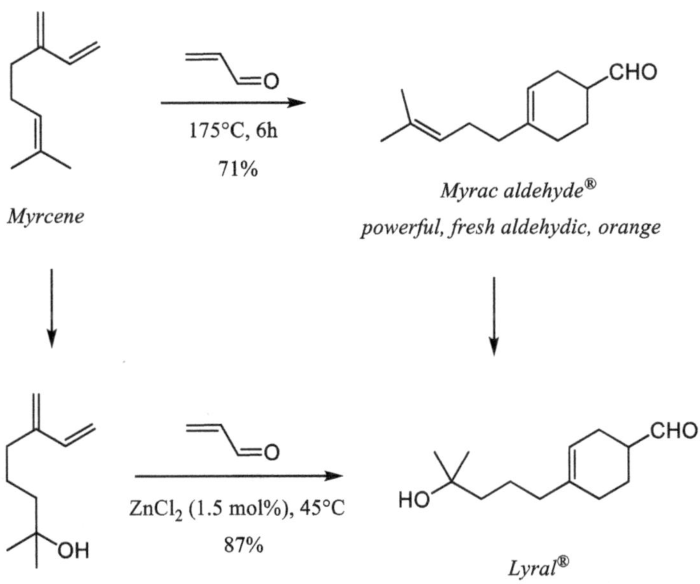

Figure 8.5. Diels–Alder preparation of Myrac aldehyde® and Lyral®.

Ambramone® is prepared from the Diels–Alder reaction of α-methylene valeraldehyde with myrcene [18]. The resulting propyl-substituted Myrac-type aldehyde intermediate is further cyclised under acidic conditions and undergoes a sequence of [1,2]-alkyl and [1,2]-hydride shifts to afford the tricyclic ketone with an elegant amber note (Figure 8.6).

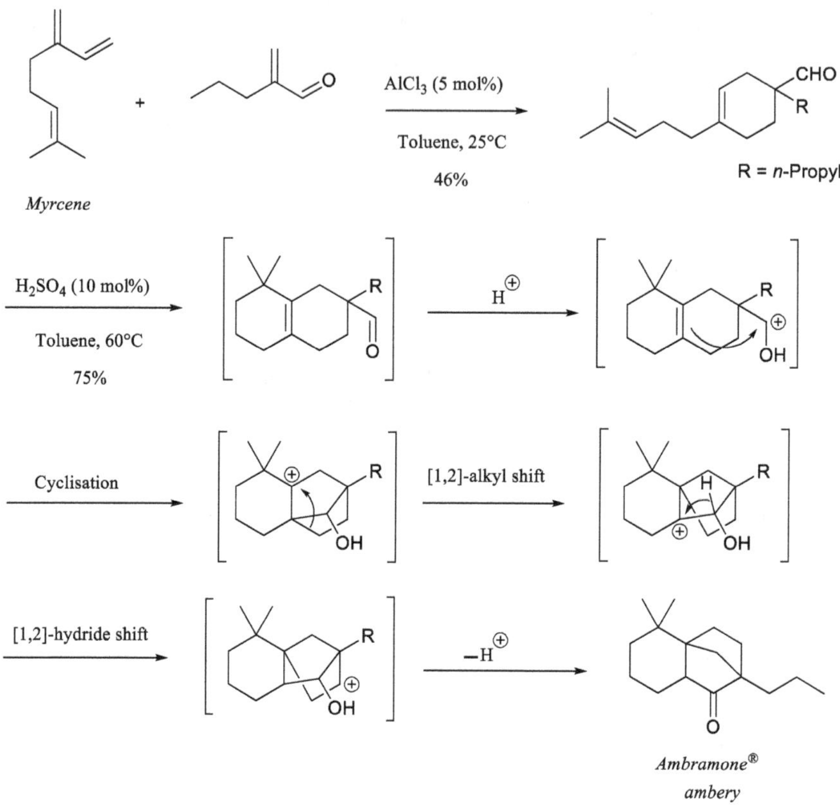

Figure 8.6. Synthesis of Ambramone®.

Iso E super® is one of the most important derivatives among the fragrant compounds prepared by a Diels–Alder approach, its annual production exceeding several hundred tons. The molecule was patented by IFF in 1975 and constitutes an essential component of the woody-ambery family [19]. Iso E super® is conveniently prepared from myrcene and 3-methylpent-3-en-2-one, affording a mixture of Diels–Alder adducts (Figure 8.7). The selectivity towards the main regioisomer is enhanced when using $AlCl_3$ catalysis, rather than thermal conditions [19]. Further cyclisation over H_3PO_4 or H_2SO_4 affords Iso E super®.

Interestingly, it was later found [20] that the woody-ambery odour was to be attributed to Iso E Super Plus®, which is present at about 5% in commercial Iso E Super®. The odour threshold of racemic Iso E Super Plus® was determined to be 0.005 ng/L air, which is 10^5 times lower than

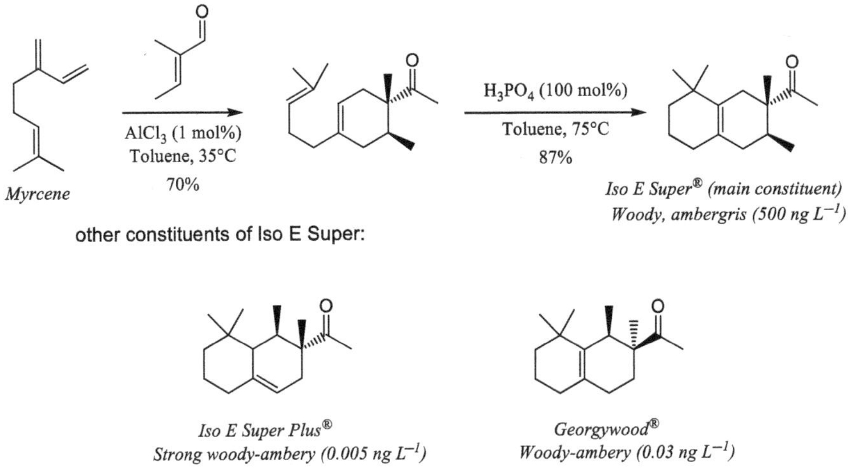

Figure 8.7. Synthesis of Iso E super® and two of its more powerful regioisomers.

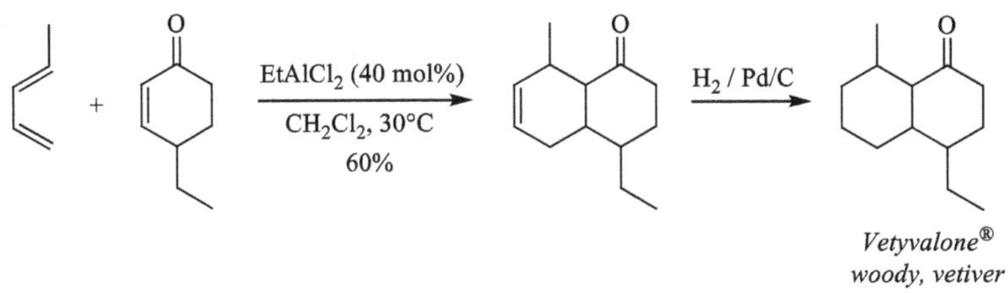

Figure 8.8. Synthesis of Vetyvalone®.

that of Iso E Super® (500 ng/L air). The formation of Iso E Super Plus® occurs during cyclisation of the Diels–Alder adduct and can be rationalised by the acid-catalysed isomerisation of the endocyclic double bond prior to cyclisation. Another isomer, Georgywood®, [21] with a close structure, was also identified in the Iso E Super® mixture and exhibits similar olfactory properties and performance with a higher odour threshold. Enantioselective syntheses of Iso E Super Plus® and Georgywood® were accomplished by Hong and Corey using an (*S*)-oxazaborolidinium-catalysed Diels–Alder reaction as a key step [22].

The vetiver odorant Vetyvalone® has been patented by Firmenich. [23] It is obtained from the reaction of piperylene with 4-ethyl-2-cyclohexen-1-one catalysed by ethyl aluminium dichloride (Figure 8.8).

8.3 Fragrant Compounds via Friedel–Crafts Reactions

Friedel–Crafts processes have been industrially set up for important compounds in the field of fragrances. Many examples concern alkylation and acylation reactions of aryl derivatives in the presence of Lewis acids, such as $AlCl_3$ [24] or protic acids, as reported for the synthesis of tetrahydronaphthalenes, [25] Herbetene® [26] or others. Thus, menthol, an extremely important flavour owing to its unique cooling properties, was prepared via alkylation of *m*-cresol and propylene [27] and further hydrogenated [28]. Lilial®, a muguet odorant now restricted because of CMR toxicity, could be prepared from aryl alkylation with *t*-butyl chloride [29]. Cyclamen aldehyde is issued from the reaction of cymene with methacrolein [30]. Galaxolide®, with its intense musk odour, is obtained by alkylation of 1,1,2,3,3-pentamethylindan with propylene oxide, [31] and other cyclic musks, such as Phantolide® and Fixolide® [32] or Celestolide®, are obtained via acylation of the corresponding indene derivatives [33, 34]. Trimofix®, a woody ambery macrocyclic ketone, is prepared from the acylation of trimethyl cyclododecatriene, a cyclic trimer of isoprene [35].

The Friedel–Crafts reaction classically requires stoichiometric amounts of Lewis acids. In the recent years, several triflate and triflimide derivatives have been reported to be efficient in catalytic amounts in these coupling reactions.

8.3.1 *Catalytic hydroarylation of olefins*

Besides the classical stoichiometric method to access fused cyclohexane structures bearing *gem*-dimethyl groups, the synthesis of this class of compounds by catalytic intramolecular cyclisations of olefin-tethered aryl derivatives has been described using only 1 mol% of $Bi(OTf)_3$ as the catalyst (Figure 8.9) [36].

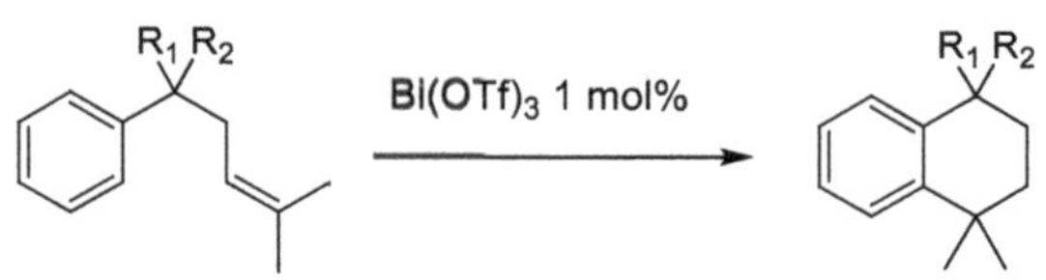

Figure 8.9. Catalytic intramolecular hydroarylation.

Figure 8.10 presents some olfactory features of these aryl-fused derivatives prepared catalytically. These compounds, bearing a *gem*-dimethyl group, generally exhibit fruity and floral notes.

8.3.2 *Tandem allylation–cyclisation of phenol derivatives*

Benzopyran structures have been prepared by intermolecular coupling reactions from phenols and allyl derivatives using 1 mol% of $In(OTf)_3$ as the catalyst through a tandem allylation–cyclisation procedure (Figure 8.11) [37].

The olfactory evaluation of some of these chroman structures is presented in Figure 8.12.

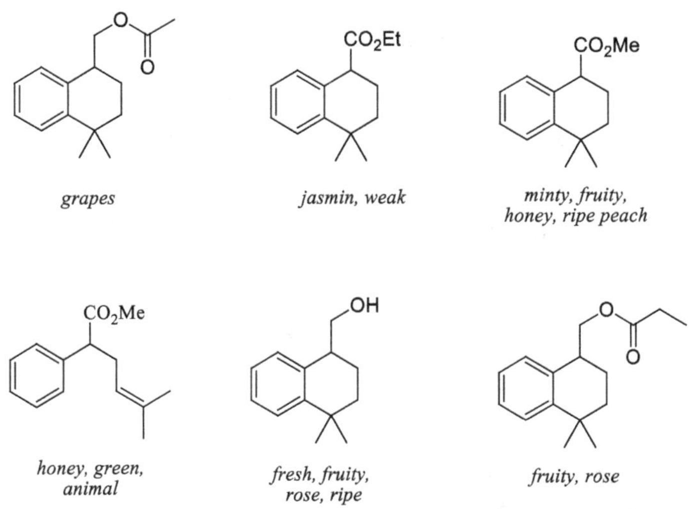

Figure 8.10. Olfactory evaluation of arene-fused cyclohexane structures.

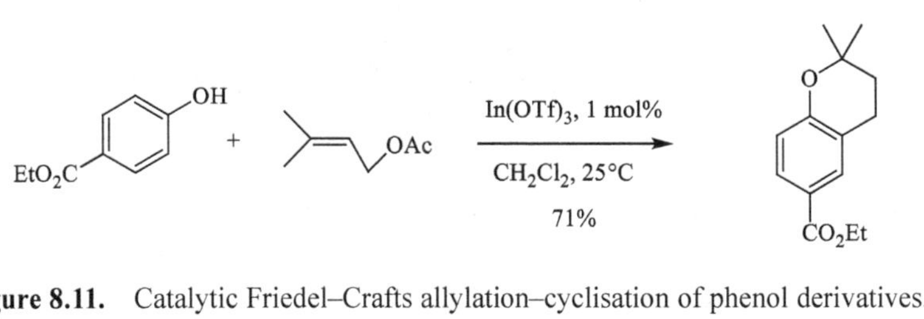

Figure 8.11. Catalytic Friedel–Crafts allylation–cyclisation of phenol derivatives.

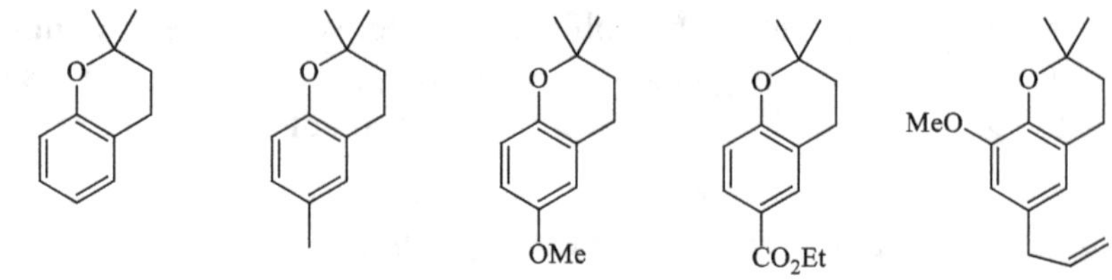

Figure 8.12. Olfactory evaluation of chroman derivatives.

8.4 Prins, Carbonyl-Ene and Related Reactions in the Fragrance Industry

8.4.1 *Odorants from the Prins reaction*

The Prins reaction was first reported in 1919 [38] and represents an often-encountered transformation in the flavour and fragrance chemistry. It provides an easy and cheap access to important fragrant materials, and its use in the fragrance industry has been recently reviewed [39]. In its simplest form, an alkene is reacted with formaldehyde to lead to a mixture of compounds mainly consisting of unsaturated alcohol, diol and 1,3-dioxane depending on the reaction conditions and stoichiometry. Some important commodity materials have been made available via this chemistry, including nopol [40] and cyclomethylene citronellol [41]. A Prins reaction was the ultimate step in the synthesis of Ambermax® [42], an elegant ambergris odorant derived from isolongifolene.

More complex structures can be achieved via multiple Prins reaction sequences on the starting substrate, giving access to molecules such as Jasmopyrane® [43], starting from 1-octene (Figure 8.13). Jasmopyrane®, with its fresh jasmine odour, is an economic substitute for jasmonates in floral compositions. Similarly, *p*-toluene sulfonic acid (PTSA)-catalysed Prins reaction on Isoelgenone afforded a mixture of compounds, among which 5-(4-methyl-3,6-dihydro-2H-5-pyranyl)-2-hexanone and 5-(4,4-dimethyl-1,3-dioxolan-5-yl)-2-hexanone were found to exhibit a pleasant plum odour (Figure 8.13) [44].

Isoprenol, obtained from isobutylene and formaldehyde, is an essential synthon for both aroma and fragrant chemistry. The Prins reaction of a suitable aldehyde and isoprenol is a method of choice for easy access to

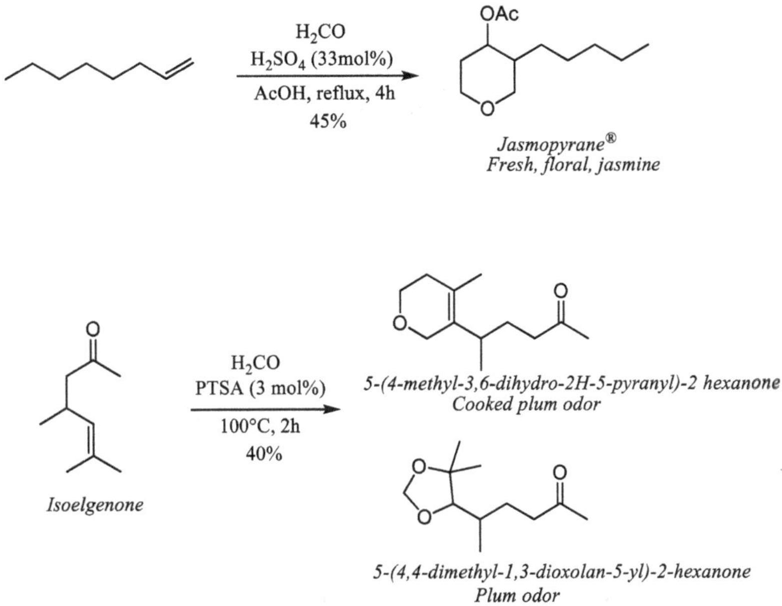

Figure 8.13. Odorants from consecutive Prins reactions.

dihydro- and tetrahydropyran-based odorants. In this respect, the high-volume fragrant material Florol® is prepared by a Prins reaction using isovaleraldehyde and isoprenol (Figure 8.14). BASF has recently filed a patent describing an improved procedure for the preparation of Florol® using strongly acidic cation exchangers, such as Amberlyst catalysts (50 w/w% Amberlyst® 131) [45].

Another important derivative is Rosyrane [46], with floral rosy notes, prepared from a *p*-toluenesulfonic acid-catalysed Prins reaction of benzaldehyde and isoprenol (PTSA 0.2 mol%, toluene, 77%). Hydrogenation over Pd/C affords Doremox® [47] as a mixture of *cis*- and *trans*-isomers (Figure 8.14). The *cis*-isomer exhibits a note of green, rose oxide type, which is quite powerful, while the *trans*-isomer has a much weaker odour, with a green, slightly minty character and a floral undertone.

An enantioselective synthesis of *cis*-Doremox® and other related 4-methyl-2-substituted tetrahydropyrans was recently reported [48]. The methodology was based on an asymmetric Prins cyclisation catalysed by a new class of confined imino-imidodiphosphate catalysts (5 mol% catalyst, cyclohexane, 69%).

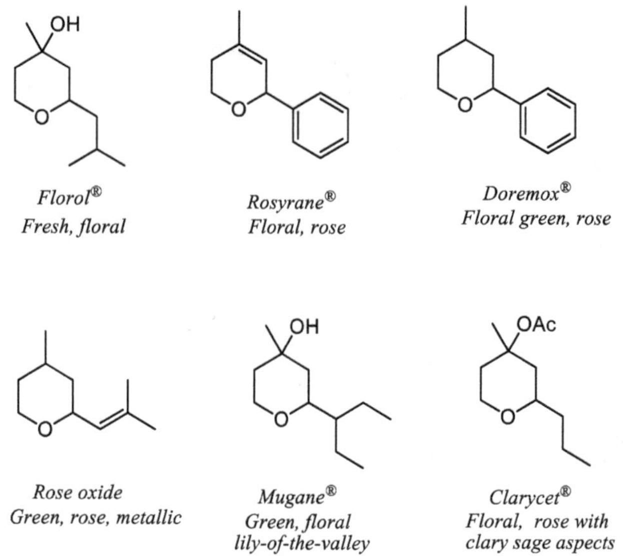

Figure 8.14. Dihydro- and tetrahydropyran-based odorants.

Mugane® is a close analogue of Florol®. It has been prepared in 64% yield from 2-ethylbutyraldehyde and isoprenol using Montmorillonite K10 as the catalyst. It exhibits a green, floral, lily-of-the-valley odour [49].

The preparation of Clarycet®, an acetylated analogue of Florol® and Mugane®, has been reported using *p*-toluenesulfonic acid on silica gel in a solvent-free process, followed by acetylation [50].

8.4.2 *Carbonyl-ene reactions*

The carbonyl-ene reaction concerns the coupling between a carbonyl derivative and an olefin to afford an unsaturated alcohol [51]. A typical example of the use of the carbonyl-ene reaction in the field of fragrances involves the synthesis of (−)-β-santalol, a highly appreciated sandalwood odorant. It was prepared in an enantioselective fashion in 2009 [52]. A key step in this synthesis was the coupling between the methylenecyclohexene derivative **1** and trimethylsilylpropynal, in the presence of a stoichiometric amount of Me_2AlCl, to afford the propargyl alcohol **2**. (−)-β-Santalol was further obtained after a few steps with 97% enantiomeric excess (Figure 8.15).

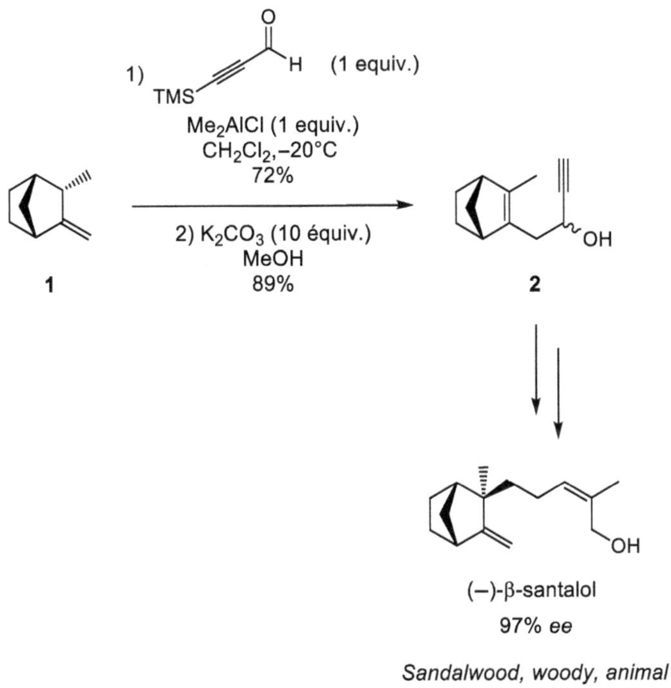

Figure 8.15. Synthesis of (−)-β-santalol.

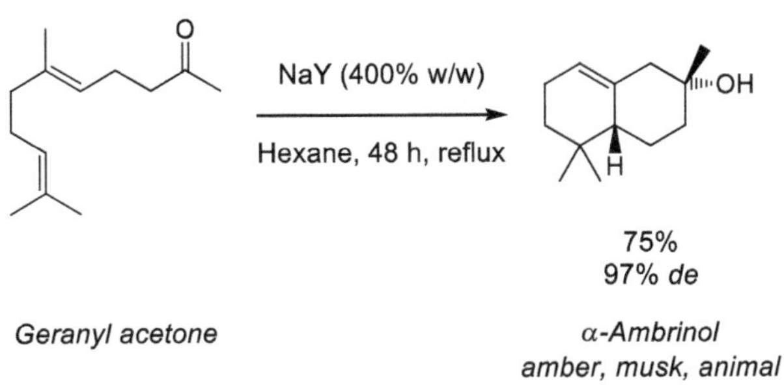

Figure 8.16. Geranyl acetone cyclisation to α-ambrinol catalysed by zeolite NaY.

α-Ambrinol (Figure 8.16) is used as a fragrance for its powerful ambery, musky, natural and animal odours. It has been synthesised by a tandem 1,5-diene cyclisation – carbonyl-ene reaction from geranyl acetone using zeolite NaY. α-Ambrinol was obtained with 75% yield and 97% diastereomeric excess [53].

The cyclisation of the same compound under Lewis acid conditions ($BF_3.OEt_2$) afforded the bicyclic ether **3** without any trace of α-ambrinol (Figure 8.17) [54].

(−)-Menthol, one of the most commonly used flavouring ingredients, is mainly obtained from *Mentha arvensis*, its essential oil containing between 73 and 86% of (−)-menthol. Due to an increase in demand, industrial preparations by both BASF and Takasago have been developed. Their processes involve a first Lewis acid-catalysed ene-cyclisation of (+)-citronellal to (−)-isopulegol, which is then hydrogenated to (−)-menthol.

Takasago's original process involved the use of $ZnBr_2$ in stoichiometric amounts, yielding (−)-isopulegol with a 100% enantiomeric excess after crystallisation at −50°C (Figure 8.18) [55]. In 2002, they released a patent involving the use of tris(2,6-diarylphenoxy)aluminium catalyst, giving a ratio of (−)-isopulegol to other isomers at above 99% in a 95% yield [56].

BASF uses an $Al(OR)_3$ Lewis acid for their ene-cyclisation, yielding (−)-isopulegol with a 100% enantiomeric excess after crystallisation [57].

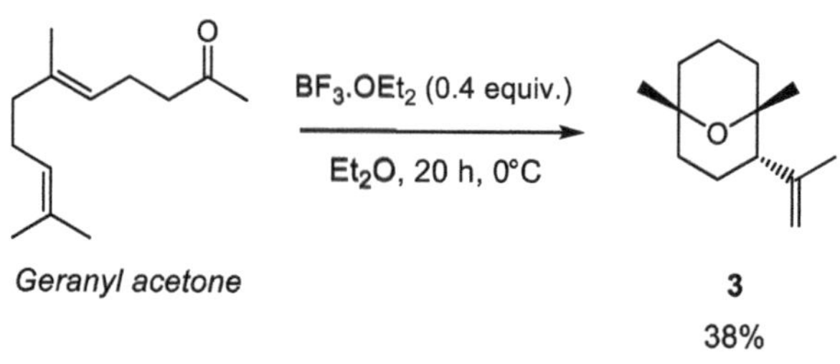

Figure 8.17. Geranyl acetone cyclisation catalysed by $BF_3.OEt_2$.

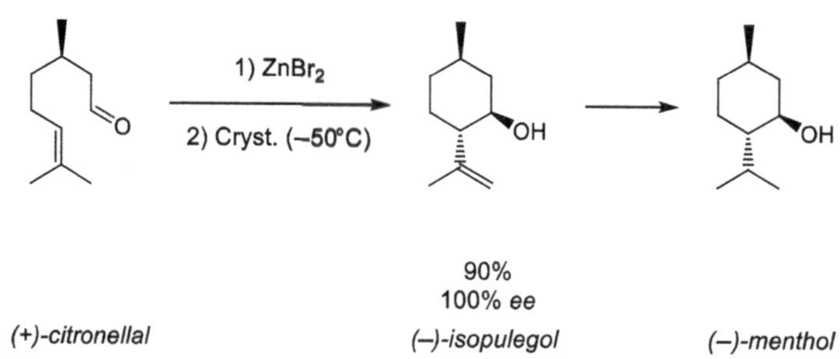

Figure 8.18. Synthesis of (−)-isopulegol by carbonyl-ene cyclisation.

The cyclisation of (+)-citronellal to (−)-isopulegol was also carried out using scandium trifluoromethanesulfonate (5 mol%) as the catalyst [58]. This ene-reaction was performed at low temperatures, and (−)-isopulegol was obtained in a 95% yield with 88% enantiomeric excess (Figure 8.19).

8.4.3 *Oxy-Cope rearrangement*

The oxy-Cope rearrangement, a variation of the Cope rearrangement, involves the skeletal reorganisation of 1,5-dienols to unsaturated carbonyl compounds [59]. In the field of fragrances, the use of a palladium (II) complex as the catalyst allowed the synthesis of a methyl analogue of the macrocycle Ambretone, affording exclusively the *E* isomer **5** from dienyl alcohol **4** (Figure 8.20) [60]. Ambretone® is known for its musky, ambery odour.

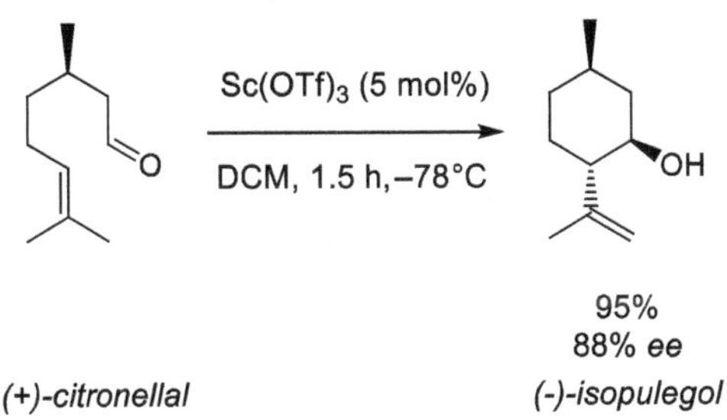

Figure 8.19. Synthesis of (−)-isopulegol catalysed by a Lewis superacid.

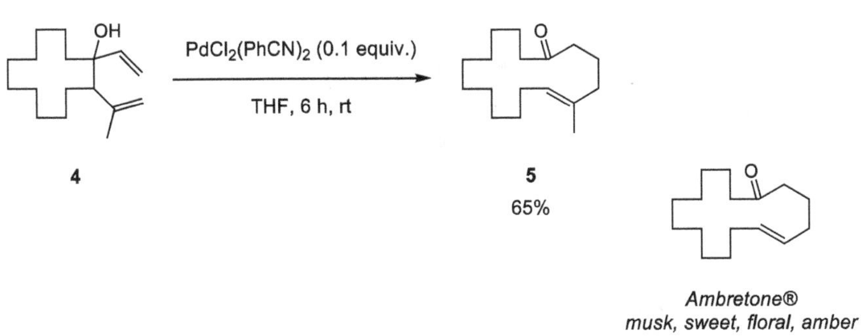

Figure 8.20. Synthesis of the methyl analogue of Ambretone® **5**.

8.5 Cyclisation of Dienes and Related Structures Catalysed by Lewis Acids

8.5.1 *Cyclisation of dienes and trienes*

The electrophilic activation of isolated double bonds followed by the intramolecular cyclisation with a second olefin affords unsaturated cyclic structures.

The cyclisation of 1,5-diene derivatives with excess H_2SO_4 has been reported for the synthesis of pseudo-ionones [61]. The use of H_3PO_4 has been used for the second step of the synthesis of Iso E Super®, a woody, floral and ambergris odorant (see Figure 8.7) [62]. The synthesis of Georgywood® could also be carried out in the presence of stoichiometric amounts of $MeAlCl_2$ as the Lewis acid (Figure 8.21) [63].

The cyclisation of 1,6-dienes has been catalysed by $Sn(NTf_2)_4$ at 2–5 mol% [64]. Highly substituted and functionalised cyclohexane structures such as **6** were obtained in good yields (Figure 8.22).

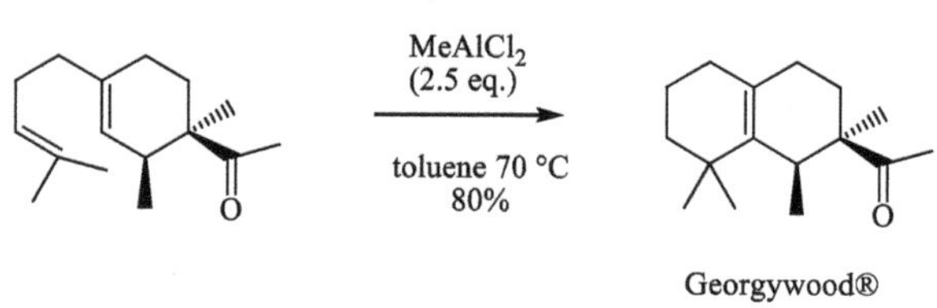

Figure 8.21. Lewis acid mediated synthesis of Georgywood® via a 1,5-diene cyclisation.

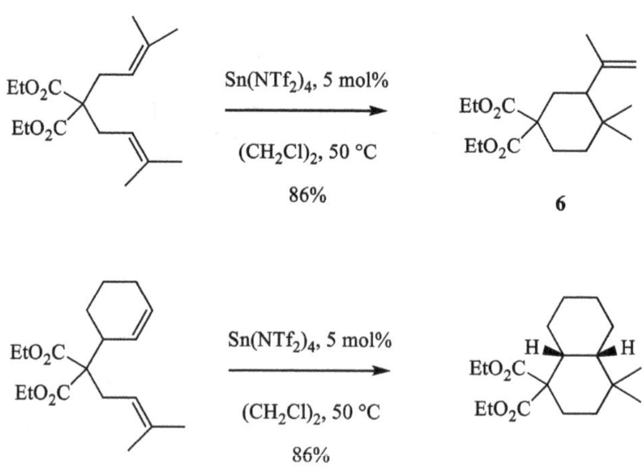

Figure 8.22. Catalytic cyclisation of 1,6-dienes to substituted cyclohexane structures.

Cyclohexane structures bearing *gem*-dimethyl groups and/or propenyl moieties are found in several interesting flavouring compounds, such as irones, ionones, damascones, carvone and isopulegol. Compound **6** presents weak orris notes. Some of its derivatives, presenting different flavouring aspects, have been reported [65, 66] as shown in (Figure 8.23).

The cyclisation of polyenes is widely found in the biosynthesis of cyclic terpenoids [67]. A Lewis acid-catalysed cyclisation of trienes has been reported with the use of $Bi(OTf)_3$ [68]. Thus, the cyclisation of triene **7** led to the bicyclic structure **8**, involving a double cyclisation and a 1,2-methyl shift (Figure 8.24).

Some bicylo[4.4.0]derivatives are well-known flavouring agents, such as Iso-E-super® and Georgywood®. The olfactory notes of derivatives of **8** and analogous compounds have been reported (Figure 8.25) [65].

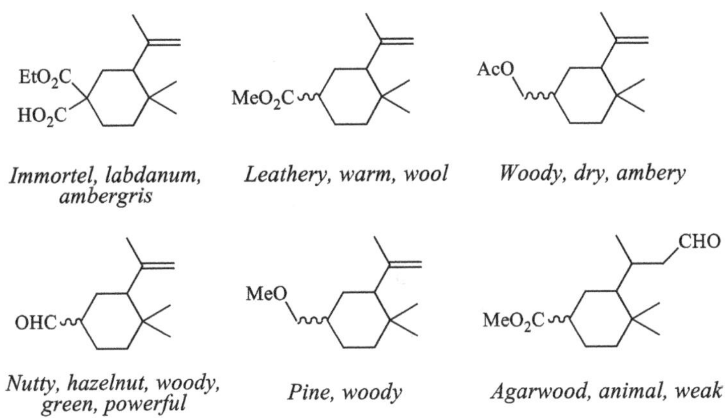

Figure 8.23. Olfactory features of *gem*-dimethylcyclohexane structures derived from **6**.

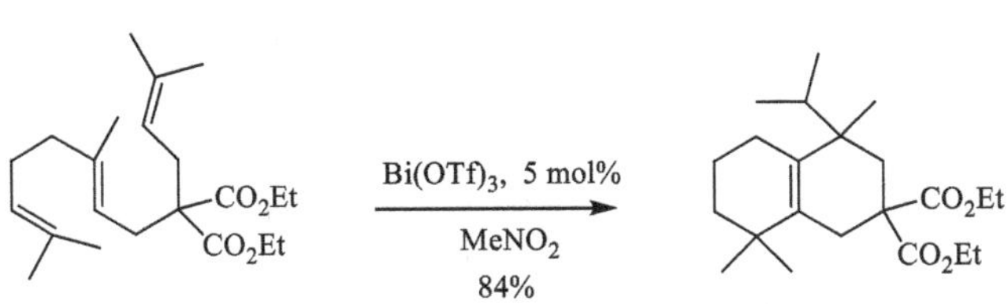

7 **8**

Figure 8.24. Catalytic tandem cyclisation of **7** with a methyl shift rearrangement.

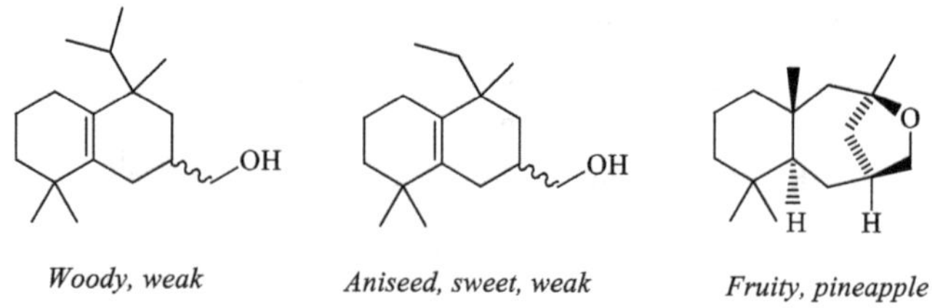

Figure 8.25. Olfactory evaluation of bicyclic compounds derived from triene cyclisations.

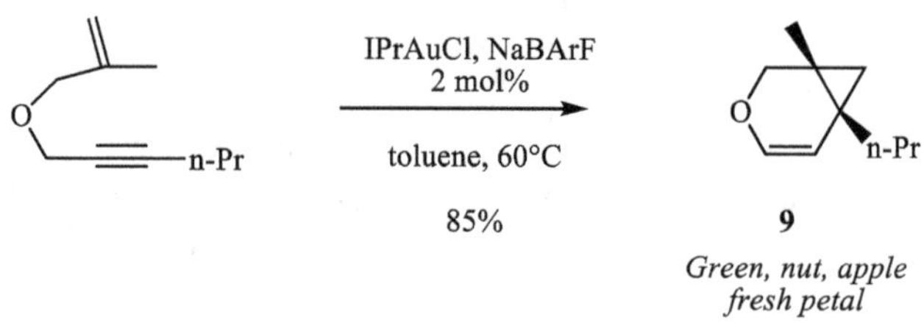

Figure 8.26. Au(I)-catalysed cycloisomerisation of 1,6-enyne ethers.

8.5.2 *Cyclisation of enynes*

Several functionalised oxa-bicyclo[4.1.0]-hept-4-ene derivatives have been efficiently prepared in an Au(I)-catalysed oxygen-tethered enyne cyclisation. An Au–carbene complex (IPrAuCl, 2 mol%) combined with $NaBAr_3F$ afforded, for example, the cyclopropane-fused derivative **9** in 85% yield (Figure 8.26) [69]. Both Au(I) and Cu(I) catalysts have also been reported for the synthesis of thujone-type compounds by cyclisation of enyne alcohol derivatives [70].

8.5.3 *Cyclisations involving enol ethers*

Spirobicyclic tetrahydrofuran and tetrahydropyran derivatives have been obtained in the $Bi(OTf)_3$-catalysed intramolecular cyclisation (at 0.1 mol%) of allenic, cyclic enol ethers [71]. The example in Figure 8.27 shows this selective cyclisation operating under mild conditions. The obtained alcohol presents fresh notes of natural leguminous and asparagus.

The tandem $Bi(OTf)_3$-catalysed double cyclisation of unsaturated hydroxy enol ethers derived from dihydrofuran has been reported for the stereocontrolled formation of bridged tricyclic structures (Figure 8.28) [72].

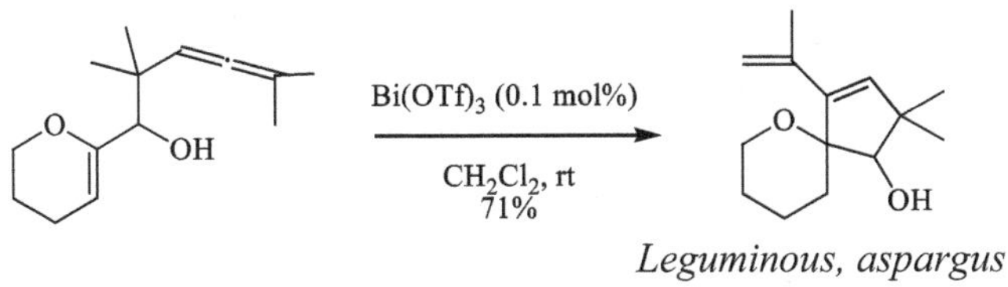

Figure 8.27. Cyclisation of an allene enol ether alcohol to a spiro ether compound.

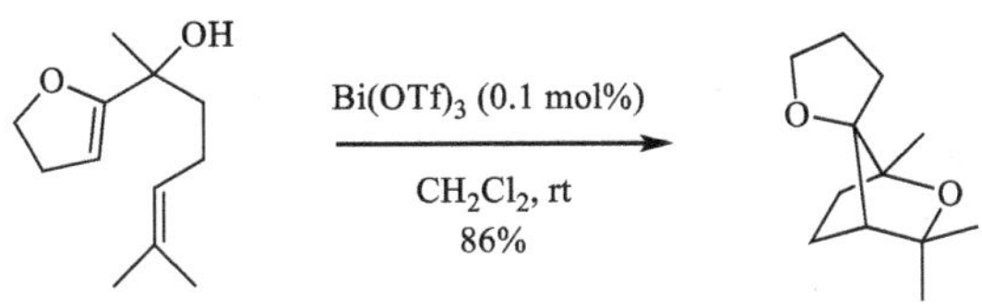

Figure 8.28. Tandem cyclisation of enol ether alcohols to tricyclic structures.

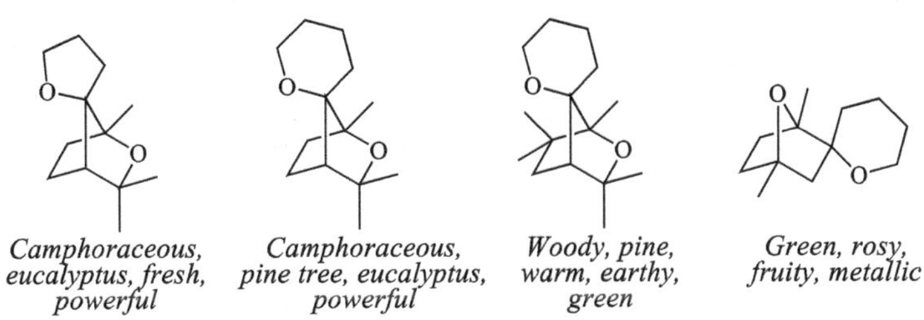

Figure 8.29. Olfactory evaluation of some spiro tricyclic diether structures.

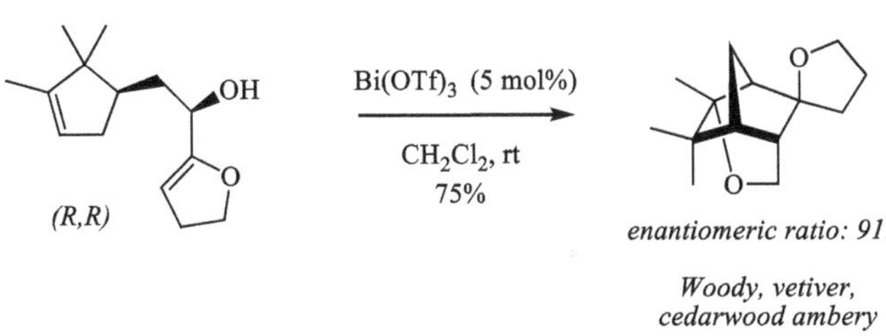

Figure 8.30. Enantioselective tandem cyclisation of a campholenic enol ether derivative.

The olfactory properties of this class of compounds have been evaluated, and generally fresh and woody notes are present (Figure 8.29) [73].

These intramolecular cyclisations have been extended to campholenic aldehyde enol ether derivatives for the preparation of complex polycyclic structures with vetiver and woody notes (Figure 8.30) [74].

8.6 Cyclisation of Unsaturated Alcohols, Carboxylic Acids and Thiols

8.6.1 *Preparation of cyclic ethers from olefinic alcohols*

A direct route to the synthesis of cyclic ethers is the cycloisomerisation of unsaturated alcohols. In the case of protic acids, the cyclisation of unsaturated alcohols has been reported in strong acid media, the protic acids being generally used in over-stoichiometric amounts or as solvents, with the possibility of rearrangements [75, 76]. Catalytic cyclisations of olefinic alcohols have been described in the presence of HOTf [77].

In the presence of Lewis acids, this hydroalkoxylation takes place in a Markovnikov-type addition [78]. The use of $AlCl_3$ and $SnCl_4$ in stoichiometric amounts or in excess has been reported for these alcohol cycloisomerisations [79].

In the field of fragrances, a typical example of Lewis acid-mediated cyclisation concerns the transformation of homofarnesol to Ambrox® using achiral or chiral $SnCl_4$-based complexes. Thus, (–)-Ambrox®, which is highly appreciated for its musky, labdanum, woody and animal notes, could be prepared in 30% yield and 42% enantiomeric excess using two equivalents of a naphthol-based chiral complex derived from $SnCl_4$ (Figure 8.31) [80, 81].

The naturally occurring spirobicyclic theaspirane presenting blackcurrant and fruity notes has been prepared from the cyclisation of the unsaturated alcohol of reduced β-ionone, under an acid-catalysed process (Figure 8.32) [82]. The four isomeric spiro ethers could be further enantioselectively prepared from dihydro-β-ionone via a chemoenzymatic synthesis [83].

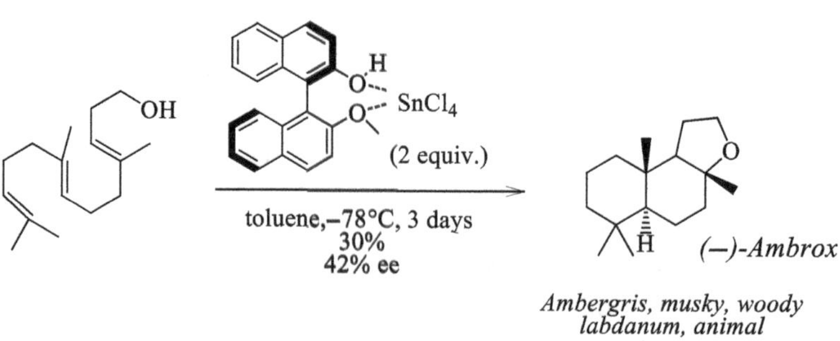

Figure 8.31. Synthesis of (–)-Ambrox® via alcohol cycloisomerisation.

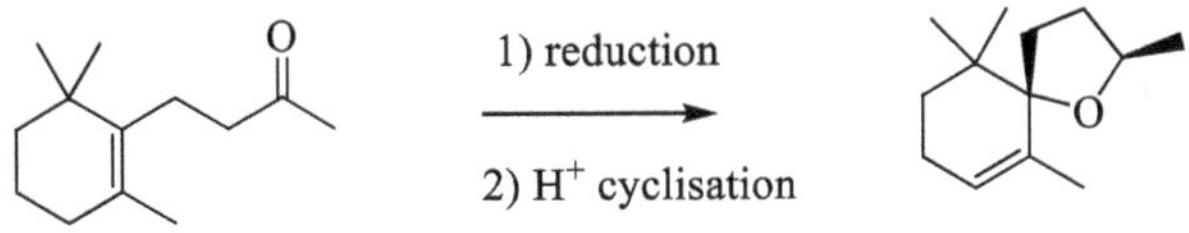

Theaspirane

Blackcurrant, fruity, woody

Figure 8.32. Cyclisation to theaspirane from reduced dihydro-β-ionone.

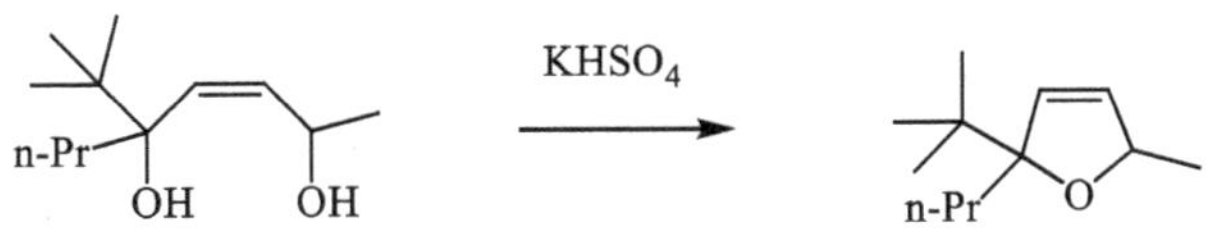

Cassyrane
Cassis, blackcurrant

Figure 8.33. Synthesis of Cassyrane®.

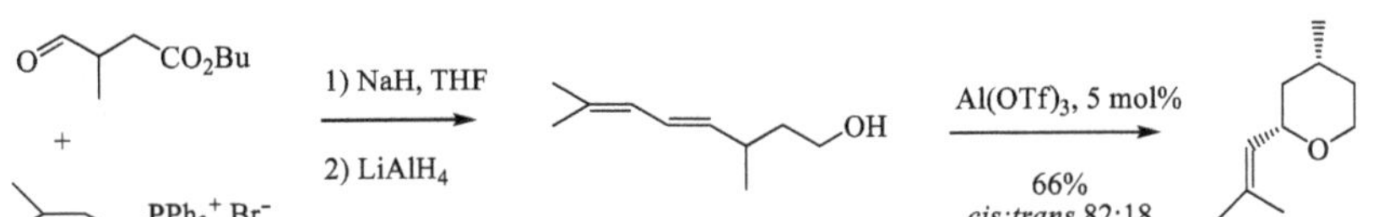

10

Rose oxide
rose, green, metallic

Figure 8.34. Lewis acid-catalysed cyclisation of alcohol **10** to rose oxide.

The dihydrofuran derivative Cassyrane®, known for its cassis and blackcurrant top notes, has been obtained via the mild acid-catalysed cyclisation of the substituted 1,4-diol derivative (Figure 8.33) [84]. The different stereoisomers of Cassyrane® could be prepared in enantiomeric form [85].

First isolated from the essential rose oil in 1959, rose oxide is a volatile cyclic ether [86]. The most appreciated enantiomer is (2*S*,4*R*)-*cis*-rose oxide (Figure 8.34), presenting pronounced floral, rose, green and metallic notes with a low threshold of 0.5 ppb [87].

Most of the syntheses of rose oxide concern citronellol oxidation (O_2, hν; Br_2) and further acid-promoted cyclisation, affording *cis/trans* mixtures [88]. Allene alcohol intermediate derivatives could also be cyclised to rose oxide by the stoichiometric use of $HgCl_2$ [89] or by $Pd(OAc)_2$-catalysed reactions [90]. The catalytic synthesis of rose oxide has been achieved via the cyclisation of dienic alcohol **10** in the presence of 5 mol% of $Al(OTf)_3$ (Figure 8.34) [91].

Structures analogous to rose oxide with different ring substitutions were prepared following the same synthetic strategy and presented mostly green and fruity notes [91].

Structure-activity studies on rose oxide led to the synthesis of 4-methyl-2-phenyltetrahydro-2*H*-pyrane (Doremox®), a heavier analogue with a higher substantivity. [47] As for rose oxide, the most appreciated isomer is (2*S*,4*R*)-Doremox®, presenting a powerful rose scent [92]. The synthesis of Doremox® has also been reported via a Prins reaction, as well as via a hetero-Diels–Alder cyclisation between benzaldehyde and isoprene, followed by double-bond reduction. [90] The direct Lewis acid-catalysed cyclisation of the styrene derivative **11** to the Doremox® structure in the presence of 5 mol% of $Al(OTf)_3$ has also been reported (Figure 8.35) [91].

The most efficient Lewis acid catalysts reported for direct and catalysed hydroalkoxylations of unsaturated alcohols are the highly electrophilic $Al(OTf)_3$ and $Sn(OTf)_4$ [93].

Rhubafuran® is a benchmark odorant of this family of cyclic ethers, with an intense fruity character of rhubarb and grapefruit notes [94]. Its synthesis proceeds through the allylation of 2-phenylpropanal, followed by its cyclisation in the presence of $Hg(OAc)_2$ and further reduction (Figure 8.36) [95, 96].

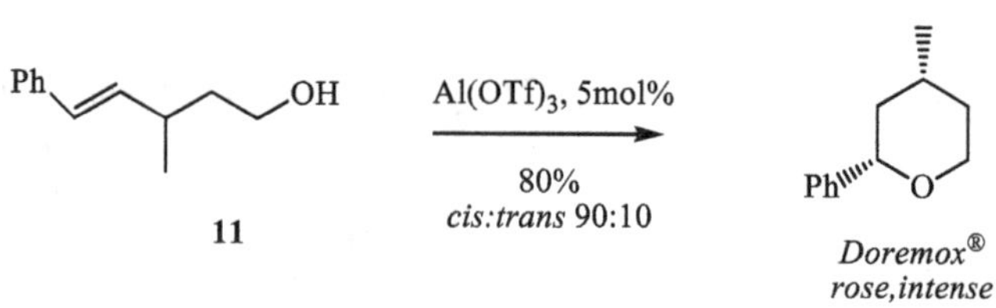

Figure 8.35. Lewis acid catalytic cyclisation of **11** to Doremox®.

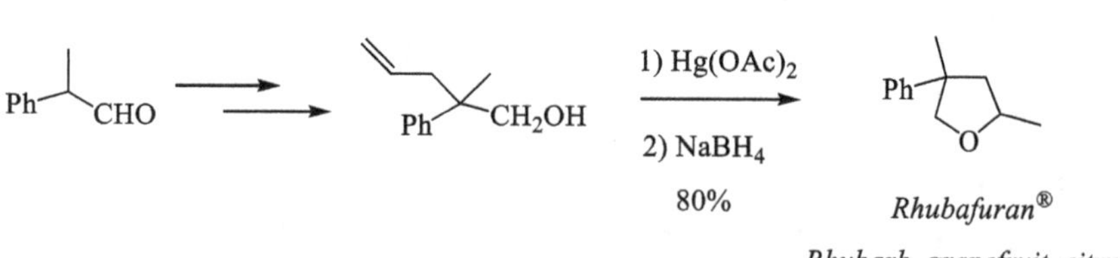

Figure 8.36. Synthesis of Rhubafuran®.

8.6.2 *Lactones by Lewis acid-catalysed cyclisation of unsaturated acids*

Lactone derivatives, in particular γ- and δ-lactones, are found in numerous natural products and food compositions and present interesting organoleptic properties [97]. Their synthesis generally proceeds through the cyclisation of the corresponding hydroxy acids.

The catalytic and regioselective cycloisomerisation of unsaturated carboxylic acids in the presence of Lewis acids has also been reported with metal triflates, in particular with $Al(OTf)_3$ and $Sn(OTf)_4$, as well as with protic superacids [98]. Thus, the easily prepared Noxolide® lactone was appreciated for its strong fresh nuts and green notes (Figure 8.37) [99].

For α-alkoxy carboxylic acids, a $Cu(OTf)_2$-catalysed isomerisation/cyclisation sequence to dioxolanones has been reported [100]. Among these compounds, phenyl dioxolanone presented a cocoa-like odour (Figure 8.38).

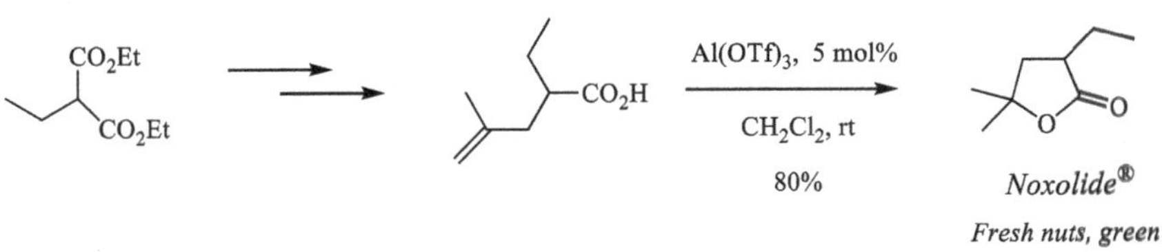

Figure 8.37. Al(III)-catalysed cyclisation to lactone Noxolide.

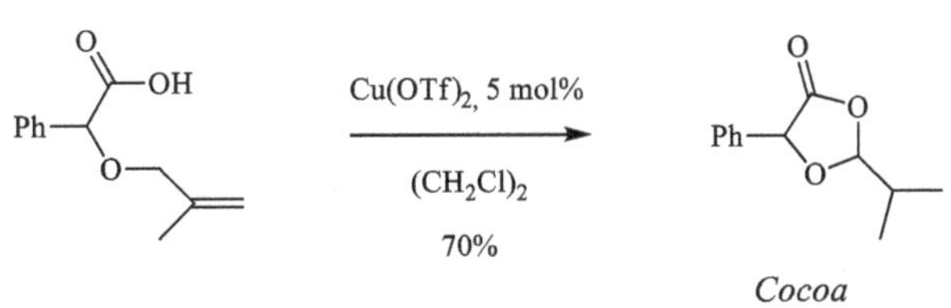

Figure 8.38. Cyclisation of unsaturated carboxylic acids to dioxolanones.

8.6.3 *Synthesis of cyclic sulphides via catalytic cyclisation of unsaturated thiols*

Thiols are among the most powerful sulphur compounds known for their organoleptic properties [101]. For the direct synthesis of thioethers from

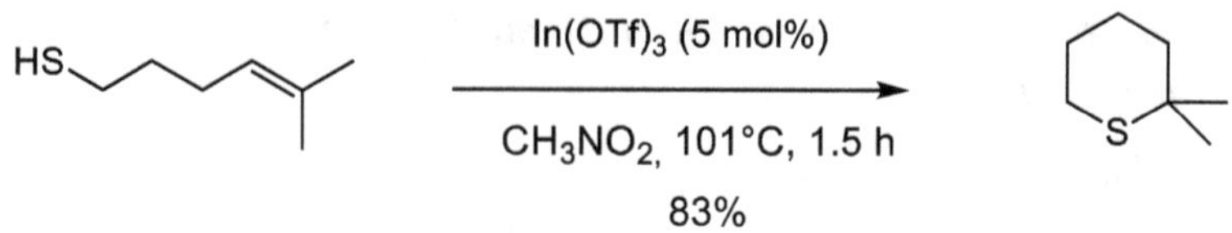

Figure 8.39. In(III)-catalysed intramolecular hydrothiolation of olefins.

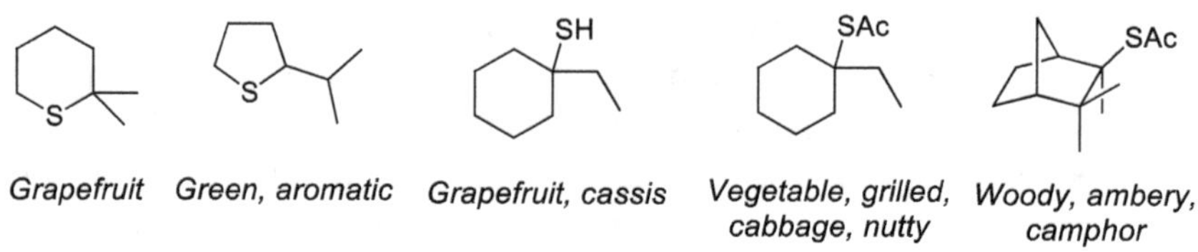

Figure 8.40. Examples of thioethers prepared by catalytic hydrothiolation.

unsaturated thiols, $In(OTf)_3$ has been developed as an efficient electrophilic catalyst, affording Markovnikov-type adducts with non-activated olefins (Figure 8.39. [102])

Some sulphur compounds prepared through this catalytic hydrothiolation procedure with their main olfactory notes are presented in (Figure 8.40).

The thiol and thioacid addition to olefins has been extended to the preparation of thio-camphene derivatives in $In(OTf)_3$-catalysed reactions under mild conditions, avoiding the classical camphene to isobornane skeleton rearrangement [103].

8.7 Conclusions

The large structural variety of molecules presenting an interest in the field of flavours and fragrances involves the use of multiple reaction types and reagents for their syntheses. Numerous examples illustrate that the earlier procedures make use of stoichiometric or even over-stoichiometric amounts of catalysts. It becomes more important nowadays to re-examine some of these classical synthetic methods, aiming to set up novel and catalytic methodologies, thereby allowing for efficient processes under mild conditions.

A more active catalyst may achieve novel transformations and allow access to new structures that are difficult to obtain by classical means.

Highly catalytic processes should involve greener chemistry with less cost, less waste, lower energy demand, simplified reaction work-up and less post-treatment of secondary products, as well as less environmental impact.

Going from stoichiometry to catalysis can be achieved, for example, by using more active and/or more selective catalytic species. This can render the transformations more efficient. Thus, for example, several acid-mediated cyclisations are still carried out using the proper acid as the solvent. It has been shown that one can use very low catalytic amounts of stronger acids for the same transformations [5, 77, 104]. Recyclable polymer-supported catalysts may also constitute a promising perspective [105].

In the context of the reactions mediated by Lewis acids, classical Lewis acids such as $AlCl_3$, BF_3, $TiCl_4$, and $ZnBr_2$ can be efficiently replaced, for example, by the corresponding metal triflates [6, 93, 106].

The possibility of recovery and reuse of catalysts also becomes an important issue. It is interesting to note that some active Lewis superacids such as metal triflates can be easily recovered and recycled [36].

Environmental, sustainability, toxicity and safety issues are to be taken into account, considering the 12 principles of green chemistry [107], particularly when setting up novel atom-economy syntheses in this creative field of new fragrances.

References

[1] C. S. Sell. (2003). *A Fragrant Introduction to Terpenoid Chemistry*. RCS Cambridge (UK).

[2] G. Frater, J. A. Bajgrowicz, and P. Kraft. (1998). Fragrance chemistry. *Tetrahedron Rep.*, 54, 7633–7703.

[3] G. Ohloff, W. Pickenhagen, and P. Kraft. (2012). *Scent and Chemistry*. Wiley-VCH, Weinheim, DE.

[4] S. Raton (1998). Heterogeneous catalysis in the fine chemical industry. *Chem. Today*, 3, 33–37.

[5] H. Yamamoto (2008). *Acid Catalysis in Modern Organic Synthesis*. Eds. Yamamoto, H. and Ishihara, K., Wiley-VCH, Weinheim, GE.

[6] S. Antoniotti, V. Dalla, and E. Duñach (2010). Metal triflimidates: Better than metal triflates as catalysts in organic synthesis: The effect of a highly delocalised counter-anion. *Angew. Chem. Int. Ed.*, 49, 7860–7888.

[7] G. A. Olah, G. K. S. Prakash, A. Molnar, and J. Sommer. (2009). *Superacid Chemistry*. Wiley, Hoboken, USA.

[8] S. Kobayashi, M. Sugiura, H. Kitagawa, Lam, and W.W.L. (2002). Rare-earth metal triflates in organic synthesis. *Chem. Rev.*, 102, 2227–2302.

[9] T. Akiyama. (2007). Stronger Bronsted acids. *Chem. Rev.*, 107, 5744–5758.

[10] O. Diels, and K. Alder. (1928). Synthesen in der hydroaromatischen Reihe. *Justus Liebig's Ann. Chem.*, 460, 98–122.

[11] J. A. Funel and S. Abele. (2013). Industrial applications of the Diels-Alder reaction. *Angew. Chem. Int. Ed.*, 52, 3822–3863.

[12] C. Vial, L. Moretti, A. Charpilloz, P. Fankhauser, and P. Fantini. (2004). *Spiro compounds as perfuming ingredients*. US 20040072721.

[13] O. Diels andK. Alder. (1927). *Verfahren zur Darstellung von Verbindungen mit hydrierten Ringsystemen*. DE 526168.

[14] A. P. Closson, A. T. Levorse, and M. G. Monteleone. (2010). *Isomers of bicyclo[2.2.1]hept-5-ene-carboxylic acid, ethyl ester and their use in perfume compositions*. US 20100092418.

[15] J. Mane, J.-J. Chanot, and M. Schroeder. (2003). *Nouveaux dérivés du norbornane et du norbornene, leur utilisation et produits parfumés les contenant*. FR 2831165.

[16] T. R. Savic-Savich and L. A. Goldblatt. (1928). *Process for producing myrcene from beta-pinene*. US 2507546.

[17] J. M. Sanders, W. L. Schreiber, and J. B. Hall. (1977). *Process for producing mixture containing 4-(4-methyl-4-hydroxyamyl)cyclohexenecarboxaldehyde, product produced, and its perfume uses*. US 4007137.

[18] P. Jaunky, J-J. Chanot, and J. Mane. (2012). *5,5-Diméthyl-2-propyl-hexahydro-2,4a-méthano-naphthalèn-l-one en tant qu'agent fragrant*. WO 2013079876.

[19] J. B. Hall and J. M. Sander. (1975). *Perfume compositions and perfurme articles containing one isomer of an octahydrotetramethyl acetonaphthone*. US 3929677.

[20] C. Nussbaumer, G. Fráter, and P. Kraft. (1999). (±)-1-[(1R*,2R*,8aS*)-1,2,3,5,6,7,8,8a-Octahydro-1,2,8,8-tetramethylnaphthalen-2-yl]ethan-1-one: Isolation and stereoselective synthesis of a powerful minor constituent of the perfumery synthetic Iso E Super®. *Helv. Chim. Acta*, 82, 1016–1024.

[21] F. Etzweiler, D. Helmlinger, C. Nussbaumer, and M. Pesaro. (1993). *Acetyl tri- and -tetramethyl-octahydronaphthalenes and fragrance compositions containing same*. US 5180709.

[22] S. Hong and E. J. Corey. (2006). Enantioselective syntheses of Georgyone, Arborone, and structural relatives. Relevance to the molecular-level understanding of olfaction. *J. Am. Chem. Soc.*, 128, 1346–1352.

[23] R. Moretti and O. Etter. (2008). *Perfuming ingredients of the woody type*. US 7378383.

[24] G. Sartori and R. Maggi. (2010). *Advances in Friedel-Crafts Reactions.* CRC Press NY.

[25] C. Fehr and J. Galindo. (1995). Process for the preparation of novel tetrahydronaphthalenes as perfume ingredients. US 5442124A.

[26] G. Frater and U. Mueller. (1988). Mechanism of the formation of herbetene from trans-didehydrobicyclofarnesol. *J. Chem. Soc. Chem. Commun.*, 1198–1200.

[27] J. Fleischer and K. Bauer. (1976). *Separating optically pure d- and l- isomers of menthol, neomenthol and isomenthol.* US3943181.

[28] W. Bierdermann, H. Köller, and K. Wedemeyer. (1978). *Process for preparing thymol.* US4086283.

[29] D. Webb. (1982). *Process for the production of phenyl substituted aldehydes and of phenyl substituted alcohols.* EP0045571.

[30] I. Scriabine. (1962). *Alkylaromatic aldehydes and their preparation.* US3023247.

[31] J. M. Sanders and L. H. Michael. (1975). *Process for production of isochromans.* US3910964.

[32] J. Stofberg, B. Van Der Wal, and K. N. Nieuwland. (1963). *Procédé de préparation de parfums musqués.* FR1392804.

[33] G. J. Muus, H. Van Essen, and W. M. Hilversum. (1959). *4-Acyl-1,1-dimethyl-6-t-butyl indanes.* US2889367.

[34] S. Kagabu and Y. Kojima. (1992). A synthesis of indane musk Celestolide. *J. Chem. Educ.*, 69, 420–421.

[35] J. Blumenthal. (1973). *Acyl-cyclododecenes, acyl-cyclododecanes and a process for producing them.* US3754036.

[36] B. Cacciuttolo, S. Poulain-Martini, F. Fontaine-Vive, M. A. Hussein Abdo, H. El-Kashef, and E. Duñach. (2014). Access to polycyclic derivatives by triflate-catalyzed intramolecular hydroarylation. *Eur. J. Org. Chem.*, 2014, 7458–7468.

[37] V. Vece, J. Ricci, S. Poulain-Martini, P. Nava, Y. Carrissan, S. Humbel, and E. Duñach. (2010). InIII catalyzed tandem C-C and C-O bond formation between phenols and allylic acetates. *Eur. J. Org. Chem.*, 2010, 2639–2648.

[38] H. J. Prins. (1919). The condensation of formaldehyde with some unsaturated compounds. *Proc. Acad. Sci.*, 22, 51–56.

[39] F. Doro, N. Akeroyd, F. Schiet, and A. Narula. (2019). The Prins reaction in the fragrance industry: 100th anniversary (1919–2019). *Angew. Chem. Int. Ed.*, 58, 7174–7179.

[40] J. P. Bain. (1946). The Reaction of β-pinene with formaldehyde. *J. Am. Chem. Soc.*, 68, 638–641.

[41] A. T. Blomquist and R. J. Himics. (1968). Terpene formaldehyde reactions. II. D-Limonene. *J. Org. Chem.*, 33, 1156–1159.

[42] J. A. Bajgrowicz and I. Frank. (2007). *Oxygen containing tri- or tetracyclic terpenoid compounds*. WO2007030963.

[43] N. P. Solov'eva, T. M. Tsirkel, I. A. Terekhina, S. A. Voitkevich, R. I. Sharapova, and E. S. Kaboshina. (1966). *3-Alkyl-4-acetoxytetrahydropyrans*. SU202964.

[44] D. Munro. (1997). *Prins reaction products of 5,7-dimethyl-oct-6-en-2-one and their use as fragrance materials*. WO1999024386.

[45] T. Stork, M. Dehn, K. Ebel, K. Beck, A. Salden, and R. Pelzer. (2017). Production of 2-substituted 4-hydroxy-4-methyltetrahydropyrans having stable quality. US 20170037021.

[46] C. Tarchini and G. E. Carouge. (1986). Dérivés pyranniques, procédé pour leur préparation et leur utilisation pour la préparation d'alcools araliphatiques. CH655932A5.

[47] H. Watkins, O. C. Liu, and J. A. Krivda. (1993). *Use of tetrahydro-4-methyl-2-phenyl-2H-pyran as perfuming ingredient*. EP0581052.

[48] L. Liu, P. S. J. Kaib, A. Tap, and B. List. (2016). A general catalytic asymmetric Prins cyclization. *J. Am. Chem. Soc.*, 138, 10822–10825.

[49] J. Mane, C. Plessis, and J.-J. Chanot. (2009). *Novel pyran derivatives, their preparation and use thereof in perfumery*. WO2009130192.

[50] A. Macedo, E. P. Wendler, A. A. Dos Santos, J. Zukerman-Schpector, and E. R. T. Tiekink. (2010). Solvent-free catalysed synthesis of tetrahydropyran odorants: The role of SiO_2. p-TSA catalyst on the Prins-cyclization reaction. *J. Braz. Chem. Soc.*, 21, 1563–1571.

[51] B. B. Snider. (1980). Lewis-acid catalyzed ene reactions. *Acc. Chem. Res.*, 13, 426–432.

[52] C. Fehr, I. Magpantay, J. Arpagaus, X. Marquet, and M. Vuagnoux. (2009). Enantioselective synthesis of (−)-*β*-santalol by a copper-catalyzed enynol cyclization-fragmentation reaction. *Angew. Chem. Int. Ed.*, 48, 7221–7223.

[53] C. Tsangarakis and M. Stratakis. (2005). Biomimetic cyclization of small terpenoids promoted by zeolite nay: Tandem formation of α-ambrinol from geranyl acetone. *Adv. Synth. Catal.*, 347, 1280–1284.

[54] K. H. Schulte-Elte and G. Ohloff. (1975). Die durch Lewis-Säuren katalysierte Cyclisierung von Geranylaceton. *Helv. Chim. Acta*, 58, 18–27.

[55] S. A. Akutagawa. (1992). Practical synthesis of (−)-menthol with the Rh-BINAP catalyst. In *Chirality in Industry*, A. N. Collins, G. N. S. and J. C., Eds.; Wiley: Tokyo, 313–323.

[56] T. Iwata, Y. Okeda, and Y. Hori. (2002). *Process for producing isopulegol*. JP 1225163.

[57] G. Heydrich. (2018). *Method for producing optically active, racemic menthol*. US 9988331.

[58] V. K. Aggarwal, G. P. Vennall, P. N. Davey, and C. Newman. (1998). Scandium trifluoromethanesulfonate, an efficient catalyst for the intermolecular carbonyl-ene reaction and the intramolecular cyclisation of citronellal. *Tetrahedron Lett.*, 39, 1997–2000.

[59] J. A. Berson and M. Jones. (1964). A synthesis of ketones by the thermal isomerization of 3-hydroxy-1,5-hexadienes. The Oxy-Cope Rearrangement. *J. Am. Chem. Soc.*, 86, 5019–5020.

[60] N. Bluthe, M. Malacria, and J. Gore. (1983). Transposition d'oxy-Cope à temperature ambiante catalysée par le dichlorure de palladium-bisbenzonitrile. *Tetrahedron Lett.*, 24, 1157–1160.

[61] Y. R. Naves and P. Bachmann. (1943). Etudes sur les matières végétales. Contribution à la connaissance des ionones. *Helv. Chim. Acta*, 2151–2165.

[62] K. G. Fahlbusch. (2007). Flavors and Fragrances. *Ullmann's Encyclopedia of Industrial Chemistry* (7th Ed.), Wiley, 45–46.

[63] G. Frater and F. Schröder (2007). Cyclisation of 1,5-dienes: An efficient synthesis of α-Georgywood. *J. Org. Chem.*, 72, 1112–1120.

[64] F. Grau, A. Heumann, and E. Duñach. (2006). Cycloisomerisation of 1,6-dienes mediated by Lewis superacid without additives. *Angew. Chem. Int. Ed.*, 45, 7285–7289.

[65] J. Godeau, F. Grau, Antoniotti, and E. Duñach. (2014). Preparation and olfactory evaluation of mono- and bicyclic compounds featuring gem-dimethylcyclohexane structures. *Flavour Fragr. J.*, 29, 59–66.

[66] F. Grau, J. C. Bayon, P. A. Aguirre, T. Parella, and E. Duñach. (2008). New aldehydes by catalytic diene cycloisomerisations. *Eur. J. Org. Chem.*, 1214–1223.

[67] K. Surendra and E. J. Corey. (2009). A short enantioselective total synthesis of the fundamental pentacyclic triterpene Lupeol. *J. Am. Chem. Soc.*, 131, 13928–13929.

[68] J. Godeau, S. Olivero, S. Antoniotti, and E. Duñach. (2011). Biomimetic cationic polyannulation reaction catalyzed by Bi(OTf)3: Cyclisation of 1,6-dienes, 1,6,10-trienes and aryl polyenes. *Org. Lett.*, 13, 3320–3323.

[69] R. Laher, C. Marin, and V. Michelet. (2020). When gold meets perfumes: Synthesis and olfactive compounds via gold-catalyzed cycloisomerization reactions. *Org. Lett.*, 22, 4058–4062.

[70] C. Fehr, M. Vuagnoux, A. Buzas, J. Arpagaus, and H. Sommer. (2011). Gold- and copper-catalyzed cycloisomerizations towards the synthesis of thujopsanone-like compounds. *Chem. Eur. J.*, 17, 6214–6220.

[71] P. Ondet, A. Joffrin, I. Diaf, G. Lemière, and E. Duñach. (2015). Cycloisomerisation of allene-enol ethers under Bi(OTf)3 catalysis. *Org. Lett.*, 17, 1002–1005.

[72] P. Ondet, L. Lempenauer, E. Duñach, and G. Lemière. (2016). Bismuth (III) triflate-catalysed tandem cyclisations towards complex polycyclic ethers. *Org. Chem. Front.*, 3, 999–1003.

[73] P. Ondet, C. Plessis, G. Lemière, and E. Duñach. (2017). Synthesis and olfactory evaluation of spiro tricyclic diether structures. *Flavour Fragr. J.*, 32, 119–125.

[74] P. Ondet, J. J. Filippi, G. Lemière, and E. Duñach. (2018). Stereocontrolled cascade cyclisation of campholenic enol ether derivatives: En route to vetiver-scented spirocycles. *Eur. J. Org. Chem.*, 980–989.

[75] R. L. Snowden, J. C. Eichenberger, S. M. Linder, P. Sonnay, C. Vial, and K. H. Schulte-Elte. (1992). Internal nucleophilic termination of biomimetic acid mediated polyene cyclisations: Stereochemical and mechanistic implications. Synthesis of (±)-Ambrox and its diastereoisomers. *J. Org. Chem.*, 57, 955–960.

[76] S. Antoniotti, S. Poulain-Martini, and E. Duñach. (2010). Electrophilic functionalization of non-activated olefins catalyzed by Lewis superacids. *Synlett*, 20, 2973–2988.

[77] L. Coulombel and E. Duñach. (2004). Triflic acid-catalysed cyclisation of unsaturated alcohols. *Green Chem.*, 6, 499–500.

[78] I. Larrosa, P. Romea, and F. Urpi. (2008). Synthesis of six-membered oxygenated heterocycles through carbon-oxygen bond-forming reactions. *Tetrahedron*, 64, 2683–2723.

[79] K. Tani and Y. Katoaka. (2001). *Catalytic Heterofunctionalisation*. Eds. Togni, A. and Grützmacher, Wiley-VCH, Weinheim, DE.

[80] K. Ishihara, S. Nakamura, and H. Yamamoto. (1999). The first enantioselective biomimetic cyclization of polyprenoids. *J. Am. Chem. Soc.*, 121, 4906–4907.

[81] K. Ishihara, H. Ishibashi, and H. Yamamoto. (2001). Enantioselective biomimetic cyclization of homo(polyprenyl)arenes. A new entry to (+)-podpcarpa-8,11,13-triene diterpenoids and (−)-tetracyclic polyprenoid of sedimentary origin. *J. Am. Chem. Soc.*, 123, 1505–1506.

[82] M. Harada, H. Matsuda, K. Ishida, and Y. Yamazaki. (2006). *Process for producing optically active theaspirane*. PCT/UP2006/317876.

[83] S. A. Abagale, C.W. Woodcock, A. M. Hooper, J. C. Caulfield, D. Withall, K. Chamberlain, S. O. Acquaah, H. Van Emden, H. Braimah, J. A. Pickett and M. A. Birkett. (2018). (2R,5S)-Theaspirane identified as the Kairomone for the Banana Weevil, Cosmopolites sordidus, from attractive senesced leaves of the host banana, Musa s. *Chem. Eur. J.*, 24, 9217–9219.

[84] P. Kraft, K. Popaj, and A. Abate. (2005). Design, synthesis and olfactory properties of 2-substituted 2-tert-butyl-5-methyl-2,5-dihydrofurans: Seco-derivatives of theaspiranes. *Synthesis*, 16, 2798–2809.

[85] H. Mouhib, W. Stahl, M. Lüthy, M. Büchel, and P. Kraft. (2011). Cassis odor through microwave eyes: Olfactory properties and gas-phase structures of all the cassyrane stereoisomers and its dihydro derivatives. *Angew. Chem. Int. Ed.*, 50, 5576–5580.

[86] C. F. Seidel and M. Stoll. (1959). Investigation of rose oxide. *Helv. Chim. Acta*, 42, 1830–1844.

[87] T. Yamamoto, H. Matsuda, Y. Utsumi, and T. Kanisawa. (2002). Synthesis and odor of optically active rose oxide. *Tetrahedron Lett.*, 43, 9077–9080.

[88] R. L. Snowden, S. M. Linder, B. L. Muller, and K. H. Schulte-Elte. (1987). *β*-Cleavage of bis(homoallylic) potassium alkoxides. Preparation of 3-hydroxypropyl and 4-hydroxybutyl propenyl ketones from γ- and δ-lactones. Synthesis of (±)-rose oxide. *Helv. Chim. Acta*, 70, 1879–1885.

[89] P. Audin, A. Doutheau, and J. Gore. (1984). Silver nitrate and mercuric salt mediated cyclisation of α-allenic alcohols. *Bull. Soc. Chim. Fr.*, 7-8, 297–306.

[90] H. Matsuda and T. Yamamoto. (1997). *Preparation of rose oxide with high content of trans-isomer*. JP 09071579.

[91] L. Coulombel, M. Weiwer, and E. Duñach. (2009). Aluminium triflate catalysed cyclisation of unsaturated alcohols: Novel synthesis of rose oxide and analogues. *Eur. J. Org. Chem.*, 5788–5795.

[92] E. Brenna, C. Fuganti, S. Ronzani, and S. Serra. (2002). Biocatalyzed preparation of optically enriched stereoisomers of 4-methyl-2-phenyl-tetrahydro-2H-pyran (Doremox®). *Can. J. Chem.*, 80, 714–723.

[93] L. Coulombel, M. Rajzmann, J. M. Pons, S. Olivero, and E. Duñach. (2006). Aluminum trifluoromethanesulfonate as an efficient catalyst for the intramolecular hydroalkoxylation of inactivated olefins: Experimental and theoretical approaches. *Chem. Eur. J.*, 12, 6356–6365.

[94] A. Abate, E. Brenna, G. Fronza, C. Fuganti, F. Gatti, S. Serra and E. Zardoni. (2004). Preparation of enantiomerically enriched isomers of odorous cyclic ethers. *Helv. Chim. Acta*, 87, 765–780.

[95] S. Murahashi, Y. Makabe, and K. Kunita. (1988). Palladium (0) catalyzed rearrangement of N-allyl enamines. *J. Org. Chem.*, 53, 4489–4495.

[96] A. J. A. Van der Weerdt, R, Plomp, and W. Apeldoorn. (1981). *Perfume compositions containing phenyltetrahydrofurans as a fragrance*. EP 49543 21-9.

[97] A. Mosandl and C. Günther. (1989). Stereoisomeric flavor compounds. Structure and properties of γ-lactones. *J. Agric. Food Chem.*, 37, 413–418.

[98] L. Coulombel and E. Duñach. (2005). Cycloisomerisation of olefinic carboxylic acids catalysed by trifluoromethanesulfonic acid. *Synth. Commun.*, 35, 153–160.

[99] J. M. E. Mane, J. C. Clinet, I. Clinet, L. Colombel, and C. Marin. (2007). *Process for the preparation of 5,5-dimethyl-3-ethyl-3,4-dihydrofuran-2-one, its olfactive characteristics and perfume compositions containing it.* FR2007-928 2906531.

[100] E. Lemière and E. Duñach. (2013). Catalytic activation of olefins by metal triflates and triflimides: Applications to fragrance chemistry. *Chem. Eur. J.*, 19, 3270–3280.

[101] S. B. Jameson. (2005). Aroma Chemicals III: Sulfur compounds in *Chemistry and Technology of Flavors and Fragrances*. Rowe, D. J. Ed. Blackwell Publ., Oxford.

[102] M. Weiwer, L. Coulombel, and E. Duñach. (2006). Regioselective indium trifluoromethanesulfonate-catalyzed hydrothiolation of non-activated olefins. *Chem. Commun.*, 2007, 332–334.

[103] M. Weiwer, X. Chaminade, J. C. Bayon, and E. Duñach. (2007). Indium trifluoromethanesulfonate-catalyzed addition of thio compounds to camphene: A novel route to 2,3,3-trimethyl-2-thiobicyclo[2.2.1]heptane derivatives. *Eur. J. Org. Chem.*, 2464–2469.

[104] J. P. Simonato. (2005). *L'Actualité Chimique*, 292, 55–56.

[105] B. Clapham, T. S. Reger, and K. D. Janda. (2001). Polymer-supported catalysis in synthetic organic chemistry. *Tetrahedron*, 57, 4637–4662.

[106] H. Yamamoto. (2000). *Lewis Acids in Organic Synthesis*, Vol. 1&2, Wiley-VCH, Weinheim.

[107] P. T. Anastas and J. C. Warner (1998). *Green Chemistry: Theory and Practice*, Oxford University Press, New York.

https://doi.org/10.1142/9781800617780_0009

Chapter 9

Metal-Catalysed Synthesis of Fragrances

Romain Laher[*,‡], Christophe Marin[†,§] and Véronique Michelet[*,¶]

[*]*Université Cote d'Azur, Institut de Chimie de Nice, Nice, France*

[†]*Expressions Parfumées, Grasse, France*

[‡]*romain.laher@unice.fr*
[§]*c.marin@epparfums.com*
[¶]*veronique.michelet@univ-cotedazur.fr*

Abstract

Historically, the fragrance industry has based its discovery of the extraction and valorisation of plants and flowers. In some cases, due to the high cost and low yield of extractions, synthetic chemists have entered the game and proposed some solutions to the fragrance industry. Among all the potential avenues of molecular chemistry, this chapter focuses on the use of transition metals, allowing access to challenging molecules with organoleptic properties. Some selected examples are proposed, including cycloadditions, reductions (C ═ C or C ═ O), syntheses of carbonyl compounds (oxidation, dehydrogenation, allylic oxidation and isomerisation), carbonylation reactions and metathesis reactions.

9.1 Introduction

Historically, the fragrance industry has based its discovery of the extraction and valorisation of plants and flowers. During the Renaissance, more particularly in the region of Grasse, perfumes entered a new era thanks to the appearance of the first synthetic perfume ingredients. The application of industrial organic chemistry has led to a significant reduction in production costs, hitherto influenced by the price of raw materials and expensive extraction processes, such as enfleurage [1].

Among the first scent molecules that appeared is 2-phenylethanol (**1**) (Figure 9.1), whose isolation from rose extract has allowed its use in floral accords, which brings a note of rose, lily of the valley, jasmine and pepper

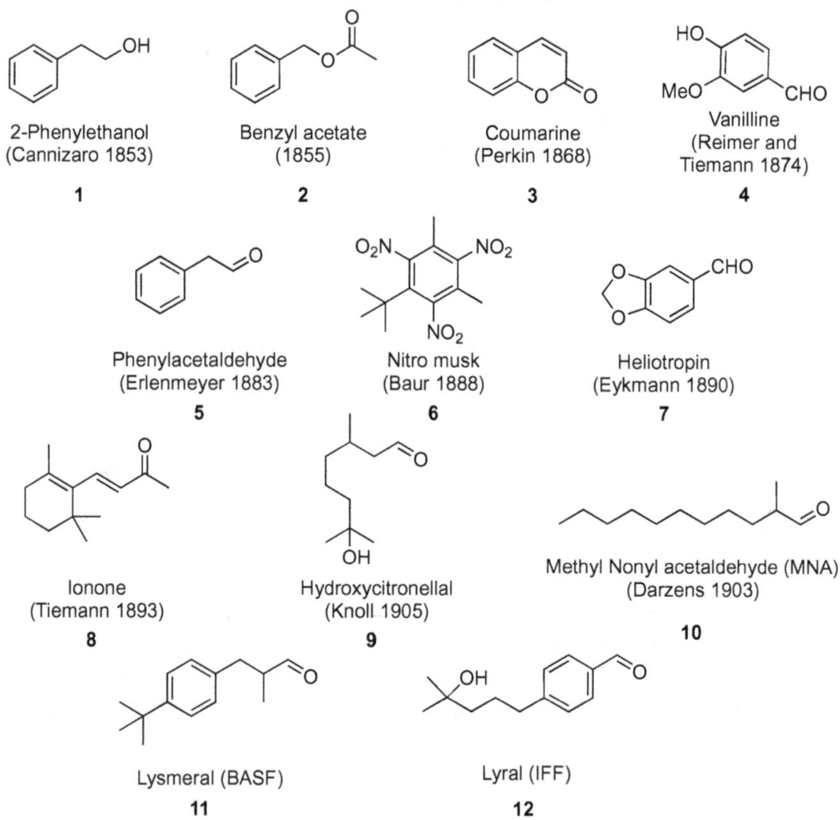

Figure 9.1. First synthetic derivatives in modern perfumery.

to the compositions. Benzyl acetate (**2**), synthesised in 1855, takes on a scent of jasmine. These two compounds appeared in 1921 in the famous "Chanel N°5" perfume. Until then extracted from tonka beans, synthetic coumarin (**3**) arrived on the market after the elucidation of its structure by Vogel in 1820, followed by its synthesis by Perkin in 1868 [2]. This period is considered to be the beginning of modern perfumery, with the use of synthesis to replace extraction. A few years later, Paul Parquet used this molecule in his Fougère Royale fragrance (1882), thus introducing to perfumery the scent family called "Fougère" in honour of its fragrance.

Synthesised from isoeugenol in 1874 by Wilhelm Haarmann and Ferdinand Tiemann, synthetic vanillin (**4**) also finds its place in the creation of Guerlain [3]. Guerlain combines it with coumarin (**3**) as well as heliotropin (**7**) for its perfume Jicky (1889). Heliotropin (**7**), whose name comes from the heliotrope flower (*Heliotropium europaeum*) from which it derives its scent of almond and vanilla, brings new notes to the catalogue of perfumers hitherto unavailable with natural molecules [3]. Nitro musk (**6**), discovered by Baur during the synthesis of TNT in 1888, was used until the 1950s in perfumes for its musky smell and very low production cost [4]. The smell of lily of the valley is a particularly sought-after odour, but it is impossible to obtain by extraction — the distillation of its bells not yielding essential oil. The synthesis of hydroxycitronellal (**9**) allowed floral accords of white flowers. Robert Bienaimé used it in 1912 to give a lily-of-the-valley facet to his floral composition, *Quelques Fleurs* (Houbigant). The use of methyl nonyl acetaldehyde (MNA) (**10**) in the composition brings amber, aldehyde, citrus and tuberose notes. MNA is still widely used in perfumery today. Ernest Beaux, in 1921, used nitrated musk (**6**) together with phenylacetaldehyde (**5**), vanillin (**4**), coumarin (**3**), β-ionone (**8**) and aldehyde C12 MNA (**10**) to prepare a perfume sample for Gabrielle "Coco" Chanel on May 5, which would later be chosen as n°5. Chanel insisted on naming her perfume after the number that she considered lucky, which is how the famous No. 5 was born. Beaux used in this composition a high concentration of linear aldehydes C10–C12, which, considered separately, bring unpleasant fatty notes but, as part of his daring mixture, led to the success for which he is known today. In 1952, Beaux designated synthetic chemistry as the future of perfumery: "It is on chemists that we will have to rely on to find new bodies from which original notes could emerge. Yes, for perfume, the future is above all in the hands of chemicals [5]." It is in this context that organic chemistry finds its place in the elegant world of perfumery. Almost

150 years after the first synthetic odorous molecule, research in perfumery shows no signs of slowing down. The major players in the industry devote a significant portion of their turnover to research. For example, Firmenich injected 10% of its turnover into research in 2018, while Givaudan invested 8.6% in the same year. The motivations for this research tend to differ from those that drove the research for original notes evoked by Beaux. Much research today aims to anticipate and replace compounds that are or will be banned due to health and/or environmental risks. The latest amendment published by the International Fragrance Association (IFRA, created in 1973) was issued in January 2020 and notably banned the use of Lysméral® (**11**) for lipstick and oral applications. In 2017, the European Commission banned the use of Lyral (**12**) due to its allergenic properties [6]. These two compounds, used since the 1960s, have a particularly interesting note of lily of the valley (Figure 9.1).

In 1987, the United Nations defined the concept of "sustainable development" in the Brundtland report. It lays the foundations for what is described today as a mode of development that meets the needs of present generations without compromising the ability of future generations to meet theirs [7]. This new method of development responds to both environmental and social issues while including the economic aspect. In the early 1990s, the chemical industry saw the birth of new means of evaluating the efficiency of its processes in relation to this context of sustainable development. Discoveries in chemistry have long been catalysed by the synthesis of novel molecular structures and the optimisation of their processes. Research in chemistry has thus made it possible to drastically reduce the number of steps in many syntheses and to improve their performance, thus allowing substantial savings for the industries in the sector [8]. In 1991, Trost added to the criteria of chemical selectivity, regio-, diastereo- and enantioselectivities, a new term for evaluating the efficiency of a chemical reaction: atom economy [9]. This term is defined as being the proportion by mass of the desired compound to the sum of the compounds obtained. Sheldon supplemented these terms by adding a concept that calculates the impact of a chemical process on the environment: the "E-Factor". The latter term calculates the share of waste generated (except water) relative to the mass of the desired product synthesised [10]. Transition metals have particularly found a place of choice in the field of catalysis. The different oxidation states which characterise them, with regard to the incomplete filling of their d atomic orbitals, make them highly reactive from a catalytic point of view. Transition metals are

heptan-2-one Pd catalysis | pentan-3-one Co catalysis | butane-1,4-diol Rh catalysis | propionic acid Ni catalysis

Figure 9.2. Examples of classic compounds produced industrially.

present in many catalytic processes developed on an industrial scale, serving as starting materials in the fragrance industry (Figure 9.2) [11].

The use of transition metals for the synthesis of compounds with interesting organoleptic properties is relatively well established and is available via different types of reactions, which we discuss in the following. In addition, the development of chiral ligands has made it possible to use them for the synthesis of enantiopure compounds, in particular high-added-value derivatives. Recently, Kraft and co-workers have reviewed the impact of the transition from today's resource-wasting petrochemical economy towards a fully renewable and biodegradable future in the fragrance sector [12].

In this chapter, we discuss the growing interest in the use of transition metals for the development of syntheses, allowing access to molecules with interesting organoleptic properties. We concentrate on cycloadditions, reductions (C = C or C = O), syntheses of carbonyl compounds (oxidation, dehydrogenation, allylic oxidation and isomerisation), carbonylation reactions and metathesis reactions. Rearrangement and cycloisomerisation reactions have been reviewed recently [13] and will not be developed in this chapter.

9.2 Diels–Alder [4 + 2] Cycloaddition Reaction

The Diels–Alder reactions are not widely known to be catalysed by transition metals; they prefer the use of poor metals such as zinc, aluminium or boron, which prove to be less expensive [14]. Nevertheless, some examples exist via titanium catalysis. The synthesis of Presantone (**15**) could thus be carried out with the use of $TiCl_4$ as a catalyst for the Diels–Alder reaction between (*E*)-3-methyl-but-3-en-2-one (**13**) and cyclopentadiene (**14**) (Figure 9.3) [15]. This compound of interest serves as a key intermediate in the synthesis of compounds with interesting olfactory properties such as Legnox® (**16**). This patent developed by the Mane company

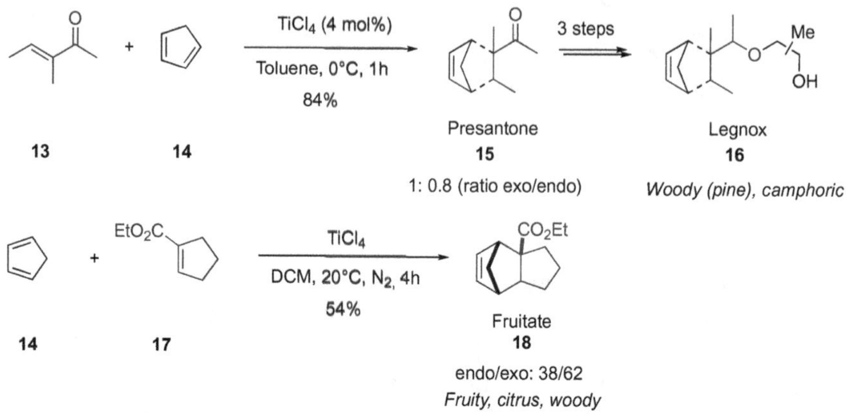

Figure 9.3. Synthesis of Presantone and Fruitate via a Diels–Alder reaction.

briefly mentions the use of methyl rhenium trioxide ($MeReO_3$) for the synthesis of Presantone on an industrial scale [16].

Another example of a Diels–Alder reaction concerns the synthesis of Fruitate® (**18**), a compound with fruity and citrus notes, developed by the company KAO (Figure 9.3). The cycloaddition between the cyclopentadiene (**14**) and the dienophile (**17**) in the presence of $TiCl_4$ is carried out at room temperature under a nitrogen atmosphere over four hours, leading to the tricyclic compound (**18**) with the exocyclic isomer as the main compound [17].

Ionone derivatives such as Iso E Super (**19**), Georgywood® (**20**) or Iso E Super Plus® (**21**) are important compounds when it comes to providing a woody note (Figure 9.4). These compounds can be found as a mixture after a Diels–Alder reaction between myrcene and 3-methyl-3-penten-2-one (**13**), followed by cyclisation in acidic condition [18]. The study of the olfactory thresholds made it possible to identify Georgywood® (**20**) and Iso E Super Plus® (**21**) as being the most powerful with very low perception thresholds of 5 and 30 pg/L, respectively.

While the method described above can lead to a mixture of several regioisomers, the [5 + 2] cycloaddition method using Wilkinson's catalyst developed by Wender [19] made it possible to carry out the [4 + 2] cycloaddition reaction (Figure 9.4) regioselectively from alkenes comprising an α-cyclopropyl unit (**23**). This unprecedented reaction allowed the synthesis of analogues of the compounds mentioned above, such as (**24**) from cyclooctenone [20]. The mechanism was described by Kraft as following

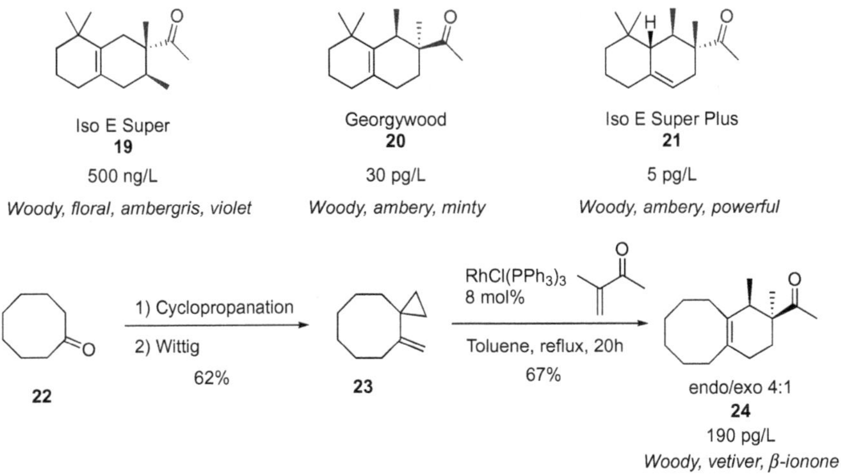

Figure 9.4. Ionone derivatives and Georgywood® analogue synthesis.

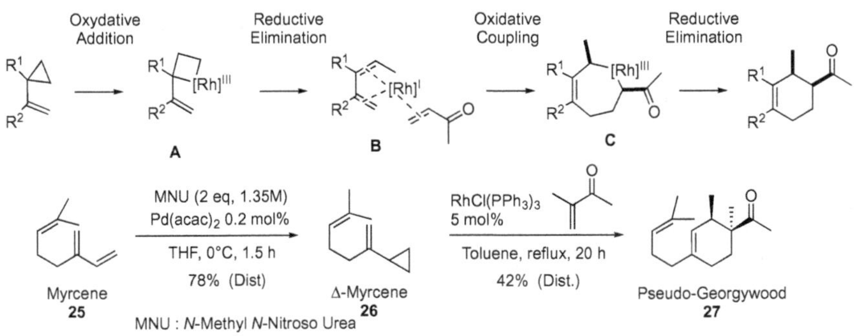

Figure 9.5. Proposed mechanism and application to the synthesis of pseudo-Georgywood.

an oxidative addition of vinyl cyclopropane to form the four-membered metallacycle [**A**] (Figure 9.5) [20, 36]. The reductive elimination of the latter leads to the formation of the diene [**B**], which, by oxidative coupling with an α, β-unsaturated ketone, generates the seven-membered metallacycle [**C**]. The latter intermediate leads to the ketone resulting from the global cycloaddition [4 + 2]. Although the intermediate **C** is not explained by Kraft, Wender's mechanistic hypotheses assuming the passage through a seven-membered metallacycle can explain the regioselectivity obtained [21]. It seems likely that a similar mechanism is responsible for the observed regioselectivity.

This method has been widely used for the synthesis of many analogues of Georgywood® (**20**) [22], and the synthesis of a precursor of this compound was the subject of a patent in 2016 by Givaudan, which reports the use of a method involving palladium-catalysed cyclopropanation of myrcene (**25**). The Δ-myrcene (**26**) thus formed reacts with the conjugated ketone in the presence of the Wilkinson catalyst complex to lead to (**27**) (Figure 9.5) [23]. Another widely used reaction for the fragrance industry is the reduction of unsaturated derivatives such as alkenes, alkynes and carbonyls.

9.3 Reduction of Triple and Double Bonds

Transition metals can be used to effect the reduction of C–C double and triple bonds (Figure 9.6). We find in particular the use of metals such as palladium, platinum and nickel but also cobalt, copper, iron and rhenium. This relatively standard reaction is widely used in perfumery to provide a larger panel of interesting olfactory compounds. For example, the heterogeneous reduction of 8-cyclohexadecenone (**28**) has been described by the company Symrise AG. This synthesis makes it possible to obtain cyclohexadecanone (**29**), a musky compound that is a minor constituent of civet absolute, marketed under the name Isomuscone® [24].

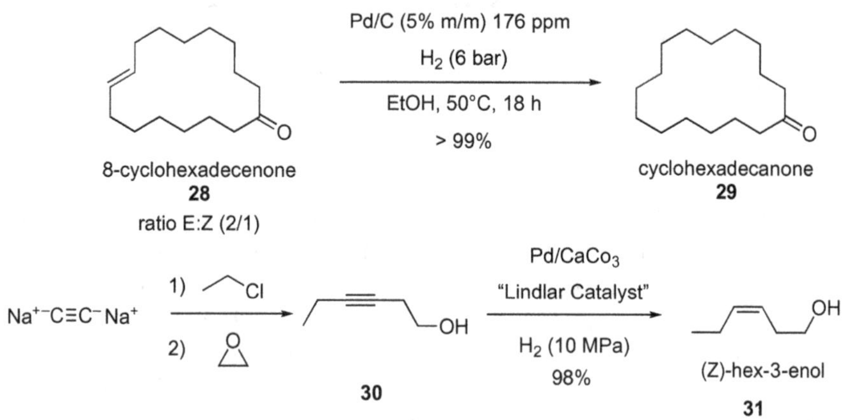

Figure 9.6. Synthesis of cyclohexadecanone and (*Z*)-hex-3-enol.

Triple-bond hydrogenation has been shown to be effective in the synthesis of (*Z*)-hex-3-enol (cis-3-hexenol) (**31**), a compound with the characteristic green notes of cut grass [24a]. A historical synthesis of this compound by the Bedoukian company is indeed based on the partial hydrogenation of the triple bond of the alcohol hex-3-yn-1-ol (**30**), the latter being obtained by monoalkylation of acetylide by ethyl chloride and its alkylation by ethylene oxide [25].

The ruthenium catalyst developed by Chaudret's team has been used for the regio- and stereoselective 1,4-hydrogenation of C–C double bonds in the synthesis of the same compound (**31**) (Figure 9.7) [26]. This synthesis has been patented from sorbitol in order to reduce the synthesis costs of (**31**), which is very present in perfumery for its green note. Sorbitol was obtained after esterification of sorbic acid into the corresponding ester (**32**), followed by reduction *via* the use of polymethoxyhydroxysilane (PMHS, low cost and non-toxic) in the presence of a catalytic amount of Zn (2-ethyl-hexanoate)$_2$ (Zn (2-EH)$_2$) in diisopropylether. In the same year, Mimoun's team reported, along with the reduction of esters, the enantioselective reduction of ketones to alcohols through the use of a modification of this catalytic system [27]. This same reduction of ester to sorbitol (**33**) was subsequently patented by the same authors for the use of a ruthenium complex [28]. The 1,4-hydrogenation of sorbol (**33**), reported by Dupau's team, turns out to be very effective here, with a selectivity greater than 99%, a yield of over 90% and a *cis*/*trans* ratio of 98/2. Dupau notes the importance of the presence of an acid catalyst to

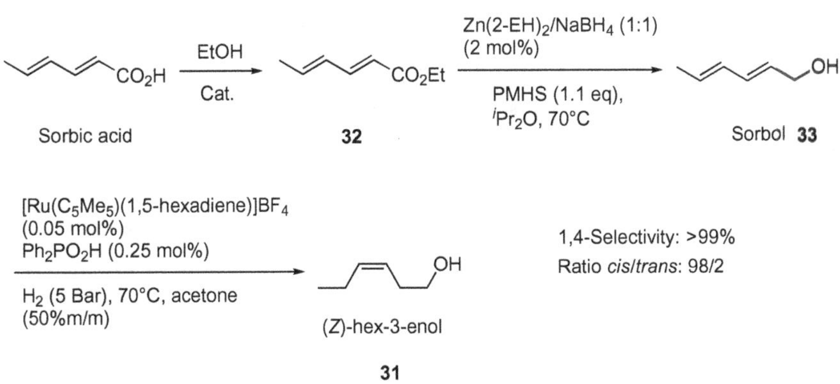

Figure 9.7. Synthesis of (*Z*)-hex-3-enol.

improve yield [29]. The second patent highlights the benefit of using acyclic dienes in improving 1,4-selectivity [30].

The reduction of double bonds in the heterogeneous phase is quite practical in its implementation, as the catalyst can be separated from the reaction mixture by filtration. However, it is not very effective in the context of enantioselective reactions. The development of a homogeneous phase catalytic system brings the possibility of controlling the stereochemistry induced during hydrogenation. The Takasago company has developed a process allowing the enantioselective hydrogenation of geraniol (**34**) into citronellol (**35**) (Figure 9.8), a compound widely used for its rosy floral notes [24a]. The reaction is carried out under mild temperature conditions at a pressure of 50 bar.

The chiral centre is formed there with an enantiomeric excess of 99%, thanks to a complex of [Ru]-(S)-BINAP. The catalytic system is very efficient with a TON of 50,000 and a TOF of 4,200 h^{-1}, allowing the company to produce 300 tons annually through this process [31]. The *E* isomer of geraniol, nerol, is hydrogenated to (*R*)-citronellol (**35**) via the use of the [Ru]-(*R*)-BINAP system.

Another elegant example of enantioselective alkene reduction was developed by the Firmenich company within the framework of the synthesis of methyl dihydrojasmonate or Hedione® (**36**), a molecule of great interest in perfumery, which constituted considerable analysis and synthesis work that started in 1957 by Demole's group (Figure 9.9) [32]. This compound present in jasmine is adorned with fruity notes and white jasmine flowers. It is important to note that the term "hedione" comes from the Greek "*Hedone*", which means "pleasure".

Further research has identified the (+)-methyl-*cis*-dihydrojasmonate isomer, Paradisone® (**37**), as the most interesting stereoisomer, with a lower perception threshold than the other three analogues (Table 9.1) [33].

The first synthesis of the diastereoisomer enriched in the most preferred *cis* (98/2) isomer was carried out in 1997 by Fehr (Firmenich)

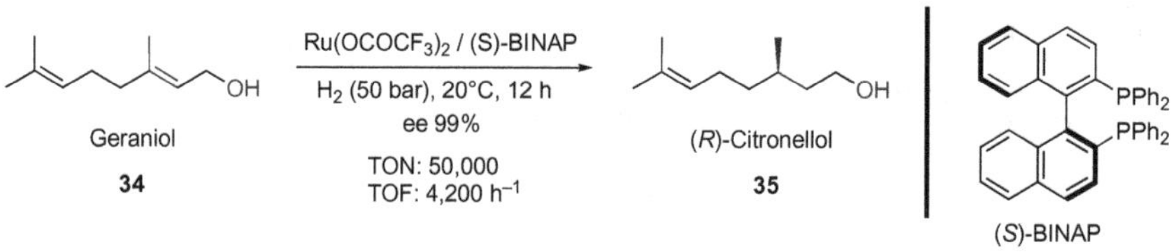

Figure 9.8. Synthesis of (*R*)-citronellol by enantioselective hydrogenation.

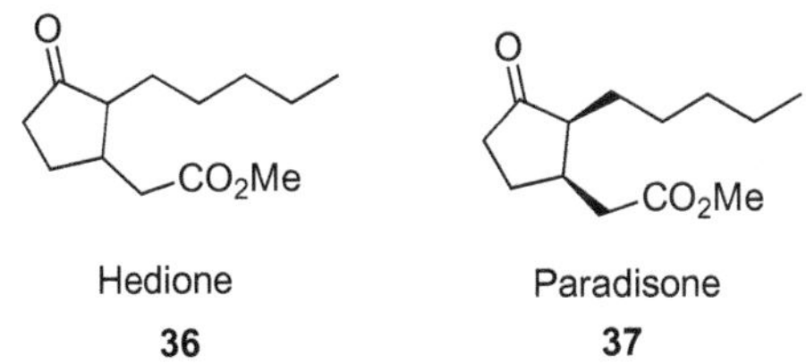

Figure 9.9. Structure of Hedione® and Paradisone®.

Table 9.1. Threshold of perception of Hedione isomers.

	Ratio cis/trans	Perception threshold (ng L^{-1})
Hedione® (Firmenich)	10:90	0.280
Kharismal® (IFF)	60:40	0.046
Hedione® HC (Firmenich)	75:25	0.037
Paradisone® (Firmenich)	94:6	0.015

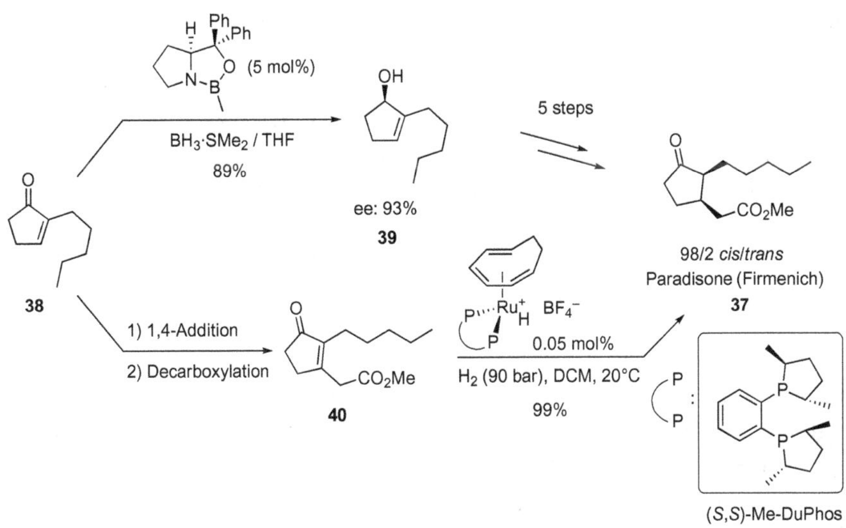

Figure 9.10. Synthesis of Paradisone®.

(Figure 9.10). The process, patented for its industrial application [34], consists of the enantioselective reduction of the *α*,*β*-unsaturated ketone (**38**) by the Corey–Bakshi–Shibata catalyst in the presence of borane in THF, resulting in alcohol (**39**) with an *ee* of 93%. It is followed by five steps: acylation, Claisen–Ireland rearrangement, epoxidation and hydride migration [35].

This elegant synthesis was superseded by the enantioselective reduction of the cyclic α, β-unsaturated ketone carried out in collaboration with Genêt's team [36]. Industrial production began in 1998 on this synthesis to produce Paradisone® at the scale of several tons per year [37]. The inexpensive cyclopentenone (**38**), obtained through the condensation of cyclopentanone with pentanal, followed by isomerisation or a Pauson–Khand reaction [38], allows the ester (**40**) to be obtained in two stages via a 1,4-addition of dimethyl malonate, followed by decarboxylation. The chemoselective and enantioselective hydrogenation of the double bond with Me-DuPhos allows the production of Paradisone® (**37**), with a yield of 99% for a *cis*/*trans* ratio of 98/2 [39]. It is also interesting to note that this last example significantly improves atom economy and lowers the E-factor compared to the version described previously.

9.4 Reduction of Carbonyl Derivatives: Synthesis of Alcohols

The reduction of the C=O bonds, mentioned above, was carried out industrially via the use of organic hydride donors, such as PMHS for ketones and aldehydes [27], or with metal hydrides, such as $LiAlH_4$ or DIBAL-H in a solution for esters [40c]. The major drawback of this type of reaction is the generation of salts, which should be neutralised or extracted at the end of the industrial process. Replacing the reducing agent with hydrogen drastically reduces the difficulties associated with these processes. The use of H_2 and a metal catalyst to replace traditional hydrides allows a marked improvement from an environmental point of view. Saudan estimates a reduction in the E-factor of the order of 300 [38] and accident risk, while improving the selectivity and efficiency of this type of reaction [41]. α-Campholenic aldehyde (**42**), derived from α-pinene (**41**), allows the synthesis by aldol condensation reaction and then reduction of alcohols (**43**), which find their place in perfumery for their woody note of sandalwood (Figure 9.11) [42].

Different structural analogues (**43a–g**) have been proposed by the industry [25, 43]. The reduction of the aldehydes and ketones functions could be carried out chemoselectively without affecting the alkene of the cyclopentene but above all without carrying out the reduction of the methylidene of Firsantol® (**43d**), for example. Furthermore, no isomerisation was observed on the stereogenic carbon. This selectivity is of great importance because the reduction of the double bond of cyclopentene results in the loss of sandalwood odour [44].

Inspired by Noyori's work on the reduction of aldehydes and ketones in homogeneous phase via the use of a ruthenium complex ([$RuCl_2$(diphosphine)(diamine)]) [45], different groups have developed their own chiral catalysts to apply them in their industrial process and thus prepare their own ingredients. This is notably the case for Firmenich, Takasago, DSM and Enantiotech (Figure 9.12) [46].

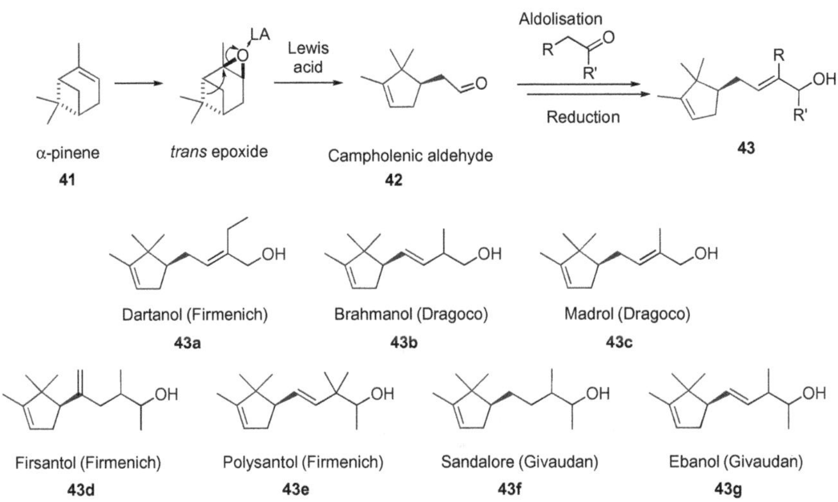

Figure 9.11. Synthesis of campholenic aldehyde derivatives.

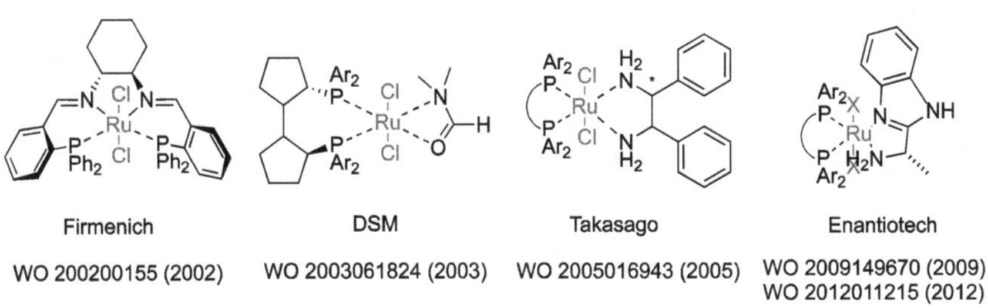

Figure 9.12. Ruthenium catalysts for the hydrogenation of carbonyl derivatives.

The synthesis of Dartanol® (**43a**) was carried out by the hydrogenation of the corresponding aldehyde (**44**) *via* the use of the cationic complex [(DPPAE)$_2$(PPh$_3$) RuCl][Cl] in the presence of a basic co-catalyst with a yield of 90% (Figure 9.13) [42].

The operating conditions were revised, allowing the improvement of the conversion (TON = 48,000) by carrying out the reaction in the presence of the complex [(en)(Xantphos)Ru(OCO^{t}Bu)$_2$] and a carboxylic acid, such as benzoic acid, in a catalytic amount [47]. The above conditions were used for the synthesis of (**43e**) from the corresponding ketone (**45**) *via* the use of the complex [(DPPAE)$_2$(PPh$_3$)RuCl][Cl]. One could note the extremely low catalytic load (S/C: 200,000), allowing the use of such a process on an industrial scale [48].

Finally, the selective hydrogenation of the prochiral face *Re* could be carried out in the presence of the complex ([RuCl$_2$(disilane)(diamine)]).

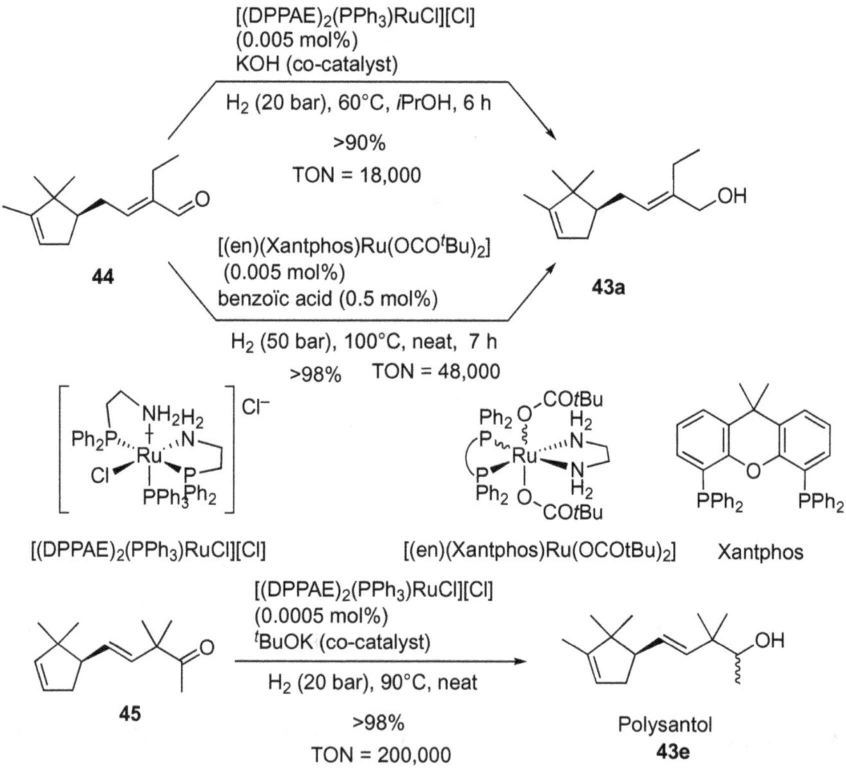

Figure 9.13. Synthesis of Dartanol® and Polysantol®.

This reaction allows the enantioselective reduction of methylated ketones (**46**) under mild conditions of temperature and at a pressure of 50 bar of hydrogen in the presence of a catalytic amount of potassium hydroxide (Figure 9.14). The Santoro group notes here the importance of the steric hindrance provided by the *t*-butyl groups carried not only by sulphur but also by aromatic groups [49].

In the same way as for aldehyde or ketone, the use of transition metals makes it possible in the reduction of esters to improve the E-factor *via* the application of hydrogenation in the homogeneous phase. The pioneering work of Milstein paved the way for the hydrogenation of esters by the application of homogeneous phase catalysis [50]. This work also allowed the replacement of much more drastic conditions for the reduction of esters with the use of heterogeneous phase catalysts. These conditions industrially required high temperatures (200–300°C) and extremely high pressures (200–300 bar) [51]. The use of a ruthenium complex has been described and developed on an industrial scale by Firmenich [38], Givaudan [52] and Takasago [53] (Figure 9.15), leading to alcohols (**48**). Nevertheless, ruthenium remains an expensive metal which should be replaced, if possible, by another metal that can also carry out this reaction, while retaining the same chemoselectivity and reactivity. Recent work by Milstein has shed light on the use of various inexpensive metals, such as cobalt and manganese as well as iron [54]. These advances have enabled Saudan's group to perform ester (**47**) reduction, leading to various compounds of interest [55], such as benzyl alcohol (**48a**), used as a solvent, or 2-phenyl-ethanol (**1**), used in large quantity for its notes of rose and honey (Figure 9.15).

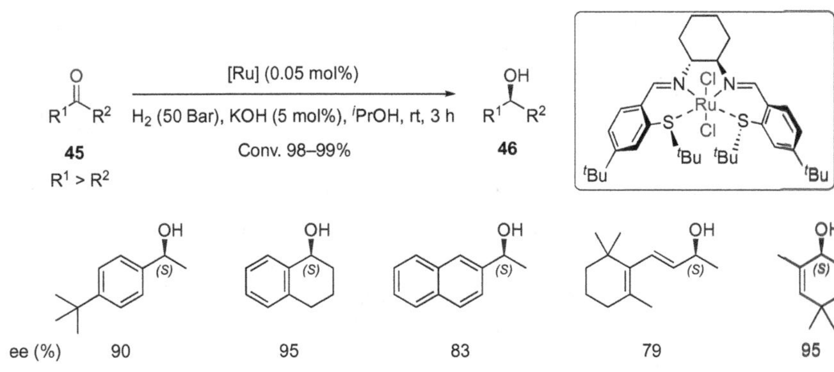

Figure 9.14. Enantioselective hydrogenation of methylated ketones.

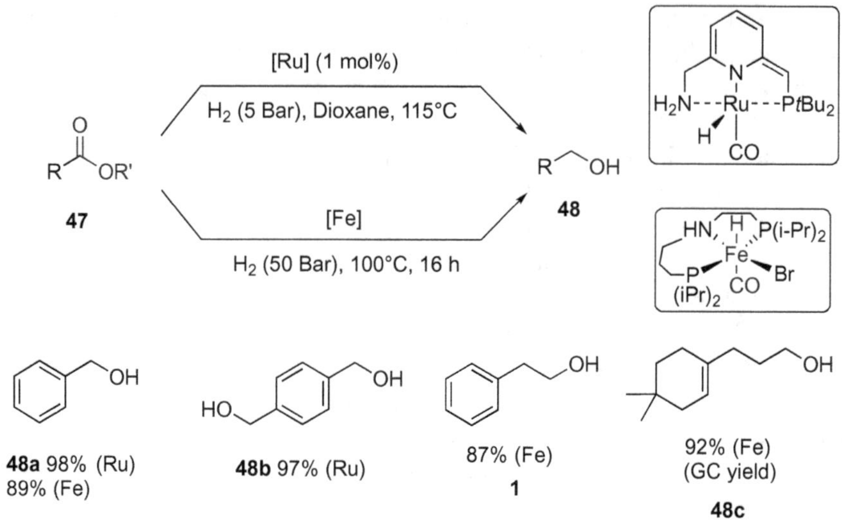

Figure 9.15. Catalytic ester hydrogenation by the ruthenium complex.

9.5 Oxidation and Allylic Isomerisation: Synthesis of Aldehyde and Ketones

Organometallic chemistry has also found applications in fragrances in the synthesis of aldehydes and ketones. Traditional stoichiometric oxidants ($KMnO_4$, $K_2Cr_2O_7$, MnO_2 and CrO_3) tend to disappear from industrial oxidation processes both for reasons of safety and atomic economy and also for the difficulties they present during processing. Molecular oxygen is the easiest and cheapest oxidant available, but other oxidants such as hydrogen peroxide or peracetic acids can still be used. Metal complexes composed of ruthenium, molybdenum, tungsten, rhenium or other transition metals are used to carry out epoxidations, oxidative cleavages, the Baeyer–Villiger reaction, the oxidation of alkanes, dehydrogenations and allylic oxidations leading to carbonyl compounds of interest for perfumery [56]. Ruthenium can be used at the laboratory scale as a strong oxidant. Ruthenium(III) chloride in the presence of sodium periodate ($NaIO_4$) is oxidised in the form of RuO_4. This compound was used by the Kraft group for the synthesis of analogues of Calone 1951® (**49**) *via* the oxidation of the methylidene function (**51**) synthesised from veratrole (**50**) in four steps (Figure 9.16). The obtained analogue (**52**) has olfactory properties close to (**49**) [57]. Kraft is interested in this study in the structure/activity aspect linking benzo[b][1,4]dioxepinones to the characteristic marine odour of Calone 1951® developed by Pfizer [58].

The disadvantage that this type of oxidation can bring lies not only in the cost of the oxidising reagent (used stoichiometrically) but also in the use of tetrachloromethane, which is prohibited for industrial use. The synthesis of citral (**56**), an important intermediate in the synthesis of other compounds for perfumery, such as pseudo-ionone (**57**), also involves a key oxidation step (Figure 9.17). The industrial process for the synthesis of citral developed by BASF is based on the vapour phase dehydrogenation of 3-methyl-but-3-en-1-ol (**53**), leading to 3-methyl-but-2-enal (**54**) after oxidation. This reaction takes place in the presence of a silver catalyst supported on

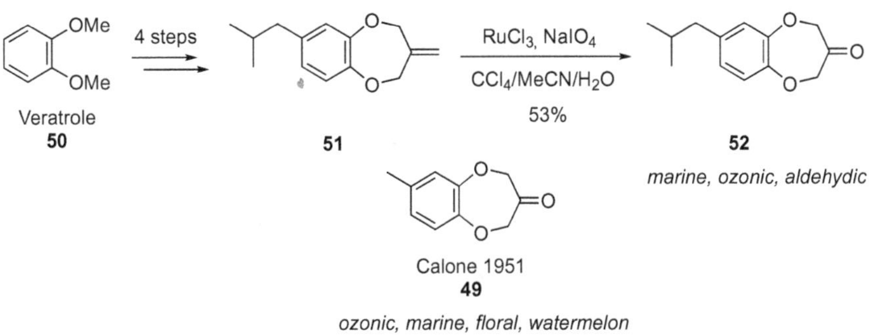

Figure 9.16. Synthesis of structural analogues of Calone 1951®.

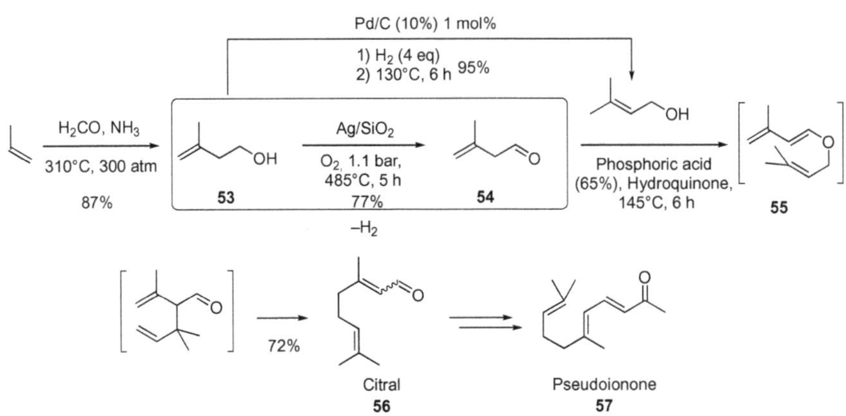

Figure 9.17. BASF process for the synthesis of citral.

SiO_2 [59]. The condensation between 3-methyl-but-2-en-1-ol formed by palladium isomerisation of (**53**) [60] and (**54**) forms the enol ether (**55**), which leads to citral (**56**) after Claisen–Cope rearrangement [61].

Other oxidation such as allylic oxidation allows the synthesis of aldehydes and ketones from the corresponding unsaturated alkene. This method can be used as an access route to the synthesis of verbenone (**58**), a 30% component of rosemary essential oil. Molecular oxygen is used in the presence of a cobalt catalyst [Co(4-methyl-pyridine)$_2$Br$_2$] to oxidise *α*-pinene (**41**) in a yield of 46% (Figure 9.18) [62]. The synthesis of verbenone (**58**) could also be carried out from verbenol (**59**) in the presence of nanoparticles of ruthenium, *tert*-hydroperoxide and the surfactant HEA$_{16}$Cl [63].

Benzylic oxidation is involved in the first step in the synthesis of aldehydes of interest, such as Lysméral® Extra (**11**), Bourgeonal® (**60**) and cyclamenaldehyde (**61**) (Figure 9.19) [63c, 64]. These compounds are used

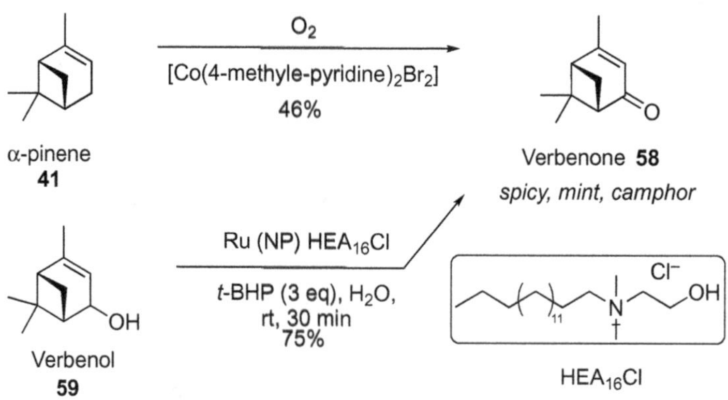

Figure 9.18. Synthesis of Verbenone.

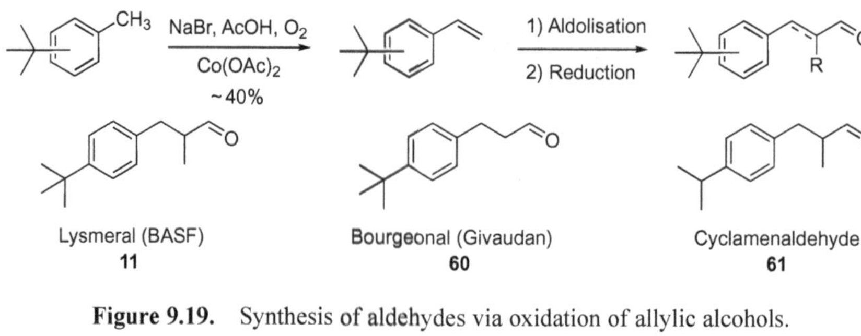

Figure 9.19. Synthesis of aldehydes via oxidation of allylic alcohols.

for their flowery notes reminiscent of the scent of lily of the valley. For example, Lysméral® was produced in this way at more than 2,500 t/year in the 2000s [65].

The oxidation step is carried out using a cobalt (II) catalyst in the presence of molecular oxygen (Figure 9.19) [66]. The process used today for the synthesis of Lysméral (**11**) has evolved: 4-*tert*-butyl-toluene is oxidised on an anode of platinum, ruthenium or chromium oxide to lead to the synthesis of dimethyl corresponding acetal. This synthesis allows the production of more than 10,000 tons of Lysméral® per year [67].

The formation of aldehydes can also be carried out *via* the isomerisation of the corresponding allyl alcohol. Aldehyde (**63**), a structural analogue of the compounds seen above (Figure 9.20), was obtained from the corresponding β,γ-unsaturated alcohol (**62**) in the presence of a ruthenium complex. This process has been patented by the Takasago company for the benefit of the synthons obtained. Citronellal (**66**) could thus be obtained from geraniol or nerol in an amount of 83% and 75%, respectively. The catalyst used also makes it possible to control the stereochemistry in a remarkable way, with *ee* greater than 99% [68].

Similarly, diethylamine-farnesyl (**67**) is isomerised to the corresponding imine and then converted to aldehyde (**68**) through acid treatment (Figure 9.21) [69]. It is interesting to note that this synthesis does not induce isomerisation other than that of the double bond in the allylic position. This reaction, carried out in the presence of a rhodium complex and a Josiphos-type ligand, is the key step towards the synthesis of Ambrox (**69**) [70], a compound with the odour of ambergris, obtained after eight steps.

A final example of isomerisation concerns the synthesis of citral (**56**). This could be synthesised from the 1,2-epoxide of myrcene (**70**) in the

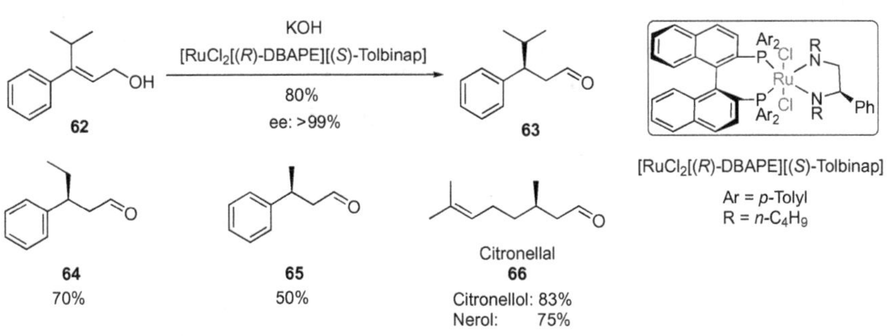

Figure 9.20. Isomerisation of unsaturated alcohols.

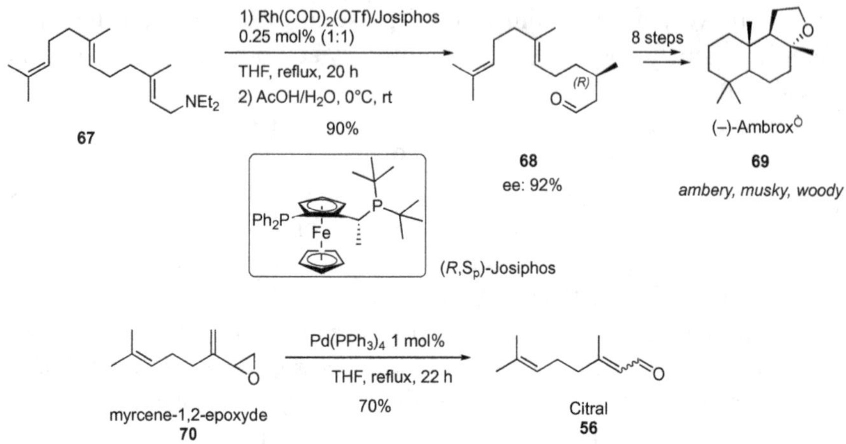

Figure 9.21. Synthesis pathway to (−)-Ambrox.

presence of a catalytic amount of palladium tetrakistriphenylphosphine at reflux of THF [71].

9.6 Carbonylation: Synthesis of Aldehyde and Ketones

The insertion of carbon monoxide is an intermediate step involved in many important industrial processes in the chemical industry. Its use allows the synthesis of carbonyl compounds (ketones and/or aldehydes) from unsaturated alkanes. The reactions resulting from this intermediate step are the starting point for the synthesis of many organic molecules of interest (alcohols, acetals, imines, etc.). The hydroformylation reaction, also referred to as the oxo process, was discovered in 1938 by Otto Roelen of the Rührchemie company (Figure 9.22) [72]. This reaction allows the synthesis of aldehydes from alkenes through the use of rhodium or cobalt complexes. Industrially, this reaction also opens up new access routes to the synthesis of approved alkenes of a carbon *via* the reduction of the aldehyde formed, followed by a dehydration reaction.

The hydroformylation reaction, leading to aldehydes, is of great importance for the synthesis of odorous compounds used in the composition of perfumes. Chanel N°5 contains, for example, dodecanal (**71**) and 2-methylundecanal (**12**), compounds that smell like orange peel.

A process for the synthesis of (**12**) consists of the hydroformylation of 1-decene (**72**), resulting in a mixture of undecanal (**73**) and 2-methyl-decanal with a ratio of 99:1 and an overall yield of 78%. A Mannich reaction is carried out on the reaction mixture by the addition of formaldehyde and di-butylamine to obtain 50% of 2-methylene-undecanal (**74**). The derivative (**12**) is obtained after hydrogenation of the mixture and distillation (Figure 9.23) [3b, 73].

The hydroformylation of 2-carene (**75**) has been shown to be a key step towards the synthesis of Spirambrene® (**78**), a compound used for its amber and woody notes (Figure 9.24) [74]. This step was performed in ethanol to drive a mixture of aldehyde (**76**) and acetal, resulting from the addition of ethanol (**77**) [75]. Notwithstanding the relatively low reactivity of the catalyst, unlikely for production on an industrial scale, this step has the advantage of using a renewable source for the synthesis of Spirambrene® (**78**). Indeed, carenes are obtained from turpentine, an oil obtained by the distillation of pine resins, for example.

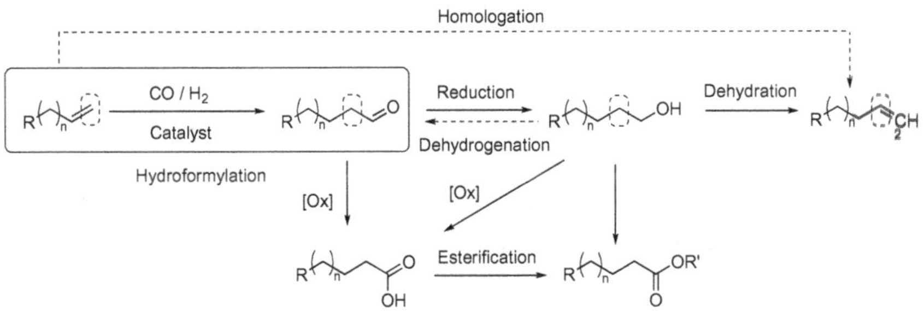

Figure 9.22. Alkene homologation via hydroformylation reaction.

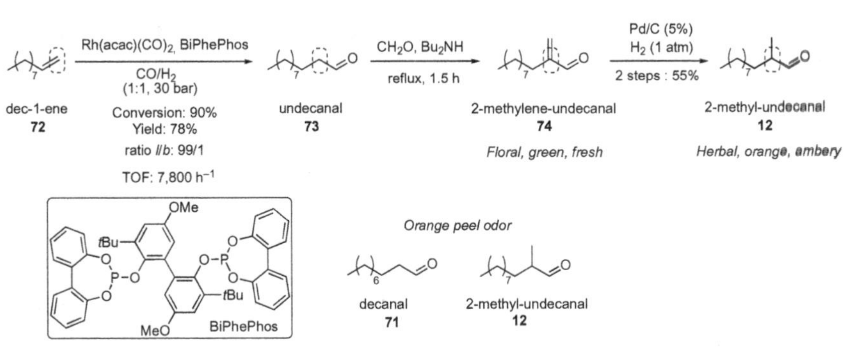

Figure 9.23. Synthesis of 2-methyl-undecanal.

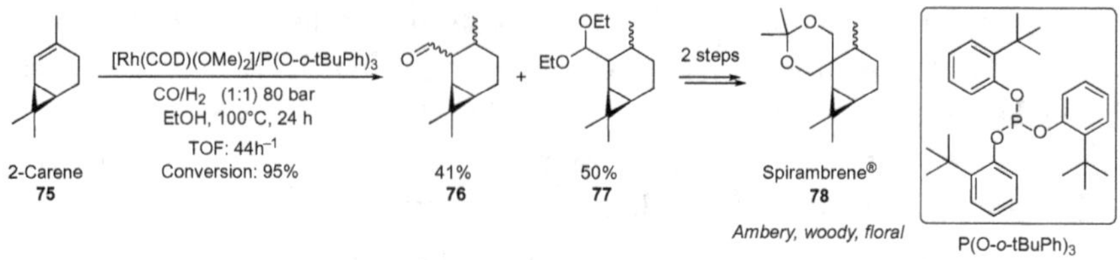

Figure 9.24. Synthesis of Spirambrene® via hydroformylation of 2-carene.

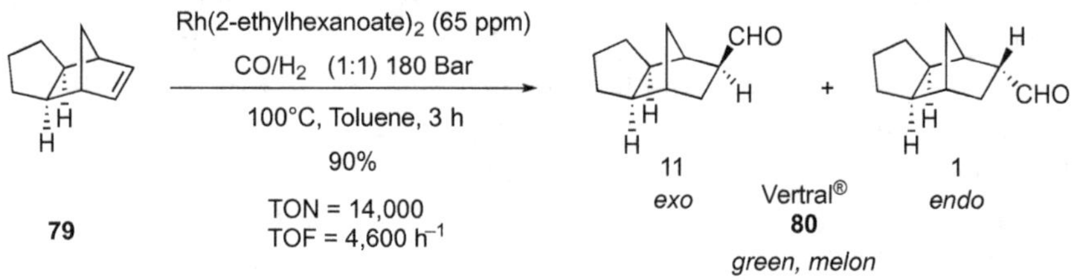

Figure 9.25. Synthesis of Vertral®.

Another elegant application of hydroformylation was carried out for the synthesis of Vertral® (**80**), a compound with green notes of melon (Figure 9.25). The hydroformylation of dihydrocyclopentadiene (**79**) in the presence of the rhodium complex [Rh(2-ethylhexanoate)$_2$] preferentially led to *exo* aldehyde with a catalytic conversion, allowing upscaling. Only 65 ppm of catalyst made it possible to obtain compound (**80**) in three hours of reaction [76].

The hydroformylation of indene was patented by BASF in 1971, and although the patent does not mention it, the use of this type of structure can be considered an olfactory analogue of Bourgeonal® (**82**) (Figure 9.26) [77]. Organoleptic evaluations made on indene aldehydes (**81**) described them as similar with their aldehyde and white flower notes [78]. This example illustrates the principle of "secochemistry", the concept of cutting bonds or creating them in order to find structures with similar properties. This technique is used in perfumery to study the importance of a cyclic structure in the organoleptic properties of the molecule [20, 79].

Although the hydroformylation reaction leads preferentially to the linear isomer, the company Chirotech has developed a chiral phosphite ligand, making it possible to obtain the compound mainly branched up to 75% in the context of the synthesis of Lysméral (**(*S*)-11**) (Figure 9.27). This elegant reaction also induces chirality with an excellent enantiomeric

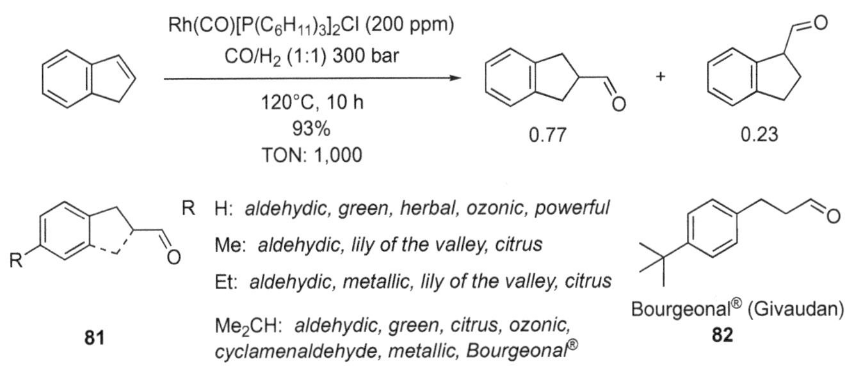

Figure 9.26. Synthesis of Bourgeonal® analogues.

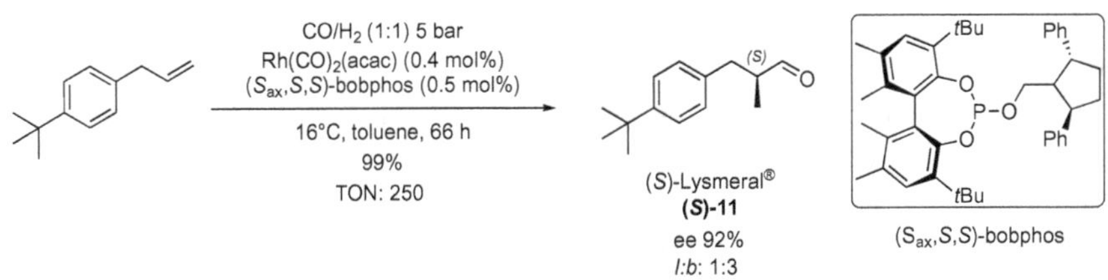

Figure 9.27. Enantioselective hydroformylation for the synthesis of (*S*)-Lysméral®.

excess of 92% [80]. This reaction was applied to the synthesis of the (*S*) enantiomer of Lysméral® (Figure 9.27), the enantiomer with the least impact on the smell of the mixture (perception threshold: <250 ng) [81]. The most potent (*R*) enantiomer (perception threshold: 1 ng) and the only enantiomer with lily-of-the-valley notes could be synthesised using the ligand (*R*ax, *R*, *R*)-bobphos [82].

Apart from hydroformylation, the insertion of CO can also lead to the production of cyclic ketones via the reaction [2 + 2 + 1] between an alkyne, an alkene and carbon monoxide. This reaction, discovered by Ullah Khand and Peter Ludwig Pauson in 1977 and now commonly referred to as the Pauson–Khand reaction, allows the synthesis of substituted cyclopentenones [83]. It has been widely used in the fragrance industry for accessing cyclopentenone derivatives. The synthesis of cyclopentenone (**86**) has been briefly mentioned as part of the process for the formation of Hedione® (**36**), where two reactions can be involved in its synthesis (Figure 9.28). Demole and Lederer patented in 1960 the synthesis of (**86**) via the reaction

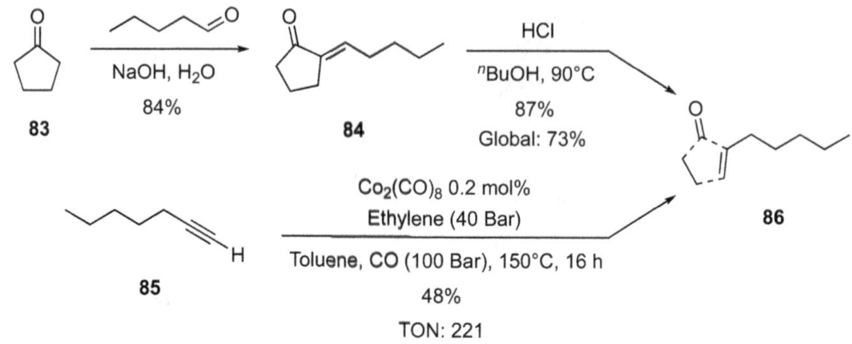

Figure 9.28. Synthesis of cyclopentenones through the Pauson–Khand reaction.

between a Grignard and an enol ether, a reaction however quite difficult to implement industrially [33b]. Chapuis reported a condensation between cyclopentanone (**83**) and pentanal, followed by the isomerisation of (**84**) [33c]. In 1990, Valentin Rautenstrauch, also a researcher in the Firmenich company, patented the synthesis of the same substrate via the application of the Pauson–Khand reaction. The reaction involves hept-1-yne (**85**) under an atmosphere of ethylene and carbon monoxide in the presence of the cobalt complex $Co_2(CO)_8$. The reaction exhibits a modest yield, which is nevertheless offset by the excellent selectivity resulting in the desired compound with 98% selectivity [39, 84].

9.7 The Metathesis Reaction: Synthesis of Alkenes and Alkynes

The metathesis reaction, whose richness and applications earned Robert H. Grubb the Nobel Prize in 2005 , allows the rearrangement of double bonds to be carried out, in a fairly general way, to generate new ones. Figure 9.29 illustrates the production of macrocycles using the ring-closing metathesis (RCM) reaction.

This reaction requires a low concentration in order to avoid the diene polymerisation reaction (ADMET stands for "acyclic diene metathesis") but also that of the macrocyclic substrate by opening and metathesis (ROMP stands for "ring-opening metathesis polymerisation").

If the metathesis reaction has been considered for the synthesis of macrocycles, it is because these are of interest in the field of olfaction [85].

Ambrettolide (**87**) is one of the natural constituents present in the seeds of ambrette (*Hibiscus abelmoschus L.*) and occurs as a 17-atom lactone [3b]. Before the Convention on International Trade in Endangered Species of Wild Fauna and Flora (CITES) protected certain species hunted for their use in perfumery, certain macrocyclic musks were used as a mixture via the incorporation of animal secretions in the compositions.

The glands of the civet (*Viverra zibetha*) produce a secretion used in dilution for its musky odour. The compounds responsible for this odour are civetone (**88**), present in an amount of 3.4%, but also cyclohexadecanone (**89**) and 6-cis-cycloheptadecenone (**90**). Finally, the male of the musk deer family (*Moschus moschiferus*) was hunted for its quality musk, of which (–)-muscone (*R* isomer) (**91**) constitutes the main scent [3b, 24a] (Figure 9.30).

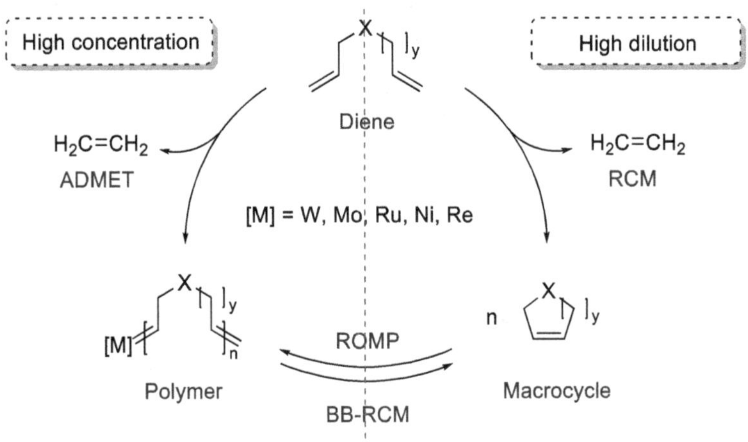

Figure 9.29. Principle of the metathesis reaction.

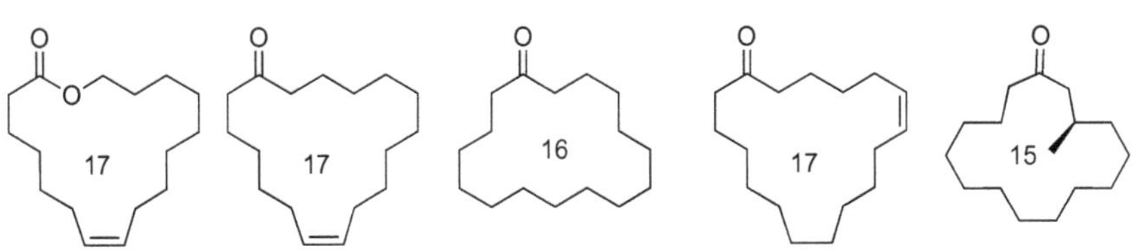

Ambrettolide **87** | Civetone **88** | cyclohexadecanone **89** | 6-*cis*-cycloheptadecenone **90** | (*R*)-Muscone **91**

Figure 9.30. Macrocyclic musks and their natural origins.

The search for new compounds has made it possible to identify new macrocyclic structures of interest for their musky note; this is the case of Cosmone® (**92**), Muscénone® (**93**), Habanolide® (**94**), Velvione® (**95**) and Globanone® (**96**) (Figure 9.31).

Each of these compounds can be synthesised through a metathesis reaction. Civetone has in particular been the subject of various studies with the aim of obtaining mainly the (*Z*) isomer. Grubbs's group reported, in 2013, *cis*-selective metathesis of various macrocycles of interest [86]. The use of a complex of ruthenium and *N*-heterocyclic ligands (NHC) with a high catalytic charge (7.5 mol%) gives access to civetone *via* the cyclising metathesis of diene (**97**) (Figure 9.32). The disadvantage of this reaction is the extremely low concentration of the substrate (3 mM), making it difficult to use this reaction on an industrial scale.

Recent progress has enabled this synthesis at a higher concentration in a commercial oil (PAO 6), of the order of 200 mM with a cyclic compound yield greater than 69%. The desired product is distilled here under reduced pressure as the reaction progresses so as to shift the equilibrium towards the formation of civetone (**88**) (Figure 9.33). Although the isomer

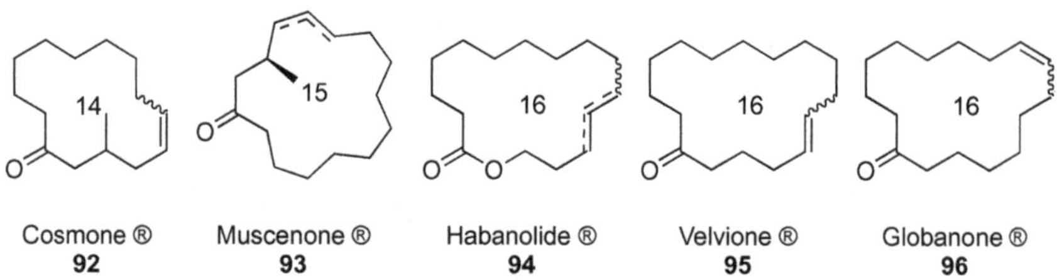

Figure 9.31. Synthetic macrocyclic musks.

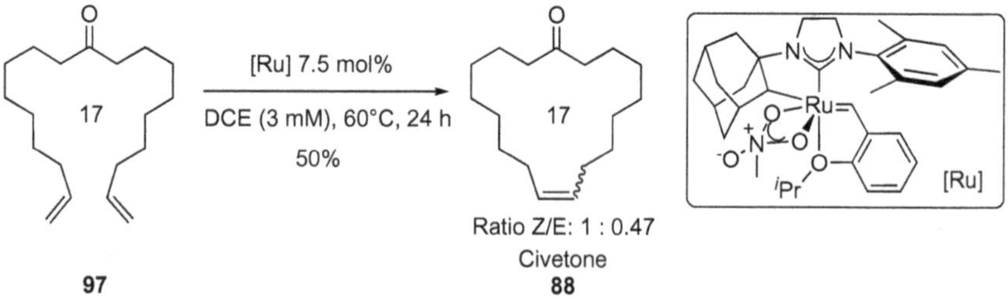

Figure 9.32. Synthesis of civetone by RCM metathesis.

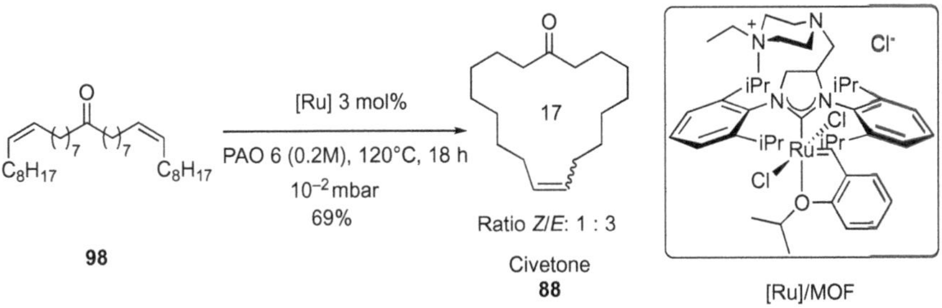

Figure 9.33. High concentration civetone cyclising metathesis.

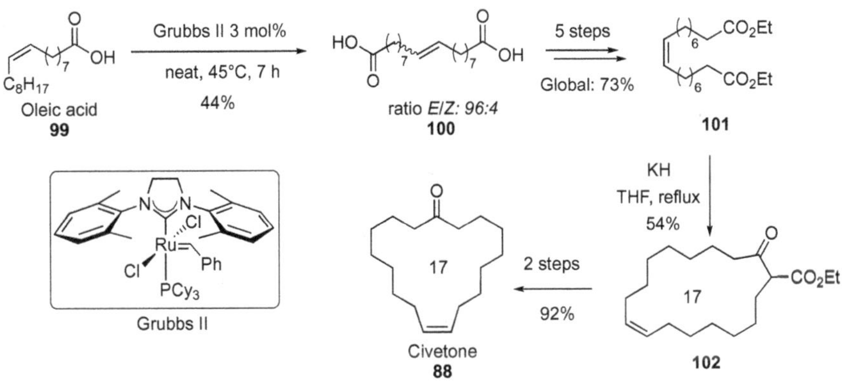

Figure 9.34. Synthesis of civetone via oleic acid metathesis.

obtained is the predominant compound (*E*), the author notes the use of a fully bio-based diene (**98**) derived from oleic acid [87].

The synthesis of bio-based precursors towards the synthesis of civetone (**88**) had also been described in 2012 for a different cyclisation strategy (Figure 9.34). Oleic acid (**99**) was engaged in an intermolecular metathesis reaction to obtain diacid (**100**). The condensation step was here carried out from the diester of configuration (*Z*) (**101**) *via* a Dieckmann reaction, leading to the keto-ester (**102**), followed by a decarboxylation to yield the desired compound [88].

The metathesis reaction has, at the present time and to our knowledge, not been able to take hold in industry intramolecularly for economic reasons related to the dilution factor, but it allows the synthesis of the starting precursors [89,90]. Givaudan has patented a similar process of using

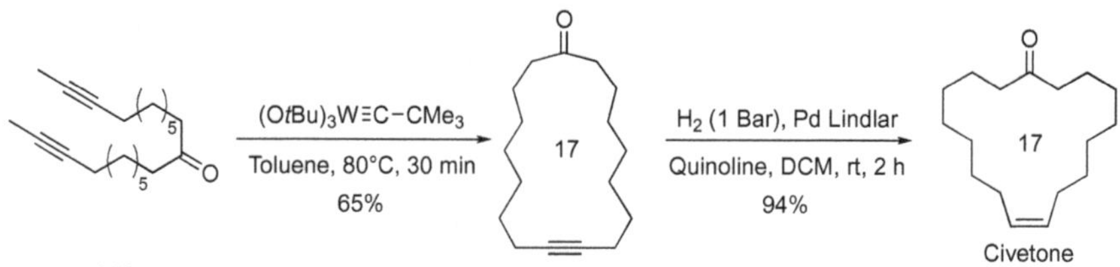

Figure 9.35. Synthesis of civetone by RCAM.

metathesis on oleic derivatives to produce a building block for use in the synthesis of Ambrettolide® (**87**) [91].

Fürstner's team showed that the metathesis reaction can as easily take place between two internal alkynes so as to obtain another internal alkyne (Figure 9.35). It uses a tungsten complex to perform the cyclising metathesis of diyne (**103**) (RCAM stands for "ring-closing alkyne metathesis") and thus obtain civetone (**88**) after partial reduction of the triple bond (**104**) to alkene. This synthesis has the advantage of being entirely selective and essentially leads to the desired (*Z*)-isomer [91].

More recently, Fürstner's group described a similar synthesis leading to the production of Muscenone® (**93**) *via* an RCAM step giving access only to the (*R*, *Z*) isomer [92].

9.8 Conclusions and Future Directions

The use of transition metals in the field of perfumes has thus become essential for manufacturers who wish to offer new innovative processes that are more respectful of the environment. Several transformations that are atom- and step-economical have been described and applied to challenging odorants. An important issue of stereoselectivity has been solved thanks to metal-mediated reactions. Nevertheless, the cost of metals and sometimes ligands slowed down the industrial commercial routes for classical ingredients. Research is still moving towards innovation as several ingredients are or will be banned in the near future. The perfumes of the future will therefore rely both on cost issues and on renewable and biodegradable ingredients. The industry is facing the degradability of each key component that is extracted and produced.

The perfumers will most probably increase their collaboration with the chemists to find solutions to replace some molecules so that the fragrance is not affected. This research will rely on artificial intelligence and on the full understanding of olfaction.

Acknowledgements

This work was supported by the University of Côte d'Azur and the Centre National de la Recherche Scientifique (CNRS). We gratefully acknowledge Expressions Parfumées for the PhD grant to R.L.

References

[1] C. Sell. (2006). *The Chemistry of Fragrances: From Perfumer to Consumer*, 2nd ed.; RSC Publ: Cambridge.

[2] (a) A. Buettner. (2017). *Springer Handbook of Odor*; Springer International Publishing: Cham.
(b) D. J. Rowe. (2005). *Chemistry and Technology of Flavors and Fragrances*; Blackwell: Oxford.

[3] (a) R. R. Calkin and J. S. Jellinek. (1994). *Perfumery: Practice and Principles*; John Wiley & Sons: New York.
(b) K.-G. Fahlbusch, F.-J. Hammerschmidt, J. Panten, W. Pickenhagen, D. Schatkowski, K. Bauer, D. Garbe and H. Surburg. (2003). Flavors and Fragrances. In *Ullmann's Encyclopedia of Industrial Chemistry*; Wiley-VCH Verlag GmbH & Co. KGaA: Weinheim, Germany.

[4] (a) A. Buettner. (2017). *Springer Handbook of Odor*; Springer International Publishing: Cham.
(b) D. J. Rowe. (2005). *Chemistry and Technology of Flavors and Fragrances*; Blackwell: Oxford.

[5] M. Edwards. (1996). *Perfume Legends: French Feminine Fragrances*; & Co. in association with HM Editions: Levallois.

[6] Rule (EU) 2017/1410 from the commission - august 2 2017 - modification of the annexes II and III of EU rule N°1223/2009 of the European Parliament and of the Council on cosmetic products.

[7] G. H. Bruntland. (1987). Report of the World Commission on Environment and Development: Our Common Future; A/42/427, p. 374.

[8] P. T. Anastas and R. H. Crabtree. (2009). *Handbook of Green Chemistry*; Wiley-VCH: Weinheim.

[9] B. Trost. (1991). The atom economy — A search for synthetic efficiency. *Science*, 254, 1471–1477.

[10] (a) R. A. Sheldon. (1992). Catalysis, the Atom Utilization Concept and Waste Minimization. In *Industrial Environmental Chemistry*, Sawyer, D. T., Martell, A. E., Eds., Springer US: Boston, MA, 99–119.

(b) R. A. Sheldon. (2007). The E Factor: Fifteen years on. *Green Chem.*, 9, 1273–1283.

[11] (a) G. D. C. Heilen, A. D. C. Nissen, O. D. C. Woertz. (1980). Process for the Preparation of Aliphatic Carbonyl Compounds. EP0013385A1.
(b) I. Omae. (2009). Characteristic reactions of group 9 transition metal compounds in organic synthesis. *Appl. Organomet. Chem.*, 23, 91–107.
(c) K. Weissermel, H.-J. Arpe. (2003). *Industrial Organic Chemistry*, 4th completely rev. ed., Wiley-VCH: Weinheim.
(d) (2000). *Ullmann's Encyclopedia of Industrial Chemistry*, 1st ed., Wiley-VCH Verlag GmbH & Co. KGaA, Ed., Wiley.
(e) G. P. Chiusoli, P. M. Maitlis. (2006). *Metal-Catalysis in Industrial Organic Processes*, RSC Pub: Cambridge.

[12] F. Elterlein, N. Bugdahn, and P. Kraft. (2024). Sniffing Out the Sustainable Future: The Renewability Revolution in Fragrance Chemistry. *Chem. Eur. J.*, 30, e202400006.

[13] P. Martinaux, R. Laher, C. Marin, and V. Michelet. (2022). Transition Metal-Catalyzed Rearrangement and CycloIsomerisation Reactions Toward Hedonic Materials. *Israel J. Chem.*, 62, e202200047.

[14] J.-A. Funel and S. Abele. (2013). Industrial Applications of the Diels–Alder Reaction. *Angew. Chem. Int. Ed.*, 52, 3822–3825.

[15] A. Muratore, E. Duñach, J.-C. Clinet, and C. Plessis. (2008). New Norbornyl Derivatives as Woody Fragrant Materials. *Chem. Biodiversity*, 5, 1099–1114.

[16] J. Mane, J.-J. Chanot, F. L. Borgne and M. Schroeder. Novel Norbornane and Norbornene Derivatives, Use Thereof and Perfumed Products Containing Same. WO2003035595A1, May 1, 2003.

[17] T. Asada, Y. Ataka, J. Etsuno and H. Takase. (1997). Tricyclocarboxylic Acid Ester and Its Production. JPH09151164A.

[18] (a) G. Fráter, U. Müller, F. Schröder. (2004). Synthesis and olfactory properties of (−)-(1R,2S)-Georgywood. *Tetrahedron: Asymmetry*, 15, 3967–3972.
(b) A. Stepanyuk, A. Kirschning. (2019). Synthetic terpenoids in the world of fragrances: Iso E Super® is the showcase. *Beilstein J. Org. Chem.*, 15, 2590–2600.

[19] P. A. Wender, H. Takahashi, B. Witulski. (1995). Transition metal catalyzed [5 + 2] cycloadditions of vinylcyclopropanes and alkynes: A homolog of the diels-alder reaction for the synthesis of seven-membered rings. *J. Am. Chem. Soc.*, 117, 4720–4721.

[20] P. Kraft. (1999). Design and Synthesis of Violet Odorants with Bicyclo[6.4.0]dodecene and Bicyclo[5.4.0]undecene Skeletons. *Synthesis*, 1999, 695–703.

[21] P. A. Wender and T. E. Smith. (1998). Transition metal-catalyzed intramolecular [4 + 2] cycloadditions: Mechanistic and synthetic investigations. *Tetrahedron*, 54, 1255–1275.
[22] P. Kraft, D. Frech, U. Müller, and G. Fráter. (2006). Synthesis and olfactory properties of spirocyclic georgywood® analogues. *Synthesis*, 2006, 2215–2223.
[23] F. Schroeder and F. Ruethi. Improvements in or Relating to Organic Compounds. US20160280615A1, September 29, 2016.
[24] (a) H. Surburg and J. Panten. (2006). *Common Fragrance and Flavor Materials: Preparation, Properties and Uses*; Wiley-VCH Verlag GmbH & Co. KGaA: Weinheim, FRG. (b) D. A. van, Dorp, R. Klok and D. H. Nugteren. (1973). New macrocyclic compounds from the secretions of the civet cat and the musk rat. *Recl. Trav. Chim. Pays-Bas*, 92, 915–928.
[25] E. Leitmannova and L. Cerveny. (2008). Common fragrance and flavor materials. *Perfum. Flavor.*, 33, 50–59.
[26] F. Bouachir, B. Chaudret, F. Dahan, F. Agbossou and I. Tkatchenko. (1991). Preparation and stoichiometric and catalytic reactivity of hydrido organometallic ruthenium complexes. X-ray crystal structure of [RuH(η.5-C8H11)2]BF4. *Organometallics*, 10, 455–462.
[27] (a) H. Mimoun. (1999). Selective reduction of carbonyl compounds by polymethylhydrosiloxane in the presence of metal hydride catalysts. *J. Org. Chem.*, 64, 2582–2589.
(b) H. Mimoun, J. Y. de Saint Laumer, L. Giannini, R. Scopelliti and C. Floriani. (1999). Enantioselective reduction of ketones by polymethylhydrosiloxane in the presence of chiral zinc catalysts. *J. Am. Chem. Soc.*, 121, 6158–6166.
[28] R. Jean-Jacques, S. Lionel, W. Patrick and D. Philippe. (2006). Hydrogenation of Esters with Ru/Tetradentate Ligands Complexes. BRPI0608686 (A2).
[29] P. Dupau and L. Bonomo. (2008). 1,4-Hydrogenation of Sorbol with Ru Complexes. WO2008120174A1.
[30] P. Dupau and L. Bonomo. (2008). 1,4-Hydrogenation of Dienes with Ru Complexes. BRPI0811413 (A2).
[31] (a) H. Takaya, T. Ohta, N. Sayo, H. Kumobayashi, S. Akutagawa, S. Inoue, I. Kasahara and R. Noyori. (1987). Enantioselective hydrogenation of allylic and homoallylic alcohols. *J. Am. Chem. Soc.*, 109, 1596–1597.
(b) N. Yamada, T. Takezawa, N. Sayo, M. Yagi, H. Kumobayashi, S. Akutagawa, H. Takaya, S. Inoue and R. Noyori. (1990). Process for Producing Optically Active Alcohols. US4962242A.
(c) Vries, J. G. de (2007). *The Handbook of Homogeneous Hydrogenation*, Elsevier, C. J., Eds.; Wiley-VCH: Weinheim.

[32] (a) E. Demole, E. Lederer and D. Mercier. (1962). Isolement et détermination de la structure du jasmonate de méthyle, constituant odorant caractéristique de l'essence de jasmin. *Helv. Chim. Acta*, 45, 675–685.
(b) D. Edouard and L. Edgar. (1964). Procédé pour la préparation de céto-esters alicycliques. CH382731A.
(c) C. Chapuis. (2012). The Jubilee of Methyl Jasmonate and Hedione®. *Helv. Chim. Acta*, 95, 1479–1511.

[33] P. Teisseire and M. Plattier. (1970). Isomer of Dihydromethyl-Jasmonate. ZA7108447 (B).

[34] C. Fehr. (1999). Epoxides, Process for Their Preparation and Their Use for the Preparation of Perfuming Ingredients. US5962706A.

[35] C. Fehr and J. Galindo. (2000). A new variant of the claisen rearrangement from malonate-derived allylic trimethylsilyl ketene acetals: Efficient, highly enantio- and diastereoselective syntheses of (+)-methyl dihydroepijasmonate and (+)-methyl epijasmonate. *Angew. Chem. Int. Ed.*, 39, 569–573.

[36] J.-P. Genêt, V. Rautenstrauch, D. A. Dobbs, K. P. M. Vanhessche, E. Brazi, J.-Y. Lenoir, J. Wiles and S. H. Bergens. (2000). Industrial synthesis of (+)-cis-methyl dihydrojasmonate by enantioselective catalytic hydrogenation; Identification of the precatalyst Ru((−)-Me-DuPHOS)(H)(η6-1,3,5-cyclooctatriene). *Angew. Chem. Int. Ed.*, 39, 1992–1995.

[37] (a) L. Saudan. (2019). 85 Years of Catalysis at Firmenich. *Chimia*, 73, 684–704.
(b) J. A. Wiles, S. H. Bergens, K. P. M. Vanhessche, D. A. Dobbs and V. Rautenstrauch. (2001). A versatile and high-yield route to active and well-defined catalysts Ru(bisphosphane)(H)(solvent)3. *Angew. Chem. Int. Ed.*, 40, 914–919.
(c) V. Rautenstrauch. (2012). Key enantioselective hydrogenation steps in the syntheses of two renin inhibitors: Missing origin and incorrect description. *Org. Process Res. Dev.*, 16, 1185–1186.
(d) V. Rautenstrauch and J.-J. Riedhauser. (1998). Process for the Preparation of (+)-(1R)-Cis-3-Oxo-2-Pentyl-1-Cyclopentaneacetic Acid. US5728866 (A).

[38] V. Rautenstrauch and W. Keim. (1993). 2-Pentyl-cyclopentenone preparation by reacting hept-1-yne with ethylene and carbon monoxide using cobalt carbonyl catalyst. CH681224 (A5).

[39] V. Rautenstrauch, R. Challand, R. Churlaud, R. H. Morris, K. Abdur-Rashid, E. Brazi and H. Mimoun. (2002). Catalytic Hydrogenation Process. JP2012087138 (A).

[40] J. Magano and J. R. Dunetz. (2012). Large-Scale Carbonyl Reductions in the Pharmaceutical Industry. *Org. Process Res. Dev.*, 16, 1156–1184.

[41] M. Beller and H. U. Blaser. (2012). *Organometallics as Catalysts in the Fine Chemical Industry*; Topics in organometallic chemistry; Springer: Berlin; New York.

[42] C. S. Sell. (2014). *Chemistry and the Sense of Smell*; Wiley: Hoboken, NJ.

[43] (a) R. E. Naipawer and W. M. Easter. (1977). 3-Methyl-5-(2,2,3-Trimethylcyclopent-3-En-1-Yl)Pentan-2-ol Compound and Perfume Compositions. US4052341A.
(b) C. Chapuis and P.-A. Blanc. (1997). Campholinic Aldehyde Derivatives, Process for Their Preparation and Their Use as Perfuming Ingredients. US5696075 (A).
(c) C. Chapuis and P.-A. Blanc. (1996). Dérivés de l'aldéhyde campholénique et leur utilisation en parfumerie. EP0694520A1.
(d) K.-H. Schulte-Elte, B. L. Muller and H. Pamingle. (1986). Hydroxylic Campholenic Aldehyde Derivatives, Their Utilization as Perfume Ingredients and Perfuming Compositions Containing Same. US4610813 (A).
(e) Chapuis, C. (1992). Alcool tertiare à sous-structure campholénique et son utilisation à titre d'ingrédient parfumant. EP0504592A1.

[44] P. Kraft, J. A. Bajgrowicz, C. Denis and G. Fráter. (2000). Odds and trends: Recent developments in the chemistry of odorants. *Angew. Chem. Int. Ed.*, 39, 2980–3010.

[45] (a) T. Ohkuma, H. Ooka, S. Hashiguchi, T. Ikariya and R. Noyori. (1995). Practical enantioselective hydrogenation of aromatic ketones. *J. Am. Chem. Soc.*, 117, 2675–2676.
(b) T. Ohkuma, H. Ooka, T. Ikariya and R. Noyori. (1995). Preferential hydrogenation of aldehydes and ketones. *J. Am. Chem. Soc.*, 117, 10417–10418.

[46] (a) V. Firmenich: Rautenstrauch, R. Churlaud, R. H. Morris and K. Abdur-Rashid. WO2002040155A1, May 23, 2002. (b) DSM: Tucker, C. E., Jiang, Q. WO2003061824A1, July 31, 2003. (c) Takasago: Mikami, K., Sayo, N. WO2005016943A1, February 24, 2005. (d) Enantiotech: WO2012041215A1, April 5, 2012.

[47] (a) P. Dupau, L. Bonomo and L. Kermorvan. (2013). Unexpected role of anionic ligands in the ruthenium-catalyzed base-free selective hydrogenation of aldehydes. *Angew. Chem. Int. Ed.*, 52, 11347–11350.
(b) P. Dupau, L. Bonomo and L. Kermorvan. (2014). Selective hydrogenation of aldehydes with Ru/bidentate ligands complexes. WO2014139854A1.

[48] (a) L. Saudan and C. Saudan. (2010). Hydrogenation of ester, ketone or aldehyde groups with ruthenium complexes having di-amine and phosphorous-nitrogen bidentate ligand. CN102227261 (A). (b) S. Lionel and S. Christophe. (2010). Hydrogenation of esters or carbonyl groups with phosphino-oxide based ruthenium complexes. CN102164673 (A).

(c) F. Santoro, L. Saudan and C. Saudan. (2012). Hydrogenation of esters or carbonyl groups with tetradentate amino/imino-thioether based ruthenium complexes. ES2624265 (T3). (d) P. Dupau. (2018). Hydrogenation towards synthetic sandalwood odorants in fragrance industry. *Helv. Chim. Acta*, 101, e1800144.

[49] (a) R. Patchett, I. Magpantay, L. Saudan, C. Schotes, A. Mezzetti and F. Santoro. (2013). Asymmetric Hydrogenation of Ketones with H2 and Ruthenium Catalysts Containing Chiral Tetradentate S2N2 Ligands. *Angew. Chem. Int. Ed.*, 52, 10352–10355. (b) F. Santoro, L. Saudan and C. Saudan. (2012). Hydrogenation of esters or carbonyl groups with tetradentate amino/imino-thioether based ruthenium complexes. WO2012084810A1.

[50] J. Zhang, G. Leitus, Y. Ben-David and D. Milstein. (2006). Efficient homogeneous catalytic hydrogenation of esters to alcohols. *Angew. Chem. Int. Ed.*, 45, 1113–1115.

[51] S. Nishimura. (2001). *Handbook of Heterogeneous Catalytic Hydrogenation for Organic Synthesis*; J. Wiley: New York.

[52] Improvements in or Relating to Organic Compounds. GB201208589D0, June 27, 2012.

[53] W. Kuriyama, Y. Ino and O. Ogata. (2013). Alcohol Production Method by Reducing Ester or Lactone with Hydrogen. US8524953B2.

[54] (a) D. Srimani, A. Mukherjee, A. F. G. Goldberg, G. Leitus, Y. Diskin-Posner, L. J. W. Shimon, Y. Ben David, and D. Milstein. (2015). Cobalt-catalyzed hydrogenation of esters to alcohols: Unexpected reactivity trend indicates ester enolate intermediacy. *Angew. Chem. Int. Ed.*, 54, 12357–12360. (b) N. A. Espinosa-Jalapa, A. Nerush, L. J. W. Shimon, G. Leitus, L. Avram, Y. Ben-David and D. Milstein. (2017). Manganese-catalyzed hydrogenation of esters to alcohols. *Chem. Eur. J.*, 23, 5934–5938. (c) T. Zell, Y. Ben-David and D. Milstein. (2014). Unprecedented iron-catalyzed ester hydrogenation. mild, selective, and efficient hydrogenation of trifluoroacetic esters to alcohols catalyzed by an iron pincer complex. *Angew. Chem.*, 126, 4773–4777.

[55] J. Quintaine and L. Saudan. (2018). Hydrogenation of Esters with Fe/Tridentate Ligands Complexes. US2018305382 (A1).

[56] L. I. Simándi. (1991). *Dioxygen Activation and Homogeneous Catalytic Oxidation: Proceedings of the Fourth International Symposium on Dioxygen Activation and Homogeneous Catalytic Oxidation*, Studies in surface science and catalysis; Elsevier: Amsterdam; New York.

[57] P. Kraft and W. Eichenberger. (2003). Conception, characterization and correlation of new marine odorants. *Eur. J. Org. Chem.*, 2003, 3735–3743.

[58] J. J. Beereboo, D. P. Cameron, and C. R. Stephens. (1970). 1,5-Benzoxepin- and -Benzodioxepin-3-Ones as Flavor and Odorant Agents. US3517031A.

[59] C. D. C. D. Dudeck, H. D. C. D. Diem, F. D. C. D. Brunnmueller, B. D. C. D. Meissner and W. D. C. D. Fliege. (1978). Preparation of 3-Alkyl-Buten-1-Als. DE2715209 (A1).

[60] H. Overwien and H. Mueller. (1972). Production of Buten-2-ol-4 Compounds. US3697580 (A).

[61] (a) P. Chabardes and J. Chazal. (1989). Procédé de préparation du citral. EP0344043A1. (b) A. Nissen, W. Rebafka and W. Aquila. (1981). Preparation of Citral. US4288636A. (c) C. Chapuis and D. Jacoby. (2001). Catalysis in the preparation of fragrances and flavours. *Appl. Catal. A: Gen.*, 221, 93–117. (d) H. Mueller, H. Overwien and H. Pommer. (1971). Production of Alk-3-en-1-ols. US3574773 (A).

[62] M. K. Lajunen, T. Maunula, and A. M. P. Koskinen. (2000). Co(II) catalysed oxidation of α-pinene by molecular oxygen. Part 2. *Tetrahedron*, 56, 8167–8171.

[63] M. Rauchdi, M. Ait Ali, A. Roucoux, and A. Denicourt-Nowicki. (2018). Novel access to verbenone via ruthenium nanoparticles-catalyzed oxidation of α-pinene in neat water. *Appl. Catal. A: Gen.*, 550, 266–273.

[64] (a) M. Dunkel, D. J. Eckhardt, and A. Stern. (1975). Hydrogenation of Cinnamaldehyde. US3520934 (A). (b) N. D. Goetz and W. D. Hoffmann. (1959). Process for the Preparation of P-Substituted Phenyl Propanols and Their Esters and of p-Substituted 3-Phenyl-2-Methyl Propanals, EP0032576 (A1). (c) M. S. Carpenter and J. W. M. Easter. (1959). 4-Tert-butyl Alpha-methyl hydrocinnamic Aldehyde. US2875131A.

[65] G. Fráter, J. A. Bajgrowicz, and P. Kraft. (1998). Fragrance chemistry. *Tetrahedron*, 54, 7633–7703.

[66] Oxidation of Alkaryl Compounds. GB1561464A, February 20, 1980.

[67] (a) F. Stecker, A. Fischer, J. Botzem, U. Griesbach, and R. Pelzer. (2012). Electrochemical Method for Producing 3 Tert-butyl Benzaldehyde-dimethylacetal. EP2411564A1. (b) S. Möhle, M. Zirbes, E. Rodrigo, T. Gieshoff, A. Wiebe and S. R. Waldvogel. (2018). Modern electrochemical aspects for the synthesis of value-added organic products. *Angew. Chem. Int. Ed.*, 57, 6018–6041.

[68] T. Ohkuma, N. Arai, and K. Matsumura. (2015). Method of Producing Optically-Active Aldehyde. US20150376102A1.

[69] (a) C. Chapuis and M. Barthe. (2000). Stereospecific Isomerisation of Allyl Amines Using Chiral Phosphorous Ligands. US6020527 (A). (b) C. Chapuis. (2000). Stereospecific Isomerisation of Allylamines with the Aid of Immobilized Phosphorated Chiral Ligands. US6350910 (B1). (c) C. Chapuis, M. Barthe and J.-Y. de Saint Laumer. (2001). Synthesis of citronellal by RhI-catalysed asymmetric isomerisation of N,N-diethyl-substituted geranyl- and nerylamines or geraniol and nerol in the presence

of chiral diphosphino ligands, under homogeneous and supported conditions. *Helv. Chim. Acta*, 84, 230–242.

[70] C. Chapuis. (2014). Enantioselective access to (−)-Ambrox® starting from *β*-farnesene. *Helv. Chim. Acta*, 97, 197–214.

[71] V. Fauchet, B. A.-S. Miguel, M. Taran and B. Delmondo. (1999). Epoxides from myrcene: New versatile tools for the synthesis of functionalized acyclic terpenoids. *Synth. Commun.*, 29, 3673–3684.

[72] O. D. A. Roelen. (1952). Process for the Production of Oxygen-Containing Compounds. DE849548C.

[73] (a) H. Bahrmann, B. Cornils, G. Diekhaus, W. Kascha and J. Weber. (1981). Process for the Production of Aldehydes. US4258214 (A). (b) Unsaturated Aldehydes and Alcohols. FR1508854 (A), (1968). (c) J. Weber and H. Springer. Process for Preparing Methyl-Nonyl-Acetaldehyde. US4263460 (A) (1981).

[74] (a) E. Giraudi. Polysubstituted Norcaranes. EP0266648 (A2).
(b) G. Rubulotta and E. A. Quadrelli. (2019). Terpenes: A Valuable Family of Compounds for the Production of Fine Chemicals. In *Studies in Surface Science and Catalysis*; Elsevier; Vol. 178, 215–229.

[75] (a) M. C. de Freitas, C. G. Vieira, E. N. dos Santos and E. V. Gusevskaya. (2013). Synthesis of Fragrance Compounds from Biorenewables: Tandem Hydroformylation–Acetalization of Bicyclic Monoterpenes. *ChemCatChem*, 5, 1884–1890. (b) E. V. Gusevskaya, J. Jiménez-Pinto and A. Börner. (2014). Hydroformylation in the Realm of Scents. *ChemCatChem*, 6, 382–411.

[76] E.-J. Brunke, C.-H. Kappey, B. Holscher and H. Struwe. (1996). 8-Exo-Formyl-2,6-Exo-Tricyclo(5.2.1.02,6)Decane, Method of Preparing It and Use of the Same. US5492887A.

[77] W. Himmele and W. Aquila. (1972). Production of 1-and-2-formylindane. DE2053736 (A1).

[78] B. Winter and S. Gallo-Flückiger. (2005). Synthesis and odor properties of substituted indane-2-carboxaldehydes. Discovery of a new floral (muguet) fragrance alcohol. *Helv. Chim. Acta*, 88, 3118–3127.

[79] (a) P. Kraft. (2008). Woody pretzels: Spirocycles from vetiver to patchouli and Georgywood®. *Chem. Biodiversity*, 5, 970–999. (b) P. Kraft, V. Di Cristofaro and S. Jordi. (2014). From Cassyrane to cashmeran – The molecular parameters of odorants. *Chem. Biodiversity*, 11, 1567–1596. (c) H. Mouhib and W. Stahl. (2014). From cats and blackcurrants: Structure and dynamics of the sulfur-containing cassis odorant cat ketone. *Chem. Biodiversity*, 11, 1554–1566. (d) P. Kraft and K. Popaj. (2006). Musk or violet? Design, synthesis and odor of seco-derivates of a musky carotol lead. *Tetrahedron*, 62, 12211–12219. (e) P. Kraft and N. Denizot.

(2013). Synthesis of a spirocyclic seco structure of the principal vetiver odorant khusimone. *Eur. J. Org. Chem.*, 2013, 49–58.

[80] G. M. Noonan, J. A. Fuentes, C. J. Cobley and M. L. Clarke. (2012). An asymmetric hydroformylation catalyst that delivers branched aldehydes from alkyl alkenes. *Angew. Chem. Int. Ed.*, 51, 2477–2480.

[81] D. Bartschat, S. Börner, A. Mosandl and J. W. Bats. (1997). Stereoisomeric flavour compounds LXXVI: Direct enantioseparation, structure elucidation and structure-function relationship of 4-tert-butyl-α-methyldihydrocinnamaldehyde. *Z. Lebensm. Unters. Forsch. A*, 205, 76–79.

[82] L. Iu, J. A. Fuentes, M. E. Janka, K. J. Fontenot and M. L. Clarke. (2019). High iso aldehyde selectivity in the hydroformylation of short-chain alkenes. *Angew. Chem. Int. Ed.*, 58, 2120–2124.

[83] P. L. Pauson and I. U. Khand. (1977). Uses of Cobalt-carbonyl acetylene complexes in organic synthesis. *Ann. N.Y. Acad. Sci.*, 295, 2–14.

[84] V. Rautenstrauch, P. Mégard, J. Cones and W. Küster. (1990). 2-pentylcyclopent-2-en-1-one by catalytic pauson-khand reaction. *Angew. Chem. Int. Ed.*, 29, 1413–1416.

[85] (a) P. Kraft and C. Berthold. (2008). (4E,8Z)-12-methyloxacyclotetradeca-4,8-dien-2-one and its 7a-homologue: Conformationally constrained double-unsaturated macrocyclic musks by ring-closing alkyne metathesis. *Synthesis*, 2008, 543–550. (b) A. S. Williams. (1999). The synthesis of macrocyclic musks. *Synthesis*, 1999, 1707–1723. (c) H. Matsuda, S. Watanabe and K. Yamamoto. (2004). New macrocyclic musk compounds. *Chem. Biodiversity*, 1, 1985–1991.

[86] (a) V. M. Marx, M. B. Herbert, B. K. Keitz and R. H. Grubbs. (2013). Stereoselective access to Z and E macrocycles by ruthenium-catalyzed Z-selective ring-closing metathesis and ethenolysis. *J. Am. Chem. Soc.*, 135, 94–97. (b) J. Morvan, A. Del Vecchio, J. Talcik, D. Bouëtard and M. Mauduit. (2023). Recent review on Z-enantioselective olefin metathesis. challenges and breakthroughs in Z-enantioselective olefin metathesis. *Eur. J. Org. Chem.*, 26, e202300671.

[87] A. Sytniczuk, M. Dąbrowski, Ł. Banach, M. Urban, S. CzarnockaŚniadała, M. Milewski, A. Kajetanowicz and K. Grela. (2018). Olefin metathesis macrocyclization at high concentration. *J. Am. Chem. Soc.*, 140, 8895–8901.

[88] H. Hagiwara, T. Adachi, T. Nakamura, T. Hoshi and T. Suzuki. (2012). Revisit to (Z)-civetone synthesis. *Nat. Prod. Commun.*, 7, 913–915.

[89] (a) A. Wartini, K. Ebel, G. Hieber and H. Weigl. (2008). Verfahren zur herstellung von makrocyclischen ketonen durch dieckmann-kondensation in der gasphase. EP1525176B1. (b) S. Shinde. (2014). Methods of

converting mixtures of palmitoleic and oleic acid esters to high value products. US8642814B1.
[90] P. N. Davey, M. A. Lovchik, A. Goeke, K. Lorincz, F. Toth and L. Ondi. (2019). Process involving cross metathesis of olefins. US20190100504A1.
[91] A. Fürstner and G. Seidel. (2000). Ring closing alkyne metathesis: Stereoselective synthesis of civetone. *J. Organomet. Chem.*, 606, 75–78.
[92] K. Lehr and A. Fürstner. (2012). A short enantioselective total synthesis of the musk odorant (R,Z)-5-muscenone via base-metal-catalysis. *Tetrahedron*, 68, 7695–7700.

https://doi.org/10.1142/9781800617780_0010

Chapter 10

Sensory Effects of COVID-19 on Perfume Creation and Perception

Stephanie Hunter[†] **and Valentina Parma**[‡]

Monell Chemical Senses Center, Philadelphia, PA, USA

[†]*shunter@monell.org*
[‡]*vparma@monell.org*

Abstract

Although we are mostly unaware of its contributions, the sense of smell plays an essential role in our everyday life. It is fundamental for our navigation of the environment, including the recognition of possible threats, it plays a function in the creation of interpersonal relationships and it is essential to food enjoyment. Losing one's sense of smell means losing access — temporarily or permanently — to all of these vital functions. With the COVID-19 pandemic, millions of people have been newly experiencing smell disorders as a consequence of SARS-CoV-2 infection. In this chapter, we explore the types of olfactory disorders that are associated with SARS-CoV-2 and their known biological bases, with specific reference to the damage caused by SARS-CoV-2 to the sustentacular cells in the olfactory epithelium. The available interventions and the implications of temporary and permanent alterations of the sense of smell will be discussed, with particular reference to the experience of and strategies available to professionals relying on their intact chemical sense for their career.

10.1 COVID-19: A Magnifying Glass on the Chemical Senses

Perfumers, trained professionals in culinary and chemosensory science and other experts in chemosensory disciplines place high value on their sense of smell. Yet, the majority of people, laymen and scientists from other disciplines alike, underappreciate it. As an example, a group of 220 individuals responding to an online survey ranked their sense of smell as less valuable compared to their sight, hearing, balance, touch and taste [1], and even their cell phones [2]. It took millions of people globally concurrently losing their sense of smell as a result of a pandemic to make the point that our sense of smell plays an important role in our everyday lives. Indeed, it isn't until people lose their sense of smell that they come to appreciate how smell impacts our eating, social and protective behaviours [2, 3].

Although the prevalence of smell impairment is likely underestimated [4], pre-COVID-19, ~12% of people in the US [5], ~20% in Spain [6] and 16% in Germany [7] reported smell loss in their lifetime. This number has significantly increased as a result of COVID-19, as about 67% of individuals with COVID-19 have smell loss as a symptom [4], although this varies based on the SARS-CoV-2 variant [4, 8–10]. While the majority of these people will regain their sense of smell within a month of losing it [11–15], about 10–15% have smell loss for more than a month and may never recover [11, 12]. Furthermore, up to 66% of those with persistent smell loss from COVID-19 develop parosmia [16, 17] (i.e. odour distortion with a known source), similarly to what other post-viral conditions indicate [18].

Losing one's sense of smell, even if temporarily, is not the only chemosensory effect that characterises COVID-19 patients. Taste the ability to perceive sweetness, saltiness, sourness, bitterness and savoury aspects of food, among a few others (fat [19] and calorie [20]), and chemesthesis (the ability to perceive spiciness, coolness and carbonation through the oral and nasal mucosa) are affected in the acute and post-acute phase [4, 16, 21, 22].

To characterise the impact of COVID-19 on chemosensation, we review the organisation of the olfactory system, provide definitions of smell disorders, outline the known and potential mechanisms of COVID-19-related smell disorders and describe how smell disorders affect behaviour, including safety, nutrition, relationships, careers and quality of life.

10.1.1 *The olfactory system*

The detection of odours from the inhaled air is mediated by the olfactory epithelium (OE), a specialised epithelium located on the nasal conchae (or turbinates) in the most posterior region of the nasal cavity [23] (Figure 10.1). The OE of mammals consists of several cell types with different functions and morphologies: the olfactory sensory neurons (OSNs), sustentacular (Sus) cells, microvillar cells (MCs) and basal cells (BCs), as well as Bowman's glands, which secrete the layer of mucus that covers the surface of the OE [24].

OSNs have the function to detect the odorants that enter the nasal cavity. They are bipolar neurons, with the cell body located in the middle portion of the epithelium, and their axons projected via the olfactory nerve onto the olfactory bulb (OB), the first region of the brain receiving the olfactory information. On the other end of their cell body, a single unbranched dendrite departs, reaching the surface of the OE, where it ends in a spherical-shaped structure called a knob from which several cilia protrude into the nasal cavity [24]. In the cilia of the OSNs, all of the proteins

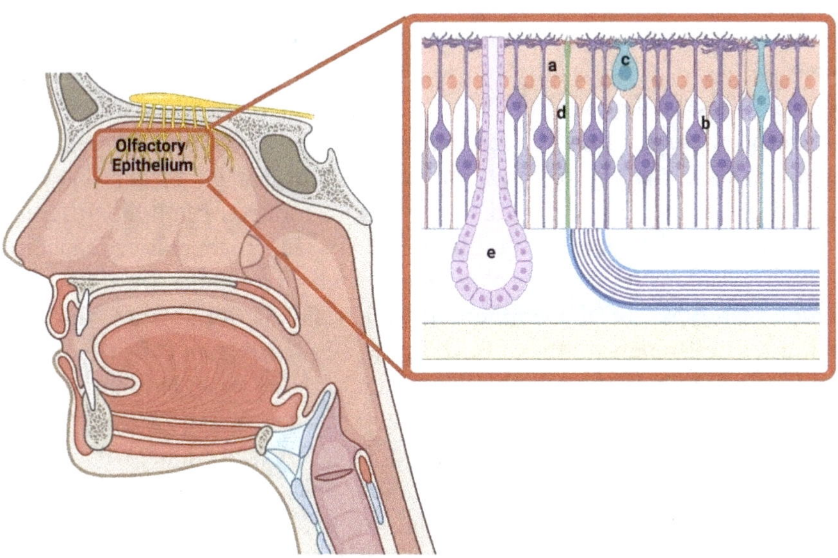

Figure 10.1. The human olfactory system. Within the olfactory epithelium, there are several cell types: (a) sustentacular cells, (b) olfactory sensory neurons, (c) microvillar cells, (d) trigeminal nerve and (e) Bowman's gland. Images created with BioRender.com.

necessary to detect odour molecules and generate a signal conveyed to the OB are expressed [25]. The detection of odorants starts with the activation of an olfactory receptor (OR), a G protein-coupled receptor [26]. Each OSN expresses only one of the hundreds (in humans, thousands for rodents) of receptor genes. All OSNs expressing the same OR are projecting to the same glomerulus in the OB, resulting in the activation of a combination of glomeruli, specific for each odour identity and intensity [26].

MCs and Sus cells are both considered supporting cells, not neuronal cells of the OE. Their function, as the name suggests, is to protect and maintain an optimal environment for the OSNs. MCs are located in the most apical portion of the OE. They present microvilli directly protruding into the nasal cavity and are suggested as key players in the protection of the OE from bacteria and chemical hazards [27–31].

Sus cells are tall and columnar glia-like cells, and they extend through the entire layer of the OE [24, 32]. Similar to MCs, Sus cells have microvilli protruding into the nasal cavity immersed in mucus. Their cell body forms a monolayer at the apical surface of the OE, wrapping the OSN dendrites and cell bodies and extending until the basal lamina [33]. Sus cells function as both epithelial (involved in secretion, endocytosis and metabolism of toxicants) and glial support cells [34–38]. They confer structural stability to the OE and regulate the extracellular ionic environment necessary for olfactory transduction and communication with neurons and basal cells [39–41]. Together with Bowman's glands, Sus cells are responsible for mucus secretion and the production of some of its components such as lysozymes, biotransformation enzymes and odorant-binding proteins [42]. In the OE, a continuous turnover of all the different cell types persists into adult life [43–46]. This regenerative capacity implies a lifelong persistence of stem cells, which can generate all the different cell types of the epithelium and maintain totipotency through unlimited self-renewal [45, 47, 48]. In the OE, these stem cells (i.e. BCs) are located at the bottom of the epithelium with the basal lamina.

The OE is also characterised by the presence of sensory trigeminal nerve endings in addition to the OSNs [49, 50] (see Figure 10.1). The trigeminal system and olfactory systems are different but often interact, as many odorants stimulate the trigeminal nerve [51]. The trigeminal system detects chemical, mechanical and thermal stimuli. It has somatosensory and nociceptive functions, which are carried out by specific subsets of nerve fibres. Chemosensory, or chemesthetic, trigeminal fibres are nociceptive free-ending fibres, which respond to a wide array of molecules.

When chemesthetic trigeminal nerve endings are stimulated, the sensations evoked are described as pungent, tingling, stinging, burning, cooling, warming, painful and irritating [49, 50, 52, 53]. To protect the airways from potentially harmful stimuli, activated trigeminal nociceptive fibres can induce different sensory-autonomic parasympathetic efferent reflexes, such as increasing excretions (e.g. nasal mucus), decreasing breathing frequency, releasing proinflammatory neuropeptides in the OE [49, 54], or coughing, sweating and sneezing [55, 56].

External and/or internal factors, such as viruses, pathogens, neurodegeneration or ageing, can affect olfactory functions and disrupt homeostasis, signalling cascades and regenerative abilities of the OE [57–60]. Based on their phenotype, it is possible to identify different types of olfactory dysfunctions.

10.1.2 *COVID-19-related mechanisms of olfactory dysfunction*

Disruption of the structure and function of the olfactory system leads to changes in the ability to perceive the intensity and the quality of odours. Anosmia is the inability to perceive odours, hyposmia is characterised by a significant reduction of olfactory function, and hyperosmia is the increased sensitivity to odours. These dysfunctions are quantitative and affect the perception of odours' intensity [61]. Parosmia and phantosmia, on the other hand, are qualitative olfactory disorders. Parosmia is a distortion of the perception of certain odours, whereas phantosmia is the sudden perception of a smell in the absence of an olfactory stimulus [61].

The pathophysiology of COVID-19 olfactory dysfunction is still not fully understood. SARS-CoV-2 is similar to other coronaviruses (i.e. SARS-CoV and MERS-CoV) in that it targets the respiratory system and infects host cells by binding to the angiotensin-converting enzyme 2 (ACE2) receptor [62–64]. Still, differently from the other known coronaviruses, SARS-CoV-2 can also affect other systems, including the gastrointestinal, cardiac and nervous systems [65]. This difference could be explained by the high affinity of SARS-CoV-2 for the ACE-2 receptor [66], which results in its ability to also infect cells with lower levels of expression of the receptor. ACE2 has been detected in the olfactory system; it is mainly expressed in the non-neuronal cell populations of the OE, such as Sus, MC and basal stem cells [67, 68].

Although Sus cells are not directly involved in the generation of the olfactory signal, they regulate several aspects of the OE homeostasis, which in turn assures the OSN function. Viral-mediated death of Sus cells could lead to changes in the ionic composition of the mucus, the prolonged presence of irritants and toxins in the mucus and the destitution of the epithelial structural integrity [69], which could lead to alterations in the normal physiology of OSNs or their death. Viral-mediated death of basal stem cells would negatively affect the OSN turnover/replacement, compromising the ability to recover from the loss of smell. Recent studies assessing samples of OE infected by SARS-CoV-2 have also shown a decrease in gene transcription of elements of the odour transduction pathway as well as changes in the nucleus architecture in the OSNs [70].

The peripheral degeneration of the OE due to the loss of Sus cells is not the only possible mechanism leading to post-viral anosmia. Due to its anatomical structure, the olfactory system is a point of access for many viruses to the central nervous system (CNS). The axons of the OSNs reach the OB by passing through the cribriform plate, a portion of the skull that separates the nasal cavity from the brain, which is perforated. Reports from hospitals and care providers show approximately 25% of patients with COVID-19 have neurological symptoms, such as vomiting, nausea and headache [71], introducing the possibility that damage to either the OB or other olfactory areas of the CNS could contribute to olfactory dysfunction. A study on transgenic mice showed that after intranasal viral application, it was possible to detect its antigens in the OB and its spreading in other areas of the CNS in the hours following the infection, reducing the neuronal density in the areas infected [72], which could compromise the olfactory functions. However, autopsies did not show any evidence of CNS damage due to SARS-CoV-2.

In most individuals, COVID-19-induced anosmia resolved within 4–6 weeks from its onset, but in approximately 10–15% of these patients, this symptom persists for six months or longer [73, 74]. Demographic and physiological factors are both considered to play a role in long-lasting anosmia. Age has been identified as a possible factor for slow or lack of recovery of the olfactory functions, with older age significantly associated with a longer prognosis. The role of inflammatory events at the level of the OE has been hypothesised as a factor contributing to long-lasting smell loss induced by COVID-19 [11, 75, 76]. Samples of olfactory mucosa from patients with persistent COVID-19-associated anosmia showed the presence of virus transcripts and infected cells, together with protracted inflammation [76].

The recovery process is accompanied by parosmia in up to 66% of those with smell loss [17]. The mechanisms for parosmia are not conclusive, although there are several hypotheses. One hypothesis suggests that parosmia is caused by newly regenerated OSNs rewiring to the wrong area in the OB, the first area in the brain processing the olfactory signal, such that an OSN with a specific OR projects to an area of the OB tuned to a different odour. Molecular triggers of parosmia have been identified, including thiols, pyrazines, disulphides and methoxypyrazines [77]. These molecules can be found in odours frequently reported to be distorted in those with parosmia, including coffee, onions, garlic, chocolate and meat. There may also be a central mechanism to parosmia, as those with hyposmia and parosmia have smaller OB volumes than those with hyposmia alone [78]. However, there is still much to be learned about the pathophysiology of parosmia.

10.1.3 *Behavioural effects of olfactory dysfunction*

10.1.3.1 *Safety*

Our sense of smell plays an important role in our safety. Foul smells can indicate that a food is toxic or spoiled prior to consumption. We also rely on our sense of smell to help protect us from hazards in our environment, such as food burning, smoke from a house fire or gas leaks. Individuals with smell disorders more frequently suffer impairments in detecting spoiled food than those without smell [79]. Those with smell loss are at an increased risk of experiencing a hazardous event because of their smell loss [80]. For example, those with anosmia are three times more likely to experience an olfactory-related hazardous event compared to those with a normal sense of smell [80]. People with congenital anosmia, who are born without a sense of smell, have an increased risk of “household accidents”, such as accidentally eating spoiled food, burning clothes while ironing, burning food and not being able to perceive smoke, compared to those who do have a sense of smell [81]. Since the COVID-19 pandemic, there have been multiple stories of families who narrowly escaped a house fire because they could not smell the smoke building up in their house due to smell loss from COVID-19 [82, 83]. Not being able to detect dangerous situations is a frequently reported fear in individuals who suffer from smell loss [84, 85], and it negatively impacts their quality of life [79]. To prevent these hazardous events, individuals that lose their sense of smell often rely on family members to tell them if a food is spoiled or not [85], avoid cooking [79] and/or ensure that smoke and carbon monoxide detectors are working.

10.1.3.2 *Nutrition*

Smell plays an integral role in food choice and eating behaviour. Aromas help us recognise that a food source is available and stimulate appetite by initiating a cephalic phase response to prepare our bodies for the incoming nutrients from food, even before the food is in our mouth [86]. The cephalic phase response includes salivation, gastrointestinal motility and exocrine and endocrine secretions to help metabolise nutrients from the food we eat [86]. For example, smelling baked goods when you walk into a mall may make your mouth water, or you may start to feel hungry and want to eat that baked good. Aromas specifically stimulate appetite for the food giving off the aroma or foods with certain characteristics (i.e. an aroma from a sweet food may stimulate appetite for other sweet foods) as opposed to food in general [87]. This is referred to as sensory-specific appetite. However, while aromas can affect food choice, they play less of a role in the amount of food actually consumed [88]. Aromas may indicate information about the energy, or caloric content, of the food prior to consumption, but this is likely due to learned associations from repeated exposure and post-ingestive consequences of the food throughout life [89–92].

Smell also contributes to the flavour and liking of foods. When food is in our mouth, the aroma of the food is perceived retronasally and combines with tastes (sweet, sour, salty, bitter and umami), textures, temperature and trigeminal sensations (i.e. burning from chilli pepper, cooling from mint) to create the flavour of the food. Flavour, which is often incorrectly referred to as the taste of foods, is consistently rated as the primary driver of food choice [93].

Due to its role in eating behaviour, it comes as no surprise that smell disorders affect one's relationship with food. The primary complaint of individuals who lose their sense of smell is that food is less flavourful and less enjoyable to eat [79]. This may result in appetite and subsequent body weight changes; however, the direction of these effects is inconsistent but may differ based on the type of smell disorder. For example, those with smell distortions, such as parosmia or phantosmia, more often report a decrease in appetite and experience weight loss, whereas those with smell loss, such as hyposmia or anosmia, are more likely to experience weight gain [94]. However, there are also many individuals with smell disorders who do not report any clinically meaningful changes in their body weight [95]. This indicates that individuals are able to cope and still consume an appropriate amount of food and beverages to maintain their energy needs. However, individuals with smell loss have a lower diet quality than normosmic individuals. Thus, smell loss typically results in consuming a less nutritious diet [96]. Individuals who lose their sense of smell often report adding sources of fats, sugars and salt to their foods to try to get some flavour and enjoyment back in the food [94, 97–99], although this varies across cultures [100]. These findings have been confirmed in long-term epidemiological studies, which report that adults with olfactory dysfunction consume diets with a higher energy density, particularly higher in saturated fats, total fat and added sugars compared to normosmic individuals [96]. Other ways that individuals with smell loss cope with food dislike include highlighting areas of flavour besides smell, including optimising the texture of foods, consuming foods at ideal temperatures and adding trigeminal components, such as chilli peppers, to their foods to enhance the flavour [79, 101]. Those with parosmia may benefit from consuming cold and/or non-aromatic foods so that no distortion is triggered and avoiding foods that trigger distorted odours [102].

10.1.3.3 *Relationships*

Scents play an important role in communication and relationships throughout the lifespan. Infants can recognise their caregivers and feel comforted by their scent [103]. In particular, the smell of their mother and their mother's breast milk can help to recognise food sources [104]. Being able to recognise the smell of the mother and the infant also helps the mother bond with her infant. Indeed, many women say that the smell of a baby is one of their favourite scents. Mothers who cannot detect their infant's odour over other infants or who are not attracted to their infant's odour have a more difficult time bonding with their child [105]. Similarly, smell loss in the mother or the infant could disrupt this bonding; however, further research is needed to address this question [106].

Scents also help with social relationships between adults. Subtle scent-based cues in body odour can help communicate emotions. Happiness, disgust and fear can all be conveyed in one's body odour

[107, 108]. In romantic relationships, scent can convey arousal and relationship status and contribute to attractiveness [109, 110]. Losing one's sense of smell can be isolating, and those with smell loss are more likely to feel lonely [111]. Feelings of isolation or loneliness may partly be due to not being able to detect odour-based communication cues, but also because of personal hygiene concerns [81, 84]. A common concern in those who lose their sense of smell is not being able to detect their own body odour [81], which may cause them to refrain from social engagements. The inability to detect their own personal hygiene may also cause individuals to feel insecure or refrain from sex. Individuals change their sexual behaviour after losing their sense of smell [112], and olfactory dysfunction is linked to decreased interest, enjoyment and participation in sex for both men and women [106].

Smell disorders can also affect relationships indirectly, due to a lack of empathy from loved ones. Unless someone has experienced a smell disorder, it is difficult to understand the impact smell disorders can have on people's lives. Their relationship to the world around them is different: they may no longer be able to smell the air after it rains, clean laundry or their significant other, or if they have parosmia, what used to be their favourite smell may now be distorted and cause them distress [113]. People with smell disorders often do not get the support they need from family, friends or significant others because their support system does not understand what they are going through or might think they are exaggerating their disorder [113]. The lack of support from their close networks is a challenging issue for those with smell disorders [113].

10.1.3.4 *Careers*

Several careers, including careers in perfumery, culinary arts, healthcare and safety, rely on having a functional sense of smell. The inability to recognise or reproduce a fragrance, creating unbalanced recipes, missing foul smells that are harbingers to diseases or missing a gas leak may end up having catastrophic outcomes for oneself and others [2, 113]. Luckily, this does not necessarily mean the individuals are no longer able to work. In a study by Haxel *et al.*, who surveyed a group of individuals whose careers rely on their sense of smell, only 5% reported no longer being able to work in their profession after losing their sense of smell [114]. However, 60% of individuals had to make special adjustments for their job, and 35% continued working without restrictions [114]. Whether the effect of smell loss from COVID-19 impacts careers differently, or whether some of these careers where smell is integral are more affected than others, is still unknown. Overall, it appears that most people can make adjustments to their jobs that focus on other aspects besides smell to be successful, especially if they are veterans in their careers. For example, veteran perfumers can use their experience to know how odours will interact to form a product and then put trust in others to be their noses [115].

10.1.3.5 *Quality of life*

Overall, the roles of olfaction on safety, nutrition, relationships and in some careers interact to make it an important contributor to one's quality of life. Individuals with smell disorders report significant impairments to their quality of life [79, 95] and have high rates of anxiety and depression [95, 102, 116]. Qualitative smell disorders, such as parosmia and phantosmia, may have a greater impact on quality of life than qualitative smell disorders, such as anosmia and hyposmia [95, 117]. There are multiple charities whose mission is to help improve the quality of life in those who lose their sense of smell. However, this issue will continue to be important as the prevalence of smell loss increases.

10.2 What is the Aftermath of COVID-19 for the Chemical Senses?

The COVID-19 pandemic has highlighted the importance our sense of smell has on our daily lives. With more people, including perfumers, now having smell disorders than ever before, more research into effective treatments and therapies to improve quality of life is needed. On the one hand, losing their sense of smell can be severely damaging to a perfumers' career, unless they are veterans in the field and are able to adapt. For those who are not able to adapt, there should be an open space to discuss problems and solutions with other perfumes without fear of not being trusted to do the job. Perfumers with smell loss may also seek disability assistance for their long-COVID-19 symptoms.

On the other hand, perfumers can use their skills to create approaches to help those with smell disorders. For example, perfumers could alter how they make perfumes to be sensitive to those with smell disorders. Many people with parosmia find perfumes to be unpleasant because some of the odour molecules in the perfumes are distorted to them. Identifying these odour molecules and removing them from the perfume could create a more pleasant experience for those with parosmia. For those with anosmia who otherwise would not be able to smell the perfumes, prioritising the trigeminal sensations could provide them with a sensory experience they otherwise would not have. This type of project could best be tackled by a team of perfumers with and without smell disorders, so that there is input on how it affects those with a smell disorder and those with normosmia alike. This has been achieved by perfumer Ugo Charron and congenital anosmic Emmanuelle Dancourt. Together, they created the perfume “Umema” based on textures, visuals and tastes that Emmanuelle loved, such as suede, flowers and umami [118]. Perfumers could also aid in developing treatments for smell disorders. Research studies are currently ongoing to understand the effectiveness of various treatments for smell disorders; however, currently, the only recommended treatment for smell disorders is smell training. Smell training includes sniffing four intense odours (rose, eucalyptus, lemon and cloves) twice a day for at least 12 weeks [119]. Unfortunately, many participants become frustrated and have a hard time adhering to the training [120]. Perfumers already practice a form of smell training as part of their career training. Their expertise could be used to communicate how smell training can help in daily life with a lay audience. Perfumers could also aid scientists in coming up with ideas to increase adherence to olfactory training for several months, which has the potential to help more people recover from their smell disorders [121].

10.3 Conclusions

The recent high rate of SARS-CoV-2 infections caused a worldwide increase in the number of people affected by smell disorders, and perfumers have not been exempt. Research to date shows the infection of sustentacular cells by SARS-CoV-2 as the primary event leading to smell loss. The mechanisms behind long-term smell loss and parosmia remain unclear, and more research on these conditions is necessary. Those experiencing long-term smell loss face an overall reduction in life quality, which includes increased mental, physical and emotional health risks due to the inability to detect smoke, gas and/or spoiled foods, changes in eating behaviours and reduced eating enjoyment, struggles with socially bonding with others and depression. Some coping strategies can help to mitigate these negative effects, but there is still a limited understanding of how deeply all these factors affect everyday life. Perfumers and other professionals who rely on their sense of smell for their careers may need to rely on others to be their nose but can adapt to still perform the job. Adapting may be easier for veterans in their field and should aid newcomers who find themselves without a sense of smell. Perfumers can aid scientists in developing coping strategies and treatments to improve quality of life in those with smell disorders.

Conflict of Interest

The authors do not declare any conflict of interest.

References

[1] J. Enoch, L. McDonald, L. Jones, P.R. Jones, D.P. Crabb. (2019). Evaluating whether sight is the most valued sense, *JAMA Ophthalmol.*, 137, 1317–1320. https://doi.org/10.1001/jamaophthalmol.2019.3537.

[2] I. Croy, S. Nordin, T. Hummel. (2014). Olfactory disorders and quality of life–an updated review, *Chem. Senses*, 39, 185–194. https://doi.org/10.1093/chemse/bjt072.

[3] L.A. Vaira, C. Gessa, G. Deiana, G. Salzano, F. Maglitto, J.R. Lechien, *et al.* (2022). The effects of persistent olfactory and gustatory dysfunctions on quality of life in long-COVID-19 patients, *Life*, 12, 141. https://doi.org/10.3390/life12020141.

[4] M.E. Hannum, V.A. Ramirez, S.J. Lipson, R.D. Herriman, A.K. Toskala, C. Lin, *et al.* (2020). Objective sensory testing methods reveal a higher prevalence of olfactory loss in COVID-19-positive patients compared to subjective methods: A systematic review and meta-analysis, *Chem. Senses*, 45, 865–874. https://doi.org/10.1093/chemse/bjaa064.

[5] H.J. Hoffman, S. Rawal, C.M. Li, V.B. Duffy. (2016). New chemosensory component in the U.S. National Health and Nutrition Examination Survey (NHANES): First-year results for measured olfactory dysfunction, *Rev. Endocr. Metab. Disord.*, 17, 221–240. https://doi.org/10.1007/s11154-016-9364-1.

[6] J. Mullol, I. Alobid, F. Mariño-Sánchez, L. Quintó, J. de Haro, M. Bernal-Sprekelsen, *et al.* (2012). Furthering the understanding of olfaction, prevalence of loss of smell and risk factors: A population-based survey

(OLFACAT study), *BMJ Open*, 2, e001256. https://doi.org/10.1136/bmjopen-2012-001256.

[7] B.N. Landis, C.G. Konnerth, T. Hummel. (2004). A study on the frequency of olfactory dysfunction, *Laryngoscope*, 114, 1764–1769. https://doi.org/10.1097/00005537-200410000-00017.

[8] K.D. Vihta, K.B. Pouwels, T.E. Peto, E. Pritchard, T. House, R. Studley, *et al.* (2022). OMICRON-associated changes in SARS-COV-2 symptoms in the United Kingdom, *medRxiv*, 2022.01.18.22269082. https://doi.org/10.1101/2022.01.18.22269082.

[9] B. Killingley, A. Mann, M. Kalinova, A. Boyers, N. Goonawardane, J. Zhou, *et al.* (2022). Safety, tolerability, and viral kinetics during SARS-CoV-2 human challenge, *Nature Portfolio.* https://doi.org/10.21203/rs.3.rs-1121993/v1.

[10] C. Menni, A.M. Valdes, L. Polidori, M. Antonelli, S. Penamakuri, A. Nogal, *et al.* (2022). Symptom prevalence, duration, and risk of hospital admission in individuals infected with SARS-CoV-2 during periods of omicron and delta variant dominance: A prospective observational study from the ZOE COVID Study, *Lancet*, 399, 1618–1624. https://doi.org/10.1016/S0140-6736(22)00327-0.

[11] C.M. Chiesa-Estomba, J.R. Lechien, T. Radulesco, J. Michel, L.J. Sowerby, C. Hopkins, *et al.* (2020). Patterns of smell recovery in 751 patients affected by the COVID-19 outbreak, *Eur. J. Neurol.*, 27, 2318–2321. https://doi.org/10.1111/ene.14440.

[12] S.T. Moein, S.M. Hashemian, P. Tabarsi, R.L. Doty. (2020). Prevalence and reversibility of smell dysfunction measured psychophysically in a cohort of COVID-19 patients, *Int. Forum Allergy Rhinol.*, 10, 1127–1135. https://doi.org/10.1002/alr.22680.

[13] E. Sakalli, D. Temirbekov, E. Bayri, E.E. Alis, S.C. Erdurak, M. Bayraktaroglu. (2020). Ear nose throat-related symptoms with a focus on loss of smell and/or taste in COVID-19 patients, *Am. J. Otolaryngol.*, 41, 102622. https://doi.org/10.1016/j.amjoto.2020.102622.

[14] L.A. Vaira, C. Hopkins, M. Petrocelli, J.R. Lechien, C.M. Chiesa-Estomba, G. Salzano, *et al.* (2020). Smell and taste recovery in coronavirus disease 2019 patients: A 60-day objective and prospective study, *J. Laryngol. Otol.*, 134, 703–709. https://doi.org/10.1017/S0022215120001826.

[15] E.R. Reiter, D.H. Coelho, Z.A. Kons, R.M. Costanzo. (2020). Subjective smell and taste changes during the COVID-19 pandemic: Short term recovery, *Am. J. Otolaryngol.*, 41, 102639. https://doi.org/10.1016/j.amjoto.2020.102639..

[16] K. Ohla, M.G. Veldhuizen, T. Green, M.E. Hannum, A.J. Bakke, S.T. Moein, *et al.* (2022). A follow-up on quantitative and qualitative

olfactory dysfunction and other symptoms in patients recovering from COVID-19 smell loss, *Rhinology.* https://doi.org/10.4193/Rhin21.415.

[17] R. Pellegrino, J.D. Mainland, C.E. Kelly, J.K. Parker, T. Hummel. (2021). Prevalence and correlates of parosmia and phantosmia among smell disorders, *Chem. Senses*, 46, bjab046. https://doi.org/10.1093/chemse/bjab046.

[18] J. Reden, H. Maroldt, A. Fritz, T. Zahnert, T. Hummel. (2007). A study on the prognostic significance of qualitative olfactory dysfunction, *Eur. Arch. Otorhinolaryngol.*, 264, 139–144. https://doi.org/10.1007/s00405-006-0157-0.

[19] P. Besnard, P. Passilly-Degrace, N.A. Khan. (2016). Taste of Fat: A Sixth Taste Modality? *Physiol. Rev.*, 96, 151–176. https://doi.org/10.1152/physrev.00002.2015.

[20] S. Iwata, R. Yoshida, Y. Ninomiya. (2014). Taste transductions in taste receptor cells: Basic tastes and moreover, *Curr. Pharm. Des.*, 20, 2684–2692. https://doi.org/10.2174/13816128113199990575.

[21] C. Cecchetto, A. Di Pizio, F. Genovese, O. Calcinoni, A. Macchi, A. Dunkel, *et al.* (2021). Assessing the extent and timing of chemosensory impairments during COVID-19 pandemic, *Sci. Rep.*, 11, 17504. https://doi.org/10.1038/s41598-021-96987-0.

[22] M.E. Hannum, R.J. Koch, V.A. Ramirez, S.S. Marks, A.K. Toskala, R.D. Herriman, *et al.* (2022). Taste loss as a distinct symptom of COVID-19: A systematic review and meta-analysis, *Chem. Senses*, 47, bjac001. https://doi.org/10.1093/chemse/bjac001.

[23] P.A. Escada, C. Lima, J. Madeira da Silva. (2009). The human olfactory mucosa, *Eur. Arch. Otorhinolaryngol.*, 266, 1675–1680. https://doi.org/10.1007/s00405-009-1073-x.

[24] E.E. Morrison, R.M. Costanzo. (1992). Morphology of olfactory epithelium in humans and other vertebrates, *Microsc. Res. Tech.*, 23, 49–61. https://doi.org/10.1002/jemt.1070230105.

[25] G.V. Ronnett, C. Moon. (2002). G proteins and olfactory signal transduction, *Annu. Rev. Physiol.*, 64, 189–222. https://doi.org/10.1146/annurev.physiol.64.082701.102219.

[26] J.D. Mainland, J.N. Lundström, J. Reisert, G. Lowe. (2014). From molecule to mind: An integrative perspective on odor intensity, *Trends Neurosci.*, 37, 443–454. https://doi.org/10.1016/j.tins.2014.05.005.

[27] R. Elsaesser, J. Paysan. (2007). The sense of smell, its signalling pathways, and the dichotomy of cilia and microvilli in olfactory sensory cells, *BMC Neurosci.*, 8, S1. https://doi.org/10.1186/1471-2202-8-S3-S1.

[28] T. Ogura, S.A. Szebenyi, K. Krosnowski, A. Sathyanesan, J. Jackson, W. Lin. (2011). Cholinergic microvillous cells in the mouse main

olfactory epithelium and effect of acetylcholine on olfactory sensory neurons and supporting cells, *J. Neurophysiol.*, 106, 1274–1287. https://doi.org/10.1152/jn.00186.2011.

[29] K. Lemons, Z. Fu, I. Aoudé, T. Ogura, J. Sun, J. Chang, *et al.* (2017). Lack of TRPM5-expressing microvillous cells in mouse main olfactory epithelium leads to impaired odor-evoked responses and olfactory-guided behavior in a challenging chemical environment, *eNeuro*, 4, ENEURO.0135-17.2017. https://doi.org/10.1523/ENEURO.0135-17.2017.

[30] Z. Fu, T. Ogura, W. Luo, W. Lin. (2018). ATP and odor mixture activate TRPM5-expressing microvillous cells and potentially induce acetylcholine release to enhance supporting cell endocytosis in mouse main olfactory epithelium, *Front. Cell. Neurosci.*, 12. https://www.frontiersin.org/articles/10.3389/fncel.2018.00071.

[31] K. Lemons, Z. Fu, T. Ogura, W. Lin. (2020). TRPM5-expressing microvillous cells regulate region-specific cell proliferation and apoptosis during chemical exposure, *Neuroscience*, 434, 171–190. https://doi.org/10.1016/j.neuroscience.2020.03.029.

[32] D.T. Moran, B.W. Jafek, P.M. Eller, J.C. Rowley 3rd. (1992). Ultrastructural histopathology of human olfactory dysfunction, *Microsc. Res. Tech.*, 23, 103–110. https://doi.org/10.1002/jemt.1070230202.

[33] F. Liang. (2020). Sustentacular cell enwrapment of olfactory receptor neuronal dendrites: An update, *Genes*, 11. https://doi.org/10.3390/genes11050493.

[34] M. Okano, S.F. Takagi. (1974). Secretion and electrogenesis of the supporting cell in the olfactory epithelium, *J. Physiol.*, 242, 353–370. https://doi.org/10.1113/jphysiol.1974.sp010711.

[35] A.R. Dahl, W.M. Hadley, F.F. Hahn, J.M. Benson, R.O. McClellan. (1982). Cytochrome P-450-dependent monooxygenases in olfactory epithelium of dogs: Possible role in tumorigenicity, *Science*, 216, 57–59. https://doi.org/10.1126/science.7063870.

[36] L.H. Bannister, H.C. Dodson. (1992). Endocytic pathways in the olfactory and vomeronasal epithelia of the mouse: Ultrastructure and uptake of tracers, *Microsc. Res. Tech.*, 23, 128–141. https://doi.org/10.1002/jemt.1070230204.

[37] E. Piras, A. Franzén, E.L. Fernández, U. Bergström, F. Raffalli-Mathieu, M. Lang, *et al.* (2003). Cell-specific expression of CYP2A5 in the mouse respiratory tract: Effects of olfactory toxicants, *J. Histochem. Cytochem.*, 51, 1545–1555. https://doi.org/10.1177/002215540305101114.

[38] T.V. Getchell. (1977). Analysis of intracellular recordings from salamander olfactory epithelium, *Brain Res.*, 123, 275–286. https://doi.org/10.1016/0006-8993(77)90479-6.

[39] W. Breipohl, H.J. Laugwitz, N. Bornfeld. (1974). Topological relations between the dendrites of olfactory sensory cells and sustentacular cells in different vertebrates. An ultrastructural study, *J. Anat.*, 117, 89–94.

[40] J.A. Rafols, T.V. Getchell. (1983). Morphological relations between the receptor neurons, sustentacular cells and Schwann cells in the olfactory mucosa of the salamander, *Anat. Rec.*, 206, 87–101. https://doi.org/10.1002/ar.1092060111.

[41] T.V. Getchell. (1986). Functional properties of vertebrate olfactory receptor neurons, *Physiol. Rev.*, 66, 772–818. https://doi.org/10.1152/physrev.1986.66.3.772.

[42] Y. Suzuki, M. Takeda, N. Obara, N. Suzuki, N. Takeichi. (2000). Olfactory epithelium consisting of supporting cells and horizontal basal cells in the posterior nasal cavity of mice, *Cell Tissue Res.*, 299, 313–325. https://doi.org/10.1007/s004419900135.

[43] P.P.C. Graziadei, G.A.M. Graziadei. (1979). Neurogenesis and neuron regeneration in the olfactory system of mammals. I. Morphological aspects of differentiation and structural organization of the olfactory sensory neurons, *J. Neurocytol.*, 8, 1–18. https://doi.org/10.1007/BF01206454.

[44] R.B. Fletcher, D. Das, L. Gadye, K.N. Street, A. Baudhuin, A. Wagner, *et al.* (2017). Deconstructing olfactory stem cell trajectories at single-cell resolution, *Cell Stem Cell*, 20, 817-830.e8. https://doi.org/10.1016/j.stem.2017.04.003.

[45] J.E. Schwob, W. Jang, E.H. Holbrook, B. Lin, D.B. Herrick, J.N. Peterson, *et al.* (2017). Stem and progenitor cells of the mammalian olfactory epithelium: Taking poietic license, *J. Comp. Neurol.*, 525, 1034–1054. https://doi.org/10.1002/cne.24105.

[46] M.A. Durante, S. Kurtenbach, Z.B. Sargi, J.W. Harbour, R. Choi, S. Kurtenbach, *et al.* (2020). Single-cell analysis of olfactory neurogenesis and differentiation in adult humans, *Nat. Neurosci.*, 23, 323–326. https://doi.org/10.1038/s41593-020-0587-9.

[47] J.E. Schwob. (2005). Restoring olfaction: A view from the olfactory epithelium, *Chem. Senses*, 30, i131–i132. https://doi.org/10.1093/chemse/bjh149.

[48] R. Abu-Dawud, N. Graffmann, S. Ferber, W. Wruck, J. Adjaye. (2018). Pluripotent stem cells: Induction and self-renewal, *Philos. Trans. R. Soc. Lond. B Biol. Sci.*, 373, 20170213. https://doi.org/10.1098/rstb.2017.0213.

[49] T.E. Finger, B. Böttger. (1993). Peripheral peptidergic fibers of the trigeminal nerve in the olfactory bulb of the rat, *J. Comp. Neurol.*, 334, 117–124. https://doi.org/10.1002/cne.903340110.

[50] M.L. Schaefer, B. Böttger, W.L. Silver, T.E. Finger. (2002). Trigeminal collaterals in the nasal epithelium and olfactory bulb: A potential route for direct modulation of olfactory information by trigeminal stimuli, *J. Comp. Neurol.*, 444, 221–226. https://doi.org/10.1002/cne.10143.

[51] J. Frasnelli, B. Schuster, T. Hummel. (2007). Interactions between olfaction and the trigeminal system: What can be learned from olfactory loss, *Cereb. Cortex*, 17, 2268–2275. https://doi.org/10.1093/cercor/bhl135.

[52] P.M. Richards, E.C. Johnson, W.L. Silver. (2010). Four irritating odorants target the trigeminal chemoreceptor TRPA1, *Chemosens. Percept.*, 3, 190–199. https://doi.org/10.1007/s12078-010-9081-1.

[53] A.I. Basbaum, D.M. Bautista, G. Scherrer, D. Julius. (2009). Cellular and molecular mechanisms of pain, *Cell*, 139, 267–284. https://doi.org/10.1016/j.cell.2009.09.028.

[54] F. Viana. (2011). Chemosensory properties of the trigeminal system, *ACS Chem. Neurosci.*, 2, 38–50. https://doi.org/10.1021/cn100102c.

[55] A.D. Mickle, A.J. Shepherd, D.P. Mohapatra. (2015). Sensory TRP channels: The key transducers of nociception and pain, *Prog. Mol. Biol. Transl. Sci.*, 131, 73–118. https://doi.org/10.1016/bs.pmbts.2015.01.002.

[56] W.L. Silver, T.E. Finger. (2009). The anatomical and electrophysiological basis of peripheral nasal trigeminal chemoreception, *Ann. N. Y. Acad. Sci.*, 1170, 202–205. https://doi.org/10.1111/j.1749-6632.2009.03894.x.

[57] M. Lad, W. Sedley, T.D. Griffiths. (2022). Sensory loss and risk of dementia, *Neuroscientist*, 10738584221126090. https://doi.org/10.1177/10738584221126090.

[58] A. Alkelai, T. Olender, C. Dode, S. Shushan, P. Tatarskyy, E. Furman-Haran, *et al.* (2017). Next-generation sequencing of patients with congenital anosmia, *Eur. J. Hum. Genet.*, 25, 1377–1387. https://doi.org/10.1038/s41431-017-0014-1.

[59] Z.M. Soler, Z.M. Patel, J.H. Turner, E.H. Holbrook. (2020). A primer on viral-associated olfactory loss in the era of COVID-19, *Int. Forum Allergy Rhinol.*, 10, 814–820. https://doi.org/10.1002/alr.22578.

[60] S.M. Bromley, R.L. Doty. (2019). Health histories and medical evaluations of patients with complaints of chemosensory dysfunction, *Handb. Clin. Neurol.*, 164, 219–227. https://doi.org/10.1016/B978-0-444-63855-7.00014-9.

[61] J. Frasnelli, T. Hummel. (2005). Olfactory dysfunction and daily life, *Eur. Arch. Otorhinolaryngol.*, 262, 231–235. https://doi.org/10.1007/s00405-004-0796-y.

[62] J. Lan, J. Ge, J. Yu, S. Shan, H. Zhou, S. Fan, *et al.* (2020). Structure of the SARS-CoV-2 spike receptor-binding domain bound to the ACE2 receptor, *Nature*, 581, 215–220. https://doi.org/10.1038/s41586-020-2180-5.

[63] I. Hamming, W. Timens, M.L.C. Bulthuis, A.T. Lely, G.J. Navis, H. van Goor. (2004). Tissue distribution of ACE2 protein, the functional receptor for SARS coronavirus. A first step in understanding SARS pathogenesis, *J. Pathol.*, 203, 631–637. https://doi.org/10.1002/path.1570.

[64] W. Li, M.J. Moore, N. Vasilieva, J. Sui, S.K. Wong, M.A. Berne, *et al.* (2003). Angiotensin-converting enzyme 2 is a functional receptor for the SARS coronavirus, *Nature*, 426, 450–454. https://doi.org/10.1038/nature02145.

[65] M. Kopańska, E. Barnaś, J. Błajda, B. Kuduk, A. Łagowska, A. Banaś-Ząbczyk. (2022). Effects of SARS-CoV-2 inflammation on selected organ systems of the human body, *Int. J. Mol. Sci.*, 23. https://doi.org/10.3390/ijms23084178.

[66] K.G. Andersen, A. Rambaut, W.I. Lipkin, E.C. Holmes, R.F. Garry. (2020). The proximal origin of SARS-CoV-2, *Nat. Med.*, 26, 450–452. https://doi.org/10.1038/s41591-020-0820-9.

[67] D.H. Brann, T. Tsukahara, C. Weinreb, M. Lipovsek, K. Van den Berge, B. Gong, *et al.* (2020). Non-neuronal expression of SARS-CoV-2 entry genes in the olfactory system suggests mechanisms underlying COVID-19-associated anosmia, *Sci. Adv.*, 6. https://doi.org/10.1126/sciadv.abc5801.

[68] K. Gupta, S.K. Mohanty, A. Mittal, S. Kalra, S. Kumar, T. Mishra, *et al.* (2021). The cellular basis of loss of smell in 2019-nCoV-infected individuals, *Brief. Bioinform.*, 22, 873–881. https://doi.org/10.1093/bib/bbaa168.

[69] A. Hernandez-Clavijo, K.Y. Gonzalez-Velandia, U. Rangaswamy, G. Guarneri, P. Boscolo-Rizzo, M. Tofanelli, *et al.* (2022). Supporting cells of the human olfactory epithelium co-express the lipid scramblase TMEM16F and ACE2 and may cause smell loss by SARS-CoV-2 spike-induced syncytia, *Cell. Physiol. Biochem.*, 56, 254–269. https://doi.org/10.33594/000000531.

[70] M. Zazhytska, A. Kodra, D.A. Hoagland, J. Frere, J.F. Fullard, H. Shayya, *et al.* (2022). Non-cell-autonomous disruption of nuclear architecture as a potential cause of COVID-19-induced anosmia, *Cell*, 185, 1052-1064.e12. https://doi.org/10.1016/j.cell.2022.01.024.

[71] A.A. Asadi-Pooya, L. Simani. (2020). Central nervous system manifestations of COVID-19: A systematic review, *J. Neurol. Sci.*, 413, 116832. https://doi.org/10.1016/j.jns.2020.116832.

[72] J. Netland, D.K. Meyerholz, S. Moore, M. Cassell, S. Perlman. (2008). Severe acute respiratory syndrome coronavirus infection causes neuronal death in the absence of encephalitis in mice transgenic for human ACE2, *J. Virol.*, 82, 7264–7275. https://doi.org/10.1128/JVI.00737-08.

[73] J.R. Lechien, C.M. Chiesa-Estomba, E. Beckers, V. Mustin, M. Ducarme, F. Journe, *et al.* (2021). Prevalence and 6-month recovery of olfactory dysfunction: A multicentre study of 1363 COVID-19 patients, *J. Intern. Med.* https://doi.org/10.1111/joim.13209.

[74] J.K. Logue, N.M. Franko, D.J. McCulloch, D. McDonald, A. Magedson, C.R. Wolf, *et al.* (2021). Sequelae in adults at 6 months after COVID-19 infection, *JAMA Netw. Open*, 4, e210830. https://doi.org/10.1001/jamanetworkopen.2021.0830.

[75] M. Gousseff, P. Penot, L. Gallay, D. Batisse, N. Benech, K. Bouiller, *et al.* (2020). Clinical recurrences of COVID-19 symptoms after recovery: Viral relapse, reinfection or inflammatory rebound? *J. Infect.*, 81, 816–846. https://doi.org/10.1016/j.jinf.2020.06.073.

[76] G.D. de Melo, F. Lazarini, S. Levallois, C. Hautefort, V. Michel, F. Larrous, *et al.* (2021). COVID-19-related anosmia is associated with viral persistence and inflammation in human olfactory epithelium and brain infection in hamsters, *Sci. Transl. Med.*, 13, eabf8396. https://doi.org/10.1126/scitranslmed.abf8396.

[77] J.K. Parker, C.E. Kelly, S.B. Gane. (2022). Insights into the molecular triggers of parosmia based on gas chromatography olfactometry, *Commun. Med.*, 2, 58. https://doi.org/10.1038/s43856-022-00112-9.

[78] A. Mueller, N.D. Abolmaali, A.R. Hakimi, T. Gloeckler, B. Herting, H. Reichmann, *et al.* (2005). Olfactory bulb volumes in patients with idiopathic Parkinson's disease a pilot study, *J. Neural Transm.*, 112, 1363–1370. https://doi.org/10.1007/s00702-005-0280-x.

[79] T. Miwa, M. Furukawa, T. Tsukatani, R.M. Costanzo, L.J. DiNardo, E.R. Reiter. (2001). Impact of olfactory impairment on quality of life and disability, *Arch. Otolaryngol. Head Neck Surg.*, 127, 497–503. https://doi.org/10.1001/archotol.127.5.497.

[80] T.S. Pence, E.R. Reiter, L.J. DiNardo, R.M. Costanzo. (2014). Risk factors for hazardous events in olfactory-impaired patients, *JAMA Otolaryngol. Head Neck Surg.*, 140, 951–955. https://doi.org/10.1001/jamaoto.2014.1675.

[81] I. Croy, S. Negoias, L. Novakova, B.N. Landis, T. Hummel. (2012). Learning about the functions of the olfactory system from people without a sense of smell, *PLoS One*, 7, e33365. https://doi.org/10.1371/journal.pone.0033365.

[82] A. Elassar. (2021). A teenager saved her family when they lost their sense of smell from Covid-19 and didn't realize their house was on fire. CNN.

[83] A. Gilbert. (2022). Parents couldn't smell smoke when their house caught fire. Their 2-year-old saved them. USA Today.

[84] E.H. Blomqvist, A. Brämerson, P. Stjärne, S. Nordin. (2004). Consequences of olfactory loss and adopted coping strategies, *Rhinology*, 42, 189–194.

[85] S. Nordin, E.H. Blomqvist, P. Olsson, P. Stjärne, A. Ehnhage. (2011). Effects of smell loss on daily life and adopted coping strategies in patients with nasal polyposis with asthma, *Acta Otolaryngol.*, 131, 826–832. https://doi.org/10.3109/00016489.2010.539625.

[86] P.A. Smeets, A. Erkner, C. de Graaf. (2010). Cephalic phase responses and appetite, *Nutr. Rev.*, 68, 643–655. https://doi.org/10.1111/j.1753-4887.2010.00334.x.
[87] H.F.A. Zoon, C. de Graaf, S. Boesveldt. (2016). Food odours direct specific appetite, *Foods*, 5, 12. https://doi.org/10.3390/foods5010012.
[88] M.G. Ramaekers, P.A. Luning, R.M.A.J. Ruijschop, C.M.M. Lakemond, J.H.F. Bult, G. Gort, *et al.* (2014). Aroma exposure time and aroma concentration in relation to satiation, *Br. J. Nutr.*, 111, 554–562. https://doi.org/10.1017/S0007114513002729.
[89] J.M. Brunstrom, G.L. Mitchell. (2007). Flavor-nutrient learning in restrained and unrestrained eaters, *Physiol. Behav.*, 90, 133–141. https://doi.org/10.1016/j.physbeh.2006.09.016.
[90] D.M. Small, M.G. Veldhuizen, J. Felsted, Y.E. Mak, F. McGlone. (2008). Separable substrates for anticipatory and consummatory food chemosensation, *Neuron*, 57, 786–797. https://doi.org/10.1016/j.neuron.2008.01.021.
[91] M.R. Yeomans, S. Boakes. (2016). That smells filling: Effects of pairings of odours with sweetness and thickness on odour perception and expected satiety, *Food Qual. Prefer.*, 54, 128–136. https://doi.org/10.1016/j.foodqual.2016.07.010.
[92] M.R. Yeomans, R. Re, M. Wickham, H. Lundholm, L. Chambers. (2016). Beyond expectations: The physiological basis of sensory enhancement of satiety, *Int. J. Obes.*, 40, 1693–1698. https://doi.org/10.1038/ijo.2016.112.
[93] International Food Information Council. (2020). 2020 Food & Health Survey.
[94] R.D. Mattes, B.J. Cowart, M.A. Schiavo, C. Arnold, B. Garrison, M.R. Kare, *et al.* (1990). Dietary evaluation of patients with smell and/or taste disorders, *Am. J. Clin. Nutr.*, 51, 233–240. https://doi.org/10.1093/ajcn/51.2.233.
[95] D.A. Deems, R.L. Doty, R.G. Settle, V. Moore-Gillon, P. Shaman, A.F. Mester, *et al.* (1991). Smell and taste disorders, a study of 750 patients from the University of Pennsylvania smell and taste center, *Arch. Otolaryngol. Head Neck Surg.*, 117, 519–528. https://doi.org/10.1001/archotol.1991.01870170065015.
[96] S. Rawal, V.B. Duffy, L. Berube, J.E. Hayes, A.K. Kant, C.M. Li, *et al.* (2021). Self-reported olfactory dysfunction and diet quality: Findings from the 2011-2014 National Health and Nutrition Examination Survey (NHANES), *Nutrients*, 13, 4561. https://doi.org/10.3390/nu13124561.
[97] R.D. Mattes, B.J. Cowart. (1994). Dietary assessment of patients with chemosensory disorders, *J. Am. Diet. Assoc.*, 94, 50–56. https://doi.org/10.1016/0002-8223(94)92041-9.

[98] K. Aschenbrenner, C. Hummel, K. Teszmer, F. Krone, T. Ishimaru, H.S. Seo, *et al.* (2008). The influence of olfactory loss on dietary behaviors, *Laryngoscope*, 118, 135–144. https://doi.org/10.1097/MLG.0b013e318155a4b9.

[99] L. Turner, P.J. Rogers. (2022). Varied effects of COVID-19 chemosensory loss and distortion on appetite: Implications for understanding motives for eating and drinking, *Foods*, 11, 607. https://doi.org/10.3390/foods11040607.

[100] R. Pellegrino, T. Hummel, R. Emrich, R. Chandra, J. Turner, T. Trone, *et al.* (2020). Cultural determinants of food attitudes in anosmic patients, *Appetite*, 147, 104563. https://doi.org/10.1016/j.appet.2019.104563.

[101] S.R. Hunter, P.H. Dalton. (2022). The need for sensory nutrition research in individuals with smell loss, *Clin. Nutr. Open Sci.* https://doi.org/10.1016/j.nutos.2022.11.002.

[102] J.K. Parker, C.E. Kelly, B.C. Smith, A.F. Kirkwood, C. Hopkins, S. Gane. (2021). Patients' perspectives on qualitative olfactory dysfunction: Thematic analysis of social media posts, *JMIR Form. Res.*, 5, e29086. https://doi.org/10.2196/29086.

[103] R.M. Sullivan, P. Toubas. (1998). Clinical usefulness of maternal odor in newborns: Soothing and feeding preparatory responses, *Neonatology*, 74, 402–408. https://doi.org/10.1159/000014061.

[104] S. Doucet, R. Soussignan, P. Sagot, B. Schaal. (2012). An overlooked aspect of the human breast: Areolar glands in relation with breastfeeding pattern, neonatal weight gain, and the dynamics of lactation, *Early Hum. Dev.*, 88, 119–128. https://doi.org/10.1016/j.earlhumdev.2011.07.020.

[105] I. Croy, T. Mohr, K. Weidner, T. Hummel, J. Junge-Hoffmeister. (2019). Mother-child bonding is associated with the maternal perception of the child's body odor, *Physiol. Behav.*, 198, 151–157. https://doi.org/10.1016/j.physbeh.2018.09.014.

[106] A. Blomkvist, M. Hofer. (2021). Olfactory impairment and close social relationships. A narrative review, *Chem. Senses*, 46, bjab037. https://doi.org/10.1093/chemse/bjab037.

[107] J.H.B. de Groot, P.A. Kirk, J.A. Gottfried. (2020). Encoding fear intensity in human sweat, *Philos. Trans. R. Soc. Lond. B Biol. Sci.*, 375, 20190271. https://doi.org/10.1098/rstb.2019.0271.

[108] D. Chen, J. Haviland-Jones. (2000). Human olfactory communication of emotion, *Percept. Mot. Skills*, 91, 771–781. https://doi.org/10.2466/pms.2000.91.3.771.

[109] R. Herz, M. Inzlicht. (2002). Sex differences in response to physical and social factors involved in human mate selection: The importance of smell for women, *Evol. Hum. Behav.*, 23.

[110] A. Wisman, I. Shrira. (2020). Sexual chemosignals: Evidence that men process olfactory signals of women's sexual arousal, *Arch. Sex. Behav.*, 49, 1505–1516. https://doi.org/10.1007/s10508-019-01588-8.

[111] V.M. Desiato, D.A. Levy, Y.J. Byun, S.A. Nguyen, Z.M. Soler, R.J. Schlosser. (2021). The prevalence of olfactory dysfunction in the general population: A systematic review and meta-analysis, *Am. J. Rhinol. Allergy*, 35, 195–205. https://doi.org/10.1177/1945892420946254.

[112] C. Merkonidis, F. Grosse, T. Ninh, C. Hummel, A. Haehner, T. Hummel. (2015). Characteristics of chemosensory disorders—results from a survey, *Eur. Arch. Otorhinolaryngol.*, 272, 1403–1416. https://doi.org/10.1007/s00405-014-3210-4.

[113] D.L. Burges Watson, M. Campbell, C. Hopkins, B. Smith, C. Kelly, V. Deary. (2021). Altered smell and taste: Anosmia, parosmia and the impact of long Covid-19, *PLoS One*, 16, e0256998. https://doi.org/10.1371/journal.pone.0256998.

[114] B.R. Haxel, A. Nisius, K. Fruth, W.J. Mann, A. Muttray. (2012). Deficits in medical counseling in olfactory dysfunction, *HNO*, 60, 432–438. https://doi.org/10.1007/s00106-011-2448-z.

[115] A. Breeden. (2021). When COVID-19 stole their smell, these experts lost much more. *The New York Times*.

[116] P. Kohli, Z.M. Soler, S.A. Nguyen, J.S. Muus, R.J. Schlosser. (2016). The association between olfaction and depression: A systematic review, *Chem. Senses*, 41, 479–486. https://doi.org/10.1093/chemse/bjw061.

[117] P. Bonfils, P. Avan, P. Faulcon, D. Malinvaud. (2005). Distorted odorant perception: Analysis of a series of 56 patients with parosmia, *Arch. Otolaryngol. Head Neck Surg.*, 131, 107–112. https://doi.org/10.1001/archotol.131.2.107.

[118] J. Mignot. (2022). Emmanuelle Dancourt and Ugo Charron: 'With Umema, we used the sense of touch to enter into olfactory territory'. *NEZ Magazine*.

[119] T. Hummel, K. Rissom, J. Reden, A. Hähner, M. Weidenbecher, K.B. Hüttenbrink. (2009). Effects of olfactory training in patients with olfactory loss, *Laryngoscope*, 119, 496–499. https://doi.org/10.1002/lary.20101.

[120] M.A. Fornazieri, E.C.D. Garcia, N.M.D. Lopes, I.N.I. Miyazawa, G.D.S. Silva, R.D.S. Monteiro, *et al.* (2020). Adherence and efficacy of olfactory training as a treatment for persistent olfactory loss, *Am. J. Rhinol. Allergy*, 34, 238–248. https://doi.org/10.1177/1945892419887895.

[121] O. Saatci, A. Altundag, O.A. Duz, T. Hummel. (2020). Olfactory training ball improves adherence and olfactory outcomes in post-infectious olfactory dysfunction, *Eur. Arch. Otorhinolaryngol.*, 277, 2125–2132. https://doi.org/10.1007/s00405-020-05939-3.

https://doi.org/10.1142/9781800617780_bmatter

Index

www.ingramcontent.com/pod-product-compliance
Lightning Source LLC
LaVergne TN
LVHW010330220926
848484LV00004BB/875